Meningiomas

Francesco Maiuri
Marialaura Del Basso De Caro
Editors

Meningiomas

From Pathology to Clinics

Springer

Editors
Francesco Maiuri
Department of Neuroscience
University of Naples Federico II
Napoli, Italy

Marialaura Del Basso De Caro
Department of Advanced Biomed
Sciences, Sect Path Anatomy
Federico II University Hospital
Napoli, Italy

ISBN 978-3-031-76679-4 ISBN 978-3-031-76680-0 (eBook)
https://doi.org/10.1007/978-3-031-76680-0

This Springer imprint is published by the registered company Springer Nature Switzerland AG
The registered company address is: Gewerbestrasse 11, 6330 Cham, Switzerland

Preface

Meningiomas are frequent intracranial and spinal tumors, often diagnosed as incidental findings due to the widespread use of imaging techniques. Although mostly benign, a consistent rate of these tumors shows invasive aspect, more aggressive histology, and recurrences. Thus, choosing the best management and predicting the outcome can be challenging.

This book originates from the extensive pathological and clinical experiences on meningiomas confirmed by many research studies and many patients operated on and followed up by our neuro-oncology team.

In the last two decades, the pathology, according to the World Health Organization, and the molecular biology of meningiomas, as for other central nervous system tumors, have undergone significantly changes and improvements. The primary goal is to correlate different pathological and biomolecular aspects with varying clinical behavior and outcome. In recent years, numerous research and clinical studies have contributed to defining the role of biomolecular profiling of meningiomas in predicting tumor aggressiveness, recurrence, and outcome, as well as the timing and type of radiation and medical therapy. This very important aspect is the main focus of this book.

The first part discusses the different features of meningiomas in various population groups, including elderly versus young individuals, males versus females, users of hormonal drugs, and individuals affected by genetic and familial syndromes.

The second part is mainly devoted to the molecular biology and pathological correlations of the meningiomas features, such as location and invasion. The pathological and biomolecular profiles of different locations, depending on meningeal embryology, result in different recurrence rates and outcome.

The third part is devoted to the imaging diagnosis, with a particular emphasis on radiomics; this technique enables the identification of noninvasive imaging biomarkers that could replace invasive pathologic or genetic meningioma profiling techniques.

The surgical management of meningiomas can be challenging due to brain and bone invasion and extensive dural infiltration. In such cases, the maximal safe resection must be balanced against the risk of neurological complications, mainly for tumors in critical areas. However, the Simpson grading and the concept of complete resection have been revised in recent years. For more invasive and high-grade meningiomas, medical therapies targeting biomolecular targets of the tumor are being more widely studied.

The last part is devoted to addressing the problem of recurrences by focusing on risk factors, different patterns of regrowth, and the best management.

This book is useful for all specialists in the neurooncological field, including neurosurgeons, radiotherapists, oncologists, and those working in the pathology and biomolecular fields. All will find important information in their daily clinical experience.

We thank all invited authors for their precious contributions, which have significantly improved the quality of this book.

Naples, Italy Francesco Maiuri
Naples, Italy Marialaura Del Basso De Caro

Contents

Part I Meningioma Pathology and Population-Based Aspects

Meningiomas in Children 3
Pietro Spennato, Lucia De Martino, Giulia Meccariello, Adriana Cristofano, Sara Lombardi, Maria Elena Errico, Nicola Onorini, Lucia Quaglietta, Vittoria Donofrio, Eugenio Covelli, and Giuseppe Cinalli

Meningiomas in the Elderly 17
Antonio Santoro, Daniele Armocida, and Luca D'Angelo

Meningiomas and Hormone-Related Clinical Conditions 25
Marialaura Del Basso De Caro, Rosa Maria Di Crescenzo, and Federico Russo

Meningiomas in Genetic and Familial Syndromes: Biomolecular Correlations 35
Marica Eoli

Part II Meningiomas: Pathological–Clinical Correlations

Meningioma Development: Meningeal Embryology and Tumorigenesis 45
Vratko Himic, Sanjeeva Jeyaretna, and Daniel M. Fountain

Molecular Biology and Genetics of Meningiomas: Classification and Diagnostic Methods 59
Luca Bertero, Felix Sahm, and Rosa Maria Di Crescenzo

Comprehensive Overview of Genetic Alterations in Meningioma . . . 65
Sybren L. N. Maas, Felix Sahm, and Tareq A. Juratli

Location of Meningiomas: Demographic—Pathological and Prognostic Correlations 75
Francesco Maiuri, Sergio Corvino, Giuseppe Mariniello, and Marialaura Del Basso De Caro

Brain–Tumor Interface, Brain Invasion, and Peritumoral Edema in Meningiomas: Pathology—Clinical Implications 89
Felix Behling

Malignant Meningiomas 99
Lorenzo Chiariotti, Rosa Della Monica, Rosa Maria Di Crescenzo, Sergio Corvino, Francesco Maiuri, and Marialaura Del Basso De Caro

Part III Meningioma Pathology and Imaging

Pathology: Imaging Correlations of Meningiomas—Imaging Mimics 109
Sirio Cocozza, Giuseppe Pontillo, Alessandra Scaravilli, Alessandra Alfieri, Donatella Franco, Renata Conforti, Ferdinando Caranci, and Andrea Elefante

Radiomics in Meningiomas: Pathological and Biomolecular Correlation 121
Lorenzo Ugga, Teresa Perillo, Serena Capasso, Davide Negroni, and Renato Cuocolo

PET and SPECT in Meningiomas: Pathological Correlation 131
Evangelos Kalogirou, Sotiria Founta, Paraskevi Papaspirou, and George A. Alexiou

Part IV Meningioma Pathology and Management

The Simpson Grading: Is It Still Valid in the Imaging Era? 143
Rupen Desai, Gülce Gel, Victor Ramzes Chavez-Herrera, and Theodore H. Schwartz

Aggressive Versus More Conservative Surgery for Invasive Meningiomas 153
Francesco Maiuri and Marialaura Del Basso De Caro

Meningioma Pathology and Radiotherapy at Initial Diagnosis 163
Giuseppe Minniti, L. Zaccaro, R. De Pietro, and C. Cattaneo

Biomolecular-Related Medical Therapy for Meningiomas 177
Alessia Pellerino, Francesco Bruno, Edoardo Pronello, Roberta Rudà, and Riccardo Soffietti

Part V Meningioma Pathology and Recurrences

Time to Recurrence of Meningiomas 189
Sergio Corvino

Topography of Meningioma Recurrences: Localized Versus Diffuse 195
Sergio Corvino and Francesco Maiuri

Management of Recurrences of Non-Skull Base Meningiomas 203
Francesco DiMeco, Edoardo Porto, and Lapo Bonosi

Management of Recurrent Skull Base Meningiomas 215

Sachin Govind, Khizar R. Nandoliya, Rushmin Khazanchi,
Rishi Jain, Timothy L. Sita, Priya Kumthekar,
and Stephen T. Magill

Multiple Reoperations for Recurrent Meningiomas 225
Francesco Maiuri and Sergio Corvino

Part I

Meningioma Pathology and Population-Based Aspects

Meningiomas in Children

Pietro Spennato, Lucia De Martino, Giulia Meccariello, Adriana Cristofano, Sara Lombardi, Maria Elena Errico, Nicola Onorini, Lucia Quaglietta, Vittoria Donofrio, Eugenio Covelli, and Giuseppe Cinalli

1 Introduction

Meningiomas are the most common brain tumors in adults, with an estimated incidence of 30% [1]. In contrast, they appear to be rare in children and adolescents [2].

Risk factors associated with the occurrence of meningiomas include genetic syndromes predisposing patients to tumors, such as neurofibromatosis type 2 (NF2) (up to 40% of pediatric meningiomas are associated with NF2 cases) (Fig. 1), Gorlin syndrome, and exposure to ionizing radiation [3, 4] (Fig. 2).

Pediatric meningiomas, in contrast to their adult counterparts, have unique characteristics. At presentation, they are often large, cystic, and in atypical locations, such as inside the ventricular system; the dural attachment, which is a common anatomo-radiological aspect of tumors originating from the meninges, may be missing. In NF2 patients, they are often located at the skull base and in the spinal cord (Figs. 1, 2 and 3). They are also aggressive and have a high recurrence rate, reflecting their histology, which is predominantly WHO grade 2 and 3, especially for clear cell and papillary variants [5].

The clinical presentation is highly variable, making the diagnosis late. Pediatric meningiomas have a better prognosis than adult meningiomas, despite their more aggressive features. Complete surgical resection is the treatment of choice for symptomatic meningiomas [6], and the extent of initial surgical resection is the strongest independent prognostic factor for child and adolescent meningioma [2, 7].

P. Spennato (✉) · G. Meccariello · N. Onorini · G. Cinalli
Department of Pediatric Neurosurgery, Santobono-Pausilipon Children's Hospital (AORN), Naples, Italy

L. De Martino · L. Quaglietta
Pediatric Neuro-Oncology, Santobono-Pausilipon Children's Hospital (AORN), Naples, Italy

A. Cristofano · E. Covelli
Pediatric Neuroradiology, Santobono-Pausilipon Children's Hospital (AORN), Naples, Italy

S. Lombardi
Department of Pediatric Neurosurgery, Santobono-Pausilipon Children's Hospital (AORN), Naples, Italy

Division of Neurosurgery, Department of Neurosciences, Reproductive and Odontostomatological Sciences, Università degli Studi di Napoli "Federico II", Naples, Italy

M. E. Errico · V. Donofrio
Pediatric Pathology, Santobono-Pausilipon Children's Hospital (AORN), Naples, Italy

F. Maiuri, M. Del Basso De Caro (eds.), *Meningiomas*,
https://doi.org/10.1007/978-3-031-76680-0_1

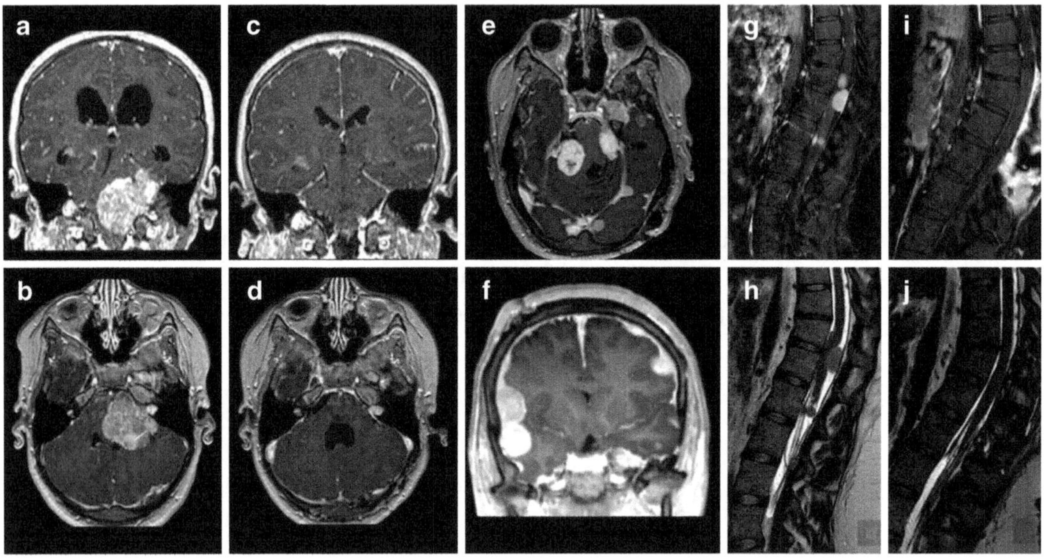

Fig. 1 (**a, b**) Left petroclival meningioma in a 4-year-old baby girl. Her clinical presentation was headache and vomiting. She underwent uneventful total resection of the lesion (transitional meningioma, grade 2) (**c**, **d**). During 14 years of follow-up, a diagnosis of NF2 was made, and the patients developed bilateral vestibular schwannomas and multiple supratentorial, infratentorial, and spinal meningiomas (**e**, **f**). At the age of 16, because of progressive paraparesis and leg pain caused by two dorsal meningiomas (**g, h**), she underwent laminectomy and gross total removal of the spinal lesions (transitional meningiomas, grade 2) (**i, j**)

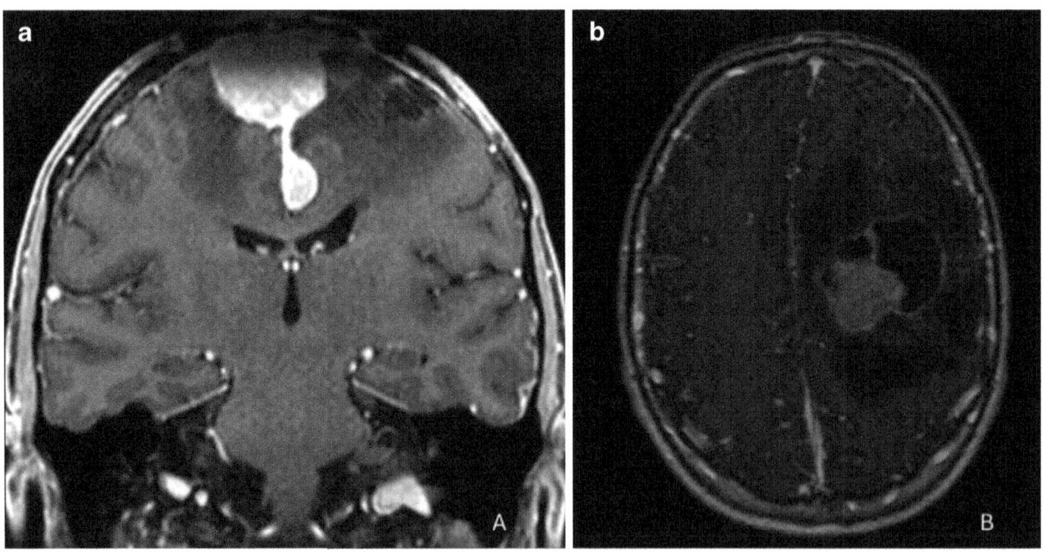

Fig. 2 Radiation-induced falcine and parasagittal meningioma (**a**) in an 18-year-old boy, who was treated 9 years previously for a supratentorial atypical teratoid/rhabdoid tumor (**b**). At that time, the patient underwent gross total removal followed by chemotherapy and radiotherapy

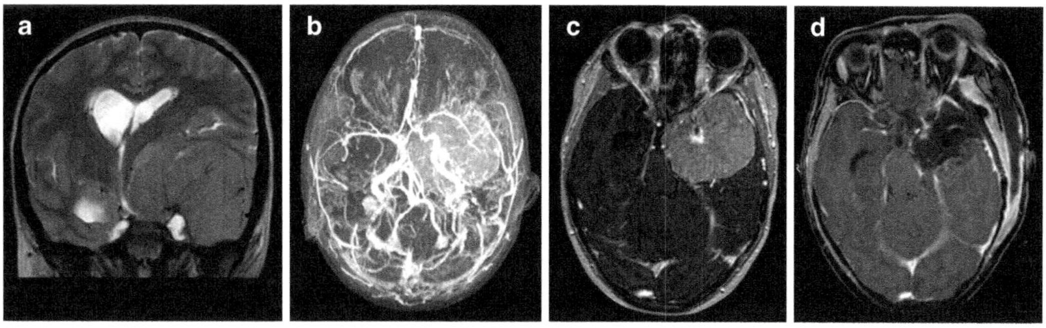

Fig. 3 Giant, highly vascularized, left sphenoid ridge meningioma in a 12-year-old boy. (**a**) Coronal T2-weighted image, (**b**) MR image, (**c**) axial contrast-enhanced T1 image. (**d**) Postoperative contrast-enhanced T1 image showing gross total removal (meningothelial meningioma, grade 2)

2 Epidemiology

The incidence of meningiomas is estimated to be 0.4–4.6% in the pediatric population [8]. They are extremely rare in infants, and their incidence increases with age [9]. The average age of onset is between 11 and 13 years. Meningiomas are more common in prepubertal males than in females, but the ratio reverses after puberty. The increased incidence with age and in postpubertal females may be explained by hormonal changes with increased estrogen exposure characteristic of this time of life [10]. Another important feature is the unusual intracranial location, such as the lateral sulcus, intraparenchymal region, or intraventricular region. The latter is more common in children (8.8–13.6%) than in adults (0.5–3%) [11]. In adults, spinal locations account for less than 5–10% of all meningiomas. In contrast, 13% of meningiomas in children/adolescents are classified as spinal, and this percentage increases in patients with NF2 [5]. In the series of Merten et al., which included 48 histologically confirmed meningiomas occurring in patients younger than 20 years, 66% of the meningiomas were supratentorial, 19% were infratentorial, 17% were intraventricular, 15% were sphenoidal, 15% were falcine and parasagittal, 15% were convex, 6% were parasellar, 15% were sphenoid ridge, 4% were other infratentorial, 12% were spinal, and 2% were orbital.

Epidemiologic studies have examined numerous potential risk factors, and the only validated associations for brain tumors to date are ionizing radiation and genetic syndromes.

Neurofibromatosis type 2 is an autosomal dominant disorder characterized by a mutation in the NF2 gene that causes loss of function of the tumor suppressor protein merlin. Meningiomas are the second most common neoplasm after vestibular schwannomas, with a pediatric incidence of 20% [12]. In contrast, only 0.35% of adult meningiomas are associated with neurofibromas [13]. The location is often atypical, with a greater prevalence of optic nerve sheath and spinal meningiomas. Furthermore, this population is more prone to develop multiple meningiomas and presents at an earlier age [14] (Fig. 2).

Another important risk factor is ionizing radiation, which causes mutations in the genome, either directly or indirectly through the formation of free radicals. Their carcinogenic effects are more pronounced in children because they are more sensitive to radiation and have more years of potential life to express the risk [10]. The incidence after exposure was 0.1% at 5 years and 1.8% at 10 years. In fact, radiation-induced (RI) meningiomas have a late onset, with a latency of 7–9 years and a mean age of 17.3 ± 3.5 years. Children with radiation exposure before the age of 5 years have a short latency period and are at increased risk for the development of anaplastic meningiomas. They do not show a sex preponderance. These patients have a worse prognosis and are more prone to recurrence [2, 7, 14].

Tumor predisposition syndromes and a history of cranial radiation also represent risk factors for the development of multiple meningiomas [6].

3 Clinical Manifestations

The clinical presentation is variable and is linked to the tumor location, mass effect, and increased intracranial pressure; the most common symptoms are headache, seizures, and focal neurological deficits, such as visual disturbances and cranial nerve palsy [15, 16]. The clinical manifestations are often nonspecific, with a mean interval of 6.5 months between symptom onset and meningioma diagnosis [16]. The most commonly reported onset symptoms and signs are increased intracranial pressure (headache, papilledema, and vomiting), which occurs in 32–62% of patients due to the rapid growth of meningiomas in childhood (Fig. 3). Seizures due to mass effects or cortical irritation and site-specific symptoms follow in frequency. When localized in the posterior fossa, they manifest with symptoms of cranial nerve palsy [7, 17]. Patients with NF2 are more likely to present with visual deficits. This may be explained by the greater incidence of optic nerve sheath meningiomas in these patients [16].

The mean interval between symptom onset and surgery is 11.6 months, with a median of 3.0 months [11].

4 Molecular Pathways

The classification of CNS tumors has undergone a vast transformation in the last few years, thanks to DNA methylation profiling. Pediatric meningiomas may also differ from their adult counterparts in terms of molecular abnormalities and methylation classes [4, 18].

In both pediatric and adult meningiomas, the most frequent genetic alteration is the monosomy of chromosome 22 [18], resulting in biallelic inactivation of the neurofibromatosis-2 (NF2) gene, located at 22q12.2 [19]. NF2 encodes merlin, an ezrin–radixin–moesin-(ERM)-related protein that plays a role as a tumor suppressor [19]. Although adult meningiomas are often sporadic tumors, children and adolescents are frequently affected by underlying oncopredisposition syndrome, mainly NF2 [20].

In non-NF2-driven pediatric meningiomas, additional pathogenic alterations have been identified. YAP1-MAML2 fusion proteins act as oncogenic drivers in the development of meningiomas, predominantly in pediatric patients, by downregulating the HIPPO pathway, which regulates the expression of genes involved in cell proliferation and apoptosis [21, 22]. A group of rhabdoid/papillary meningiomas share loss-of-function (LoF) mutations in BRCA1-associated protein-1 (BAP1) and PBRM1 [23, 24]. The clear cell meningioma subtype, which is mostly encountered in children, adolescents, and young adults, is associated with biallelic alterations in SMARCE1, which is located on chromosome 17q21. SMARCE1 encodes the protein BAF57, an ATP-dependent component of the chromatin remodeling complex SWI/SNF, which is involved in chromatin remodeling and gene transcription [25]. While adult meningiomas may harbor TRAF7, AKT1, KLF4, SMO, and PIK3CA alterations, these mutations are rare in the pediatric population [18, 26].

Recently, pediatric meningiomas have been divided into three different epigenetic subgroups (1, 2a, and 2b), while in adults, 6 prognostic subgroups (benign-1, benign-2, benign-3, intermediate-A, intermediate-B, and malignant) have been identified [27]. For pediatric meningiomas, subgroup 1 included almost exclusively clear cell meningiomas due to alterations of the SMARCE1 tumor suppressor gene and, less frequently, allelic losses at chromosomes 19 and 22. Subgroup 2A is dominated by atypical WHO grade 2 meningiomas with chromosome 22 loss and can be described as the NF2-driven subgroup of pediatric meningiomas. In subgroup 2B, a mixture of histological variants can be observed, including all rhabdoid meningiomas of WHO grade 3 [18].

5 Oncopredisposing Syndromes

Pediatric meningiomas, as for many other malignancies, may be caused by an underlying oncopredisposition syndrome. According to a recent literature review [28], 8% and 19% of pediatric patients are affected by neurofibromatosis type 1 (NF1) and type 2 (NF2), respectively. Indeed, historically, the first condition associated with meningiomas was germinal haploinsufficiency of NF2 [9, 23, 29]. The incidence of NF2 is approximately 1 in 25,000, with a penetrance of 95% [12]. The most characteristic feature of this disease is the presence of bilateral vestibular schwannomas (occurring in up to 90% of patients until the age of 20); however, meningiomas occur in approximately 50% of patients (20% of children). In contrast to sporadic meningiomas, NF2-related meningiomas are often grade 2 or 3, have a worse prognosis, and have a higher rate of recurrence [12]. Unlike NF2, NF1 is thought to be unrelated to the risk of meningioma; however, a recent meta-analysis by Kotecha suggested an NF1-related meningioma incidence of 3.4% [7]. Meningiomas may be observed in people with schwannomatosis. This rare disorder, with a reported incidence ranging from 1 in 40,000 to 1 in 1.7 million, is characterized by multiple schwannomas without vestibular schwannomas or any other stigmata of NF2. Meningioma is associated with schwannomatosis in approximately 5% of cases [6]. Interestingly, schwannomatosis is genetically related to NF2 due to variants in either INI1 (SMARCB1) or LZTR1, which are located on chromosome 22. In contrast, NF2 is unrelated to the more common NF1 gene, which is due to pathogenic variants of the tumor suppressor NF1 gene on chromosome 17 [12].

Gorlin syndrome, or multiple basal cell carcinoma syndrome, is an autosomal dominant inherited familial syndrome associated with meningiomas. Individuals with Gorlin syndrome are also at increased risk of medulloblastoma, the treatment of which may then dispose of radiation-induced meningioma [30].

Recently, several other genes, including SMARCB1, SMARCE1, BAP1, PTCH1, SUFU, PDGFB, CREBBP, WRN, and MEN1, have been associated with a predisposition to meningioma [30, 31]. Moreover, a genotype/phenotype has been established for some of them. The SMARCE1 gene is involved in the development of clear cell meningioma subtypes. SMARCE1 exhibits allelic heterogeneity: germline heterozygous pathogenic missense variants lead to Coffin-Siris syndrome, whereas germline heterozygous pathogenic LoF variants increase familial susceptibility to clear cell meningioma. Although rare, a subset of pediatric rhabdoid/papillary meningiomas, including tumors of different organs and lineages (mesotheliomas, uveal and cutaneous melanomas, and clear cell carcinoma of the kidney), may indicate the presence of BAP1 tumor predisposition syndrome [30, 31].

6 Histopathology

Meningiomas are most likely derived from arachnoid cap cells, which share many cytological features. They are composed of cells with indistinct cellular borders, eosinophilic cytoplasm, and round to ovoid nuclei that may aggregate in syncytia, form whorls, and show collagen deposition and/or laminated, concentric calcifications. These neoplasms may harbor a broad spectrum of morphological features (intracellular lumina, heterotopic mesenchymal differentiation in the form of adipose tissue, cartilage, bone, and/or myxoid matrix, rhabdoid morphology, cobweb-like appearance, etc.), with frequent overlap between two or more patterns. The fifth edition of the 2021 WHO Classification of Tumors of CNS recognizes 15 subtypes: meningothelial, fibrous, transitional, psammomatous, angiomatous, microcystic, secretory, lymphoplasmacyte-rich, metaplastic, chordoid, clear cell, rhabdoid, papillary, atypical, and anaplastic meningioma, all of which may be present in children [32]. In adults, meningothelial, transitional, and fibroblastic meningiomas are the most common variants; in addition, clear cell and papillary meningiomas occur more frequently in younger patients (Fig. 4).

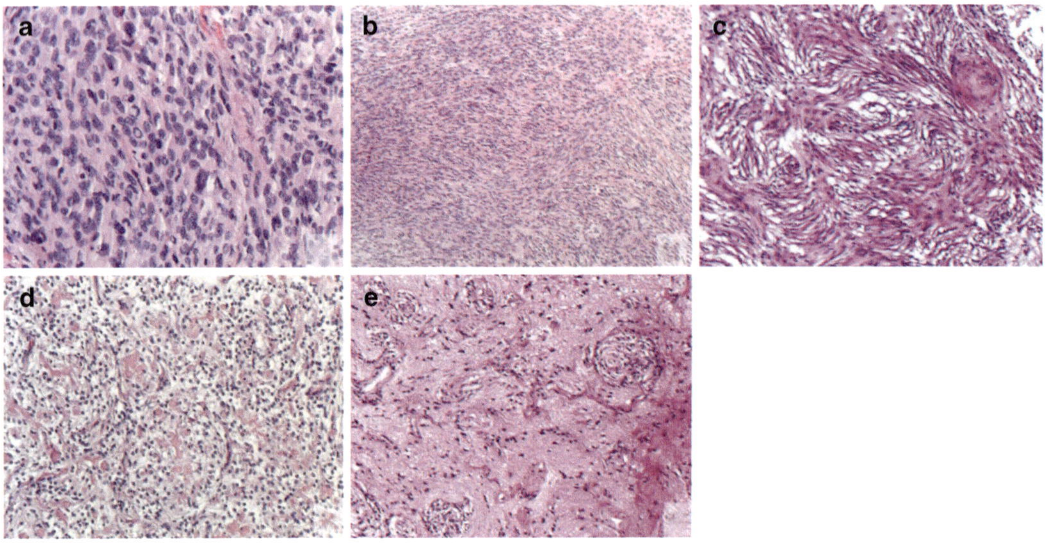

Fig. 4 Histological aspects of the most common pediatric variants. (**a**) Meningothelial meningioma: It consists of syncytium-like aggregates of cells with a classic meningothelial appearance, indistinct cell borders and oval nuclei, often demonstrating nuclear grooves, clearing, and intranuclear cytoplasmic pseudoinclusions (Ematossilina-Eosina-EE, 400×). (**b**) Fibrous meningioma: It consists of broad, intersecting fascicles of spindle cells, with variable collagen content and often a storiform pattern (EE, 100×). (**c**) Transitional meningioma: It has features of both meningothelial and fibrous type, with the formation of whorls and psammoma bodies (EE, 400×). (**d**) Clear cell meningioma is characterized by sheets of cells with round nuclei and clear cytoplasm caused by PAS-positive and diastase-sensitive glycogen content and by prominent perivascular and interstitial collagen (EE, 200×). SMARCE1 biallelic alterations, typically found in this variant, can be highlighted by the loss of nuclear expression of the protein by immunohistochemistry. (**e**) Papillary meningioma. It comprises meningothelial cells surrounding fibrovascular cores in a pseudopapillary or pseudorosette-like pattern (EE, 400×)

The meningiomas are categorized into three grades. According to the WHO CNS fifth, any subtype can be assigned to grade 2 or 3 based on specific morphologic and/or molecular diagnostic parameters: 4–19 mitoses/10HPF, brain invasion, or three of the following: increased cellularity, small cells with a high N:C ratio, prominent nucleoli, sheeting, and foci of necrosis for grade 2; >20 mitoses/10HPF, or frank anaplasia, or TERT promoter mutation and/or homozygous deletion of *CDKN2A* and/or *CDKN2B for grade 3 (anaplastic) tumors*. However, clear cell subtype and chordoid meningioma, which are historically associated with a greater likelihood of recurrence, can be assigned to grade 2, regardless of these criteria [32].

Given the rarity of pediatric meningiomas, it is not yet clear whether these criteria can be translated to children [4]. Moreover, some of the molecular alterations reported in adult meningiomas, particularly TERT promoter mutation and H3K27me3 loss of nuclear immunohistochemical expression, which are associated with high grade and poor prognosis, seem to be very rare in children [15].

Generally, the diagnosis of classic meningioma is relatively straightforward; however, given the broad range of morphologic appearances, some tumors should be differentiated from metastatic and primary neoplasms involving the dura. Immunohistochemical positivity for SSTR2a (*somatostatin receptor 2A*) can assist in identifying doubtful cases, while EMA staining can be focal or even absent, especially in fibrous and higher-grade subtypes. In the pediatric population, the potential mimickers of meningiomas are primary meningeal melanocytic tumors and mesenchymal neoplasms. In particular, schwannomas may simulate fibrous variants, especially in patients with NF2, due to the presence of spindle cells and S100 immunoreactivity, but unlike meningiomas, schwannomas are negative for

SSTR2 and EMA. Hemangioblastoma may resemble microcystic meningioma; abundant pericellular reticulin and immunoreactivity for inhibin resolve this issue. Among meningeal sarcomas, *FET::CREB*-fused tumors can mimic high-grade meningiomas, showing syncytial or sheet-like growth, cellular whorls and amianthoid-type fibers, and EMA immunoreactivity. Additional immunohistochemical staining for CD99 and desmin and specific molecular alterations, along with a distinct DNA methylation profile, can identify this new entity. *CNS tumor with BCOR internal tandem duplication* has been identified as a potential mimetic for papillary meningiomas but shows a characteristic immunophenotype (strong expression of BCOR and focal positivity for Olig2 or GFAP) and molecular signature.

Another topic is meningioangiomatosis, which is characterized by the rare proliferation of small blood vessels ensheated by meningothelial and spindle fibroblast-like cells, with variable calcification and psammoma bodies; this disease involves the leptomeninges and superficial cortex and shows perivascular spread along the Virchow–Robin spaces. Recent reports suggest that meningioangiomatosis could represent a potential specific pattern of tumor infiltration [33]. It should not be confused with brain invasion and is considered a grade 2 defining pattern.

7 Radiology

Pediatric meningiomas are often very large masses (>5 cm) with well-defined morphology [34] (Figs. 3, 5, and 6) that arise from the dura mater, the pia mater cells within the choroid plexus, or the arachnoid of the Sylvian fissures and grow quickly [34–36]. They are mostly found in the supratentorial space, but childhood and

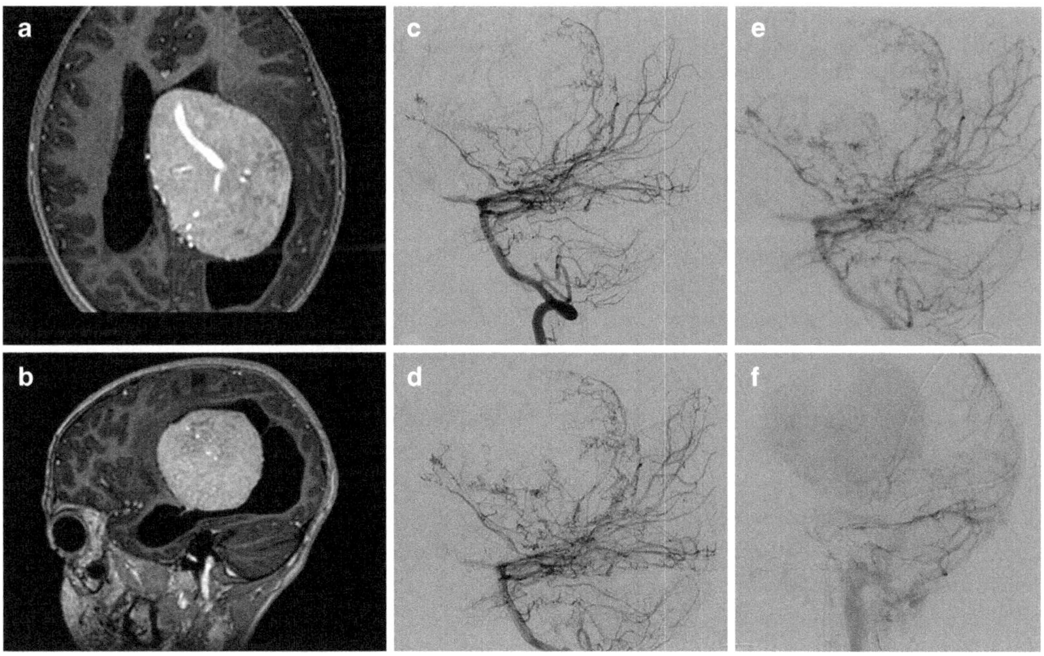

Fig. 5 Presurgery MR scan of sporadic intraventricular meningioma in an 8-year-old female child. Axial (**a**) and sagittal (**b**) MPR reconstructions of 3D-T1-weighted postcontrast sequences and (**c–f**) digital subtraction angiography (DSA) images. MR image (**a**, **b**) showing a well-defined intraventricular enhancing mass occupying the left lateral ventricle; note the intralesional hypertrophic drainage vein, compression and displacement of neighboring brain structures, obstructive hydrocephalus, and intracranial hypertension. DSA (**c–f**) revealed the so-called mother-in-law sign due to early lesion contrast enhancement from feeding arteries of the left posterior choroidal artery and continuing mass opacification after the venous phase

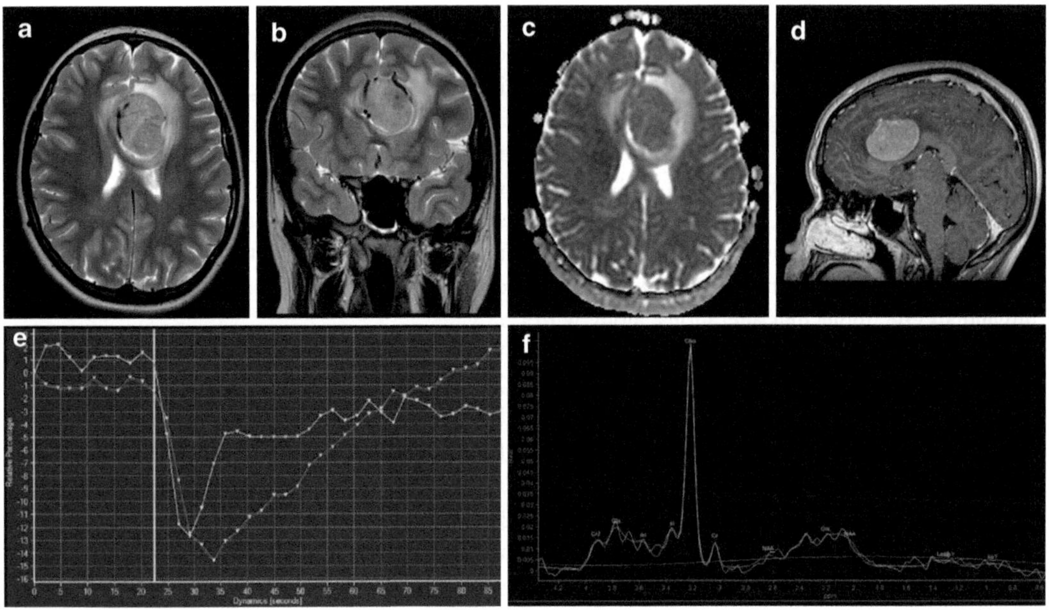

Fig. 6 Presurgery MR scan of sporadic interhemispheric meningioma in a 14-year-old girl. Axial (**a**) and coronal (**b**) T2-weighted sequences, ADC map (**c**), sagittal MPR reconstruction of 3D-T1-weighted postcontrast sequence (**d**), time–intensity curves of dynamic contrast-enhanced (DSC) MR perfusion (**e**), and short-echo-time proton MR spectroscopy (**f**). The extra-axial interhemispheric mass is localized between the anterior fossae, has regular margins, and is characterized by a hyperintense signal on T2-weighted images (**a, b**), a low apparent diffusion coefficient (ADC) (**c**), and intense and homogeneous contrast enhancement (**d**); note the CSF/vascular cleft at the interface between the meningioma and the neighboring brain, the white matter buckling sign, and perilesional edema (**a, b**). On time–intensity curves of dynamic contrast-enhanced (DSC) MR perfusion (**e**), the relaxivity did not return to baseline after the gadolinium bolus passed through the tumor (blue curve). Short-echo-time proton MR spectroscopy (**f**) revealed elevated choline with diminished creatine-phosphocreatine, NAA, and myo-inositol peaks and a small glutamine/glutamate peak centered between 2.2 and 2.4 ppm

adolescent meningiomas are often encountered at unusual sites (e.g., intraventricular (Fig. 5), Sylvian cistern (Fig. 3), planum sphenoidale, tentorium, craniocervical junction (Fig. 1), and skull base). Although they are classified as extra-axial masses, in the pediatric population, they may also have an intraparenchymal location [16, 35, 36].

On CT imaging, pediatric meningiomas could be associated with hyperostosis and/or intratumoral calcifications. All pediatric meningeal-based tumors are iso- to hyperdense compared with cortical gray matter on precontrast CT, and the solid portion of the tumor shows intense contrast enhancement [37].

On MR images, pediatric meningeal tumors show isointensity to gray matter and hypointensity to white matter on T1-weighted images; variable signal intensity to gray matter and white matter on fluid attenuated inversion recovery (FLAIR)- and T2-weighted images; and marked enhancement of the tumoral solid tissue after intravenous gadolinium injection (Figs. 5 and 6). Sometimes they are strongly heterogeneous on both pre- and postgadolinium sequences due to calcified components, cystic and necrotic areas, or regions of fibrosis and sclerosis within the mass [34–37]. Hyperostosis can be demonstrated on MR as a low signal intensity of thickening on T1-weighted and T2-weighted images at the calvarial interface. Conversely, intratumoral calcification can be large or tiny, and depending on the calcium concentration, it shows variable signal intensity on noncontrast T1-weighted images.

Cystic and necrotic changes are more frequent in pediatric patients than in adults, possibly because of their larger size and rapid growth. Usually, cystic components have the same signal intensity as cerebrospinal fluid (CSF). Like cystic parts, necrotic areas demonstrate prolonged T1 and T2 relaxation times compared to those of the solid portion of the tumor. Fibrosis and sclerosis appear hypointense on pre- and postgadolinium T1-weighted images but also marked hypointensity on T2-weighted images [34–37].

On diffusion-weighted imaging (DWI), meningiomas show a decreased apparent diffusion coefficient (ADC), but either the distinction between meningiomas from other neoplasms and tumor-like lesions or the differentiation between benign and atypical/malignant meningiomas is not always reliable [36].

Perfusion weighted imaging (PWI) (Fig. 6) may be helpful for differentiating meningeal tumors from intraparenchymal tumors due to the absence of the blood–brain barrier and pooling of the contrast agent in the interstitium in cases of meningiomas and from other extra-axial masses, such as schwannomas and paragangliomas, due to differences in vascularity and permeability. Moreover, compared with normal brain parenchyma, meningiomas tend to have a larger blood volume [35].

Long-echo-time proton MR spectroscopy (Fig. 6) in meningiomas revealed a normal neoplastic pattern of elevated choline with diminished creatine-phosphocreatine and *N*-acetylaspartate (NAA) levels. Short-echo-time spectra show a decrease in myo-inositol and an increase in alanine; the latter seems to be a specific meningioma metabolite if it is present; similarly, glutamine/glutamate has been recognized as a potential supplementary metabolite for identifying meningiomas, although it is not specific. Lactate and lipid detection cannot always be used as proof of a malignancy, although they are more frequently observed in nonbenign meningiomas [35].

Likely in adult meningiomas with typical locations, the presence of CSF/vascular clefts should be evaluated by imaging. The cleft is an entrapment of the CSF or cerebral cortical vessels found at the interface between the meningioma and the underlying brain and is useful for distinguishing an extra-axial meningioma from an intra-axial lesion. A vascular cleft interface is seen as a signal flow void or an increased or decreased signal intensity on either T1-weighted or T2-weighted images depending on whether the structure is arterial or venous and on the flow rate. The CSF cleft interface has a typical CSF signal intensity pattern [38].

The dural tail sign (DTS) is described as thickening of the dura mater adjacent to an intracranial neoplasm on contrast-enhanced imaging due to neoplastic infiltration away from the lesion; it is associated with meningiomas, but it is not specific. The lack of dural attachment in childhood and adolescence meningiomas represents another characteristic finding that makes differential diagnosis challenging [34, 36].

Intraventricular meningiomas are observed more commonly in children, arising from either the tela choroidea or from the choroid plexus, and spinal meningioma and optic nerve sheath meningioma are more frequent in patients with NF2 [16, 35].

High heterogeneity, reduced diffusivity, and poorly defined margins could suggest sarcomatous degeneration or meningiomas with higher degrees of malignancy, but these features are not reliable [35].

Digital subtraction angiography (DSA) is helpful for identifying the feeding arteries of a meningioma, distinguishing between the pial and dural blood supplies, and identifying associated vascular malformations (Fig. 5). Typically, the angiographic pattern of a meningioma is described as the mother-in-law sign because of its early enhancement during the arterial phase and its persistent opacified appearance after the venous phase [38].

8 Management

Surgical excision is generally considered the mainstay of treatment for both adult and pediatric meningiomas. According to the 2021 EANO guidelines, the key principle in meningioma surgery is "maximum safe resection with low morbidity and preservation of neurological function" [39]. This approach is particularly important in children, as studies have shown that the extent of initial resection is the single most defining prognostic factor in pediatric meningiomas [14, 40], justifying aggressive resection involving the dura and infiltrated or hyperostotic bone. Therefore, observation is generally not an option. Short-term observation might be considered in specific situations, such as optimizing clinical conditions (e.g., hydrocephalus, seizures) or for asymptomatic, small tumors. Since the number of incidental meningiomas has increased over time because of the broad availability of neuroimaging [41], a conservative approach is acceptable in asymptomatic young adults, as recommended for adults with meningioma according to the European Association of Neuro-Oncology (EANO) guidelines. These patients can be managed by an initial observation interval of 6 months and then annually for indolent tumors [6]. Observation is not recommended for children under the age of 3 years since they are frequently affected by more aggressive histology and may require additional treatment [7]. Long-term observation may also be justified in NF2 patients with asymptomatic, small meningiomas [42]. This is because tumors in these patients tend to grow more slowly, exhibit a saltatory growth pattern, and show a lower propensity for brain invasion. However, the appearance of clinical symptoms, signs, or peritumoral edema on MRI scans could indicate a more aggressive tumor, necessitating prompt surgical resection [43].

Pediatric neurosurgery presents some additional challenges compared to adult neurosurgery. Children, especially younger ones, are more vulnerable to blood loss, hypothermia, and the main complications of prolonged surgery. General measures to reduce perioperative complications should always be considered. Ambient temperature during surgery should always be elevated, and intravenous fluid should be warmed. Heat loss should be avoided using convective warming blankets and warm solutions for scabbing. Crossmatched whole blood must be collected in the operating room before surgery, and multiple intravenous access routes should always be available for rapid transfusion. When blood loss is important, equivalent to the circulating blood volume, supplemental platelets and plasma should be administered to optimize coagulation. Positioning is an additional key point during pediatric surgery. Rigid pin fixation should be avoided under 2 years of age because of the immature infant skull, and nonrigid fixation, such as a cushioned horseshoe head frame, should be favored [44].

Surgeons should always be aware of strategies to reduce operative time or blood loss. Preoperative embolization in adult meningiomas is quite common and aims to reduce vascular flow to the tumors, allowing simpler resection. In pediatric tumors, preoperative embolization has been successfully described [45] and should be considered to reduce blood loss, particularly in younger patients who have a smaller circulating volume; nevertheless, the small caliber of vessels in younger children can make the procedure challenging, and the risk of morbidity should always be considered.

Pediatric meningiomas present several unique challenges compared to those in adults, which can impact surgical resection. In children, these tumors tend to be large at presentation, exhibit locally aggressive behavior, and are frequently located in unusual sites, such as inside the ventricular system. Gross total resection of intraventricular meningiomas can be difficult to achieve, increasing the risk of recurrence. Moreover, the lesion can block cerebrospinal fluid circulation, leading to hydrocephalus and increased intracranial pressure. In this setting, some authors propose preoperative permanent or transient CSF diversion and antiedema measures to reduce morbidity [46]. Conversely, CSF diversion can cause infratentorial complications such as upward tumor herniation or intratumoral hemorrhage. In intraventricular locations, bleeding control can be a challenge. To minimize blood loss and reduce surgical time, the use of the Aquamantys® system has been proposed [47]. It is a bipolar coagulation device that combines radiofrequencies and saline solution to denature collagen fibers to obtain hemostatic sealing. It was found to be efficient in controlling diffuse bleeding in large tumors, especially when a clear bleeding point cannot be identified [47].

Compared to their adult counterparts, pediatric meningiomas have a greater recurrence rate, mostly in cases of partial resection. Some authors reported a greater histological subtype compared to the initial histology when relapsing tumors were resected, demonstrating a tendency toward malignant transformation in cases of recurrence [46]. This finding supports the importance of gross total resection as a first-line treatment and highlights the importance of histological assessment at each surgery.

Given the frailty of younger patients, the higher recurrence rate, and the evidence that gross total resection is stronger than histological grading in defining prognosis, in large tumors, staged resection can be a valid option to enhance the rate of resection, reducing intraoperative morbidity. Redo surgery should also be favored in the case of regrowing tumors to avoid or at least delay radiation therapy, reducing the toxicity to the growing brain and the risk of secondary tumors [42, 46].

For spinal meningiomas, surgical resection is the main treatment option. Neurophysiological monitoring is an essential tool to avoid neurological sequelae. Total resection should be attempted because it is associated with a good prognosis. Describing their experience with pediatric spinal meningiomas, Wang et al. reported that some pathological subtypes are prone to cause firm adhesion; this feature, together with location, size, and blood loss, can interfere with gross total resection [45].

9 Adjuvant Therapies

The management of meningioma in children and adolescents must be carefully considered on a case-by-case basis. In 2020, Szychot et al. published guidelines for the treatment of meningioma in children, teenagers, and young adults [6]. Currently, the only effective adjuvant therapy for aggressive tumors that are not completely eradicated in children is radiotherapy [6]. The radiotherapy modality of choice should be conventionally fractionated radiotherapy. Stereotactic radiosurgery should be reserved for exceptional cases in which the tumor or tumor bed is less than 2.5–3 cm and appropriately located. In general, all children and young adults up to approximately 25 years of age should be referred for proton therapy, which has the capability to reduce the exposure of nontarget tissues to radiotherapy and reduce the late effects of treatment. WHO grade 1 and 2 meningiomas should be treated with doses of 50–55 Gy in conventional fractionation, while doses up to 60 Gy should be considered for WHO grade 3 meningiomas and for some WHO grade 2 tumors (especially those with residual gross tumors). In children and young adults, such higher doses should preferably be given in 1.8 Gy fractions (i.e., 59.4 Gy in 33 fractions, rather than 60 Gy in 30 fractions as is used in older adults) [6].

Chemotherapy, hormonal therapy, and targeted therapies have been explored for refractory meningiomas. However, limited data are available regarding the use of systemic chemotherapy in the pediatric population. Although there have been some reports of modest benefits from chemotherapy and, more recently, from interferon alpha-2b, therapeutic options remain limited [48].

10 Outcome

Meningiomas are slow-growing tumors with generally favorable prognoses. Despite their benign nature, some studies have suggested shorter survival times in pediatric meningiomas patients than in adults. This is associated with factors such as NF2 gene mutations, a higher rate of malignant transformation, a larger tumor size at diagnosis, and a lack of established adjuvant therapies [49]. Additionally, pediatric meningiomas tend to be locally aggressive and recur more often, especially after partial resection, although recurrences can occur even after complete removal.

In 2011, Kotecha and coworkers published the largest meta-analysis available on meningiomas in children and adolescents, showing that patients treated with gross total resection (GTR) have a better chance of recurrence-free survival (RFS) and overall survival than those who received subtotal resection, regardless of tumor WHO grade, even if the magnitude of the difference is lower with higher WHO grade [7]. Therefore, factors that make GTR difficult, such as skull base location, can affect patient prognosis. Additionally, predisposing factors such as radiation exposure and NF2 mutations have a greater impact on outcome than the tumor grade itself.

In NF2-related meningiomas, overall survival and RFS change over time. During the first 10 years, sporadic meningiomas seem to have a worse prognosis due to their greater tendency to invade brain tissue. However, after 10 years, the trend reversed, and NF2-related meningiomas had a poorer prognosis. This can be explained by the fact that NF2 meningiomas, although less aggressive individually, are often multiple and are associated with other morbidities, leading to a less favorable long-term outcome [7].

References

1. Ostrom QT, Gittleman H, Liao P, Vecchione-Koval T, Wolinsky Y, Kruchko C, et al. CBTRUS statistical report: primary brain and other central nervous system tumors diagnosed in the United States in 2010–2014. Neuro Oncol. 2017;19(Suppl 5):v1–88.
2. Azmitia L, Taylor G, Massimi L, Visocchi M. Pediatric meningiomas: current insights on pathogenesis and management. In: Visocchi M, editor. The funnel: from the skull base to the sacrum, Acta neurochirurgica supplement, vol. 135. Cham: Springer International Publishing; 2023. p. 69–74. Disponibile su: https://link.springer.com/10.1007/978-3-031-36084-8_12.
3. Narang A, Maheshwari C, Aggarwal V, Bansal P, Singh P. Gorlin-Goltz syndrome with intracranial meningioma: case report and review of literature. World Neurosurg. 2020;133:324–30.
4. Toland A, McNulty SN, Pekmezci M, Evenson M, Huntoon K, Pierson CR, et al. Pediatric meningioma: a clinicopathologic and molecular study with potential grading implications. Brain Pathol. 2020;30(6):1134–43.
5. Dudley RWR, Torok MR, Randall S, Béland B, Handler MH, Mulcahy-Levy JM, et al. Pediatric versus adult meningioma: comparison of epidemiology, treatments, and outcomes using the surveillance, epidemiology, and end results database. J Neuro Oncol. 2018;137(3):621–9.
6. Szychot E, Goodden J, Whitfield G, Curry S. Children's Cancer and Leukaemia Group (CCLG): review and guidelines for the management of meningioma in children, teenagers and young adults. Br J Neurosurg. 2020;34(2):142–53.
7. Kotecha RS, Pascoe EM, Rushing EJ, Rorke-Adams LB, Zwerdling T, Gao X, et al. Meningiomas in children and adolescents: a meta-analysis of individual patient data. Lancet Oncol. 2011;12(13):1229–39.
8. Gao X, Zhang R, Mao Y, Wang Y. Childhood and juvenile meningiomas. Childs Nerv Syst. 2009;25(12):1571–80.
9. Perry A, Dehner LP. Meningeal tumors of childhood and infancy. An update and literature review. Brain Pathol. 2003;13(3):386–408.
10. Ostrom QT, Fahmideh MA, Cote DJ, Muskens IS, Schraw JM, Scheurer ME, et al. Risk factors for childhood and adult primary brain tumors. Neuro Oncol. 2019;21(11):1357–75.
11. He W, Liu Z, Teng H, Tang L, Jie D, Duan Z, et al. Pediatric meningiomas: 10-year experience with 39 patients. J Neuro Oncol. 2020;149(3):543–53.
12. Bachir S, Shah S, Shapiro S, Koehler A, Mahammedi A, Samy RN, et al. Neurofibromatosis type 2 (NF2) and the implications for vestibular schwannoma and meningioma pathogenesis. Int J Mol Sci. 2021;22(2):690.
13. Wilson CB. Meningiomas: genetics, malignancy, and the role of radiation in induction and treatment: The Richard C. Schneider Lecture. J Neurosurg. 1994;81(5):666–75.
14. Kotecha RS, Junckerstorff RC, Lee S, Cole CH, Gottardo NG. Pediatric meningioma: current approaches and future direction. J Neuro Oncol. 2011;104(1):1–10.
15. Tauziède-Espariat A, Pfister SM, Mawrin C, Sahm F. Pediatric meningiomas: a literature review and diagnostic update. Neuro Oncol Adv. 2023;5(Suppl 1):i105–11.
16. Wagener K, Beckhaus J, Boekhoff S, Friedrich C, Müller HL. Sporadic and neurofibromatosis type 2-associated meningioma in children and adolescents. J Neuro Oncol. 2023;163(3):555–63.
17. Opoku I, Yang L, Sun P, Zhou M, Liu Y, Ren J, et al. Pediatric cerebral meningioma: a single-center study with 10 children not associated with neurofibromatosis type 2 and literature review. Pediatr Neurosurg. 2022;57(6):422–33.
18. Kirches E, Sahm F, Korshunov A, Bluecher C, Waldt N, Kropf S, et al. Molecular profiling of pediatric meningiomas shows tumor characteristics distinct from adult meningiomas. Acta Neuropathol. 2021;142(5):873–86.
19. Smith MJ, Higgs JE, Bowers NL, Halliday D, Paterson J, Gillespie J, et al. Cranial meningiomas in 411 neurofibromatosis type 2 (NF2) patients with proven gene mutations: clear positional effect of mutations, but absence of female severity effect on age at onset. J Med Genet. 2011;48(4):261–5.
20. Perry A, Giannini C, Raghavan R, Scheithauer BW, Banerjee R, Margraf L, et al. Aggressive phenotypic and genotypic features in pediatric and NF2-associated meningiomas: a clinicopathologic study of 53 cases. J Neuropathol Exp Neurol. 2001;60(10):994–1003.

21. Schieffer KM, Agarwal V, LaHaye S, Miller KE, Koboldt DC, Lichtenberg T, et al. YAP1-FAM118B fusion defines a rare subset of childhood and young adulthood meningiomas. Am J Surg Pathol. 2021;45(3):329–40.
22. Sievers P, Chiang J, Schrimpf D, Stichel D, Paramasivam N, Sill M, et al. YAP1-fusions in pediatric NF2-wildtype meningioma. Acta Neuropathol. 2020;139(1):215–8.
23. Goutagny S, Kalamarides M. Meningiomas and neurofibromatosis. J Neuro Oncol. 2010;99(3):341–7.
24. Williams EA, Santagata S, Wakimoto H, Shankar GM, Barker FG, Sharaf R, et al. Distinct genomic subclasses of high-grade/progressive meningiomas: NF2-associated, NF2-exclusive, and NF2-agnostic. Acta Neuropathol Commun. 2020;8(1):171.
25. Smith MJ, Ahn S, Lee JI, Bulman M, du Plessis D, Suh YL. SMARCE1 mutation screening in classification of clear cell meningiomas. Histopathology. 2017;70(5):814–20.
26. Battu S, Kumar A, Pathak P, Purkait S, Dhawan L, Sharma MC, et al. Clinicopathological and molecular characteristics of pediatric meningiomas. Neuropathology. 2018;38(1):22–33.
27. Sahm F, Schrimpf D, Stichel D, Jones DTW, Hielscher T, Schefzyk S, et al. DNA methylation-based classification and grading system for meningioma: a multicentre, retrospective analysis. Lancet Oncol. 2017;18(5):682–94.
28. Santana-González A, Pérez-Ramírez M, Ramírez-Reyes G, Sánchez-Rodríguez G, Ferman-Cano F, López-Aguilar E, et al. Management experience of pediatric meningioma in a Mexican reference center. Ann Diagn Pathol. 2021;55:151811.
29. Evans DGR, Watson C, King A, Wallace AJ, Baser ME. Multiple meningiomas: differential involvement of the NF2 gene in children and adults. J Med Genet. 2005;42(1):45–8.
30. Kerr K, Qualmann K, Esquenazi Y, Hagan J, Kim DH. Familial syndromes involving meningiomas provide mechanistic insight into sporadic disease. Neurosurgery. 2018;83(6):1107–18.
31. Lee YS, Lee YS. Molecular characteristics of meningiomas. J Pathol Transl Med. 2020;54(1):45–63.
32. Louis DN, Perry A, Wesseling P, Brat DJ, Cree IA, Figarella-Branger D, et al. The 2021 WHO classification of tumors of the central nervous system: a summary. Neuro Oncol. 2021;23(8):1231–51.
33. Roux A, Zanello M, Mancusi RL, Still MEH, Nascimento FA, Tauziede-Espariat A, et al. Meningioangiomatosis: multimodal analysis and insights from a systematic review. Neurology. 2021;96(6):274–86.
34. Darling CF, Byrd SE, Reyes-Mugica M, Tomita T, Osborn RE, Radkowski MA, et al. MR of pediatric intracranial meningiomas. AJNR Am J Neuroradiol. 1994;15(3):435–44.
35. Barkovich AJ, Raybaud C. Pediatric neuroimaging. 6th ed. Philadelphia: Wolters Kluwer; 2019. 1259 p
36. Pinto PS, Huisman TAGM, Ahn E, Jordan LC, Burger P, Cohen KJ, et al. Magnetic resonance imaging features of meningiomas in children and young adults: a retrospective analysis. J Neuroradiol. 2012;39(4):218–26.
37. Jhaveri MD, Salzman KL, Osborn AG. Diagnostic imaging. Brain. 4th ed. Philadelphia, PA: Elsevier; 2021.
38. Alruwaili AA, De Jesus O. Meningioma. In: StatPearls. Treasure Island, FL: StatPearls Publishing; 2024. Disponibile su: http://www.ncbi.nlm.nih.gov/books/NBK560538/.
39. Goldbrunner R, Stavrinou P, Jenkinson MD, Sahm F, Mawrin C, Weber DC, et al. EANO guideline on the diagnosis and management of meningiomas. Neuro Oncol. 2021;23(11):1821–34.
40. Jain S, Muzumdar D, Shah A, Goel A. A treatise on pediatric meningiomas: single-center retrospective cohort experience and review of literature. J Pediatr Neurosci. 2020;15(3):238.
41. Håberg AK, Hammer TA, Kvistad KA, Rydland J, Müller TB, Eikenes L, et al. Incidental intracranial findings and their clinical impact; the HUNT MRI study in a general population of 1006 participants between 50-66 Years. PLoS One. 2016;11(3):e0151080.
42. Baumgartner JE, Sorenson JM. Meningioma in the pediatric population. J Neuro Oncol. 1996;29(3):223–8.
43. Kotch C, Brosius SN, De Raedt T, Fisher MJ. Updates in the management of central and peripheral nervous system tumors among patients with neurofibromatosis type 1 and neurofibromatosis type 2. Pediatr Neurosurg. 2023;58(5):267–80.
44. Spennato P, Nicosia G, Quaglietta L, Donofrio V, Mirone G, Di Martino G, et al. Posterior fossa tumors in infants and neonates. Childs Nerv Syst. 2015;31(10):1751–72.
45. Wang HH, Luo CB, Guo WY, Wu HM, Lirng JF, Wong TT, et al. Preoperative embolization of hypervascular pediatric brain tumors: evaluation of technical safety and outcome. Childs Nerv Syst. 2013;29(11):2043–9.
46. Arivazhagan A, Devi BI, Kolluri SVR, Abraham RG, Sampath S, Chandramouli BA. Pediatric intracranial meningiomas—do they differ from their counterparts in adults? Pediatr Neurosurg. 2008;44(1):43–8.
47. Vitulli F, Spennato P, Di Costanzo M, Cinalli MA, Onorini N, Mirone G, et al. Hemostasis in a giant intraventricular tumor using a saline-cooled radiofrequency bipolar coagulator: technical note. Childs Nerv Syst. 2023;39(8):2181–5.
48. Kaba SE, DeMonte F, Bruner JM, Kyritsis AP, Jaeckle KA, Levin V, et al. The treatment of recurrent unresectable and malignant meningiomas with interferon alpha-2B. Neurosurgery. 1997;40(2):271–5.
49. Amirjamshidi A, Mehrazin M, Abbassioun K. Meningiomas of the central nervous system occurring below the age of 17: report of 24 cases not associated with neurofibromatosis and review of literature. Childs Nerv Syst. 2000;16(7):406–15.

Meningiomas in the Elderly

Antonio Santoro, Daniele Armocida,
and Luca D'Angelo

1 Introduction

Meningiomas may occur at any age, but they are more common between the ages of 60 and 70, and the rate of occurrence increases with age [1–3]. Advanced imaging technologies and longer lifespans have led to the identification of numerous intracranial meningiomas (IM) in elderly patients [4, 5]. Elderly patients are typically defined as those over the age of 65 years [6–10].

While the main objective of surgical treatment is to remove as much tumor as possible with minimal harm and mortality, reports in existing literature have shown a wide range of mortality and complication rates in patients over 65 years old who have undergone surgery for IMs [6–9, 11–14]. There is currently no consensus on the best approach to surgical treatment of IMs in elderly patients, and the factors that predict the outcome of surgery for these patients are not well understood [13]. This raises the question of whether surgery, as a first treatment option, is as effective for older patients as it is for younger adults and what the postoperative clinical outcomes [15] will be.

Several authors have analyzed the literature to determine the morbidity and mortality rates of elderly patients who have undergone surgery for IMs, identifying clinical, radiological, biological, and surgical variables in their analysis. Previous studies [9, 12, 16] have attempted to identify and define prognostic factors for surgical outcomes in elderly patients, but the information is limited due to varying study designs and different indications for surgery across neurosurgical centers, leading to incomplete and conflicting data.

2 Literature Review

While surgery is the preferred treatment option for younger patients, it is a subject of ongoing discussions for elderly patients, particularly those with pre-existing health conditions [17]. The mortality rate following meningioma surgery in elderly patients varies widely in the literature, ranging from 0% to 20%, while in younger patients, it ranges from 0% to 6% [18–22].

Recent studies have investigated the risks associated with meningioma surgery in elderly patients; however, most of these studies focused exclusively on elderly patients and did not include control groups of younger patients [23,

A. Santoro · L. D'Angelo (✉)
Human Neurosciences Department, Neurosurgery Division, "Sapienza" University, Rome, Italy
e-mail: lucadangelo_80@hotmal.com

D. Armocida
Neuroscience Department "Rita Levi Montalcini", Neurosurgery Unit, Università degli studi di Torino, Turin, Italy

Experimental Neurosurgery Unit, IRCCS "Neuromed", Pozzilli, Italy

F. Maiuri, M. Del Basso De Caro (eds.), *Meningiomas*,
https://doi.org/10.1007/978-3-031-76680-0_2

24]. Conversely, the literature contains a limited number of series comparing meningioma surgery outcomes in elderly and younger patients, with most of these series containing fewer than 100 subjects [19, 25–29]. Bateman et al. presented the largest series of patients with intracranial meningiomas, which included 2304 elderly and 6557 young patients from a nationwide inpatient sample database [8]; however, this series did not provide information on the neurological status of the patients or the characteristics of each patient's tumor.

A more recent study by Zhao et al. [30], which specifically focused on preoperative status and neurological symptoms, confirmed the tendency of the younger group to have a better outcome than the elderly group. The most complete meta-analysis published in 2021 by Rafiq et al. [31] defined that the elderly group had a higher risk of being in ASA classes 3–5 in comparison to the young group, and the higher mortality rate observed in the elderly is likely secondary to the higher incidence of comorbidities. Preoperative comorbidity is a factor that can be easily evaluated from the scoring systems we routinely use for preoperative risk evaluation, and commonly encountered systemic factors include cardiopulmonary conditions that preclude anesthesia, thrombotic conditions that cause either venous thrombosis or bleeding tendencies, and poor vascular condition [32, 33]. Specifically, Bartek et al. demonstrated that preoperative condition, surgical duration longer than 4 h, and age influenced the outcome, while Simpson grades had less impact [34]. In other studies, Chen et al. [5] concluded that poor preoperative condition and neurological deficits were associated with poorer outcomes, whereas tumor location, size, and Simpson grades were not significantly related to the outcome [9].

The research on the risk of meningiomas in older adults appears to be inconclusive, unlike in other areas of study [17, 21, 35], this inconsistency can be attributed to the numerous variables that can be included or excluded in the analyses of various clinical studies. One of the factors contributing to the conflicting results regarding the risk of meningiomas in elderly patients is the exclusion of patients with high ASA grades from surgery in many centers; however, Nakamura et al. [21] and Li et al. [27] reported no deaths in their elderly and young cohort groups, respectively. It is important to note that Nakamura's study only included 21 patients in the elderly group, and no patients with an ASA score of 4 or higher were included in the analysis [21]. Similarly, Li et al. only investigated patients with low ASA scores (ASA 1 and 2) in their cohort [27]. Moreover, as life expectancy increases, it is difficult to assume that a 65-year-old patient has the same risks as an 80-year-old patient, even though they belong to the same age group. Ahmeti et al. [17], for instance, found that patients over 80 years old had the highest mortality rate (7 patients (17%)). The conflicting results in studies on the risk of meningiomas in older adults may be attributed to the large number of variables that can be included or excluded in the analyses of various clinical studies. For instance, patients with high ASA grades are often excluded from surgery in many centers, which may contribute to the inconsistent findings. Additionally, increased life expectancy makes it difficult to assume that patients belonging to the same age group have the same risks. A study by Ahmeti et al. [17] found that patients over 80 years old had the highest mortality rate.

3 Sapienza Experience Among Meningiomas in the Elderly

Recently, the present authors conducted a multicenter observational study aimed at identifying the risk factors associated with post-surgical outcomes in elderly patients with IM [36]. The unique aspect of this study was the comparison between two groups with an identical American Society of Anesthesiologists (ASA) score, revealing that elderly patients face a greater risk of complications and have a more difficult recovery [5–8, 31, 37]. The reason why surgery in older adults is considered riskier is due to the consideration of all other factors that may influence the outcome. Several other factors, including tumor parameters such as size, location, presence and severity of edema, radiological features, extension, grading [15, 38–40], and operative approach [36], were found to be predictive of outcome. Although operative complications, such as intracranial hematoma, hydrocephalus, and focal neurologic deterioration, are strongly related to surgical strategy and skill, non-operative complications, such as infection, atelectasis, and deep venous thrombosis, are closely linked to patients' overall health and underlying disease [41] and are correlated with age.

Intracranial meningioma resection outcomes in elderly patients did not show a more frequent location, larger tumor size, or higher edema volume compared to younger patients, although their performance statuses were significantly lower [42]. However, when elderly patients have no neurological deficits and a preoperative KPS score >70, their postoperative outcomes are favorable [43]. For patients with low preoperative KPS scores or neurological deficits, additional factors such as preoperative general health condition, tumor characteristics, and patient perspectives should be taken into account when making a decision regarding surgery [44]. Although new scoring systems may provide useful information for determining the optimal treatment for IMs in the elderly [15], surgical outcomes are more influenced by the difficulties encountered during surgery, the technology available in the operating room, and the surgeon's experience [4, 15, 45].

3.1 Dedicated Surgical Variables for Elderly Patients

It is widely accepted that younger and older adults have different ASA classifications and preoperative Karnofsky performance status (KPS), but there is no significant difference in mortality, complication rate, and long-term outcome between them after surgery for IM [15, 46]. Zhao et al. [30] found that operative complications, surgically challenging location, and diameter significantly affected the final outcome. Despite the risk of operating on patients with high prognostic risks due to comorbidities, there may be some surgical factors that could favor surgery for elderly patients [47]. For instance, the presence of some degree of cortical atrophy could facilitate surgical access and a lower risk of preoperative deficits for the same tumor volumes. As people age, it is inevitable that the brain parenchyma will undergo atrophy [48, 49]; some pathological studies have shown widening of the sulcus, particularly in the frontal and parietal lobes [49]. The potential for atrophy to create additional volume is approximately 102 cm^3 more in people over 60 years old compared to those younger than 40 years old [50]. This additional space created by parenchymal atrophy could compensate for the mass effect of the meningioma, thereby delaying the onset of symptoms from the tumor. Although most patients with intracranial meningioma present with tumor-related symptoms, the number of asymptomatic patients in both elderly and younger cohorts ranges from 5% to 35.2% and 2% to 6.5%, respectively [19].

A study demonstrated that a brain with atrophy could provide a protective effect, even in cases of acute mass effect, such as ischemic stroke [51]. In a study conducted by Zhao et al. [30], elderly patients had a significantly larger tumor diameter than the younger group, but there were no significant differences in preoperative symptoms between the two groups, indicating that elderly individuals may be insensitive to symptoms.

Conversely, elderly patients tend to have a higher comorbidity rate than younger adults, and long-standing conditions like hypertension and diabetes can cause the intracranial vessels and brain parenchyma to become more fragile [26]. Additionally, the dura of elderly patients may have more adhesions, making it more susceptible to tearing. The location and size of a tumor can also influence the complexity of surgery.

Surgery is considered the most critical factor influencing postoperative outcomes. In terms of safety, subtotal resection to preserve eloquent neurological and neurovascular structures intact may be an acceptable strategy for elderly patients, given their shorter life expectancy. This approach can alleviate symptoms of intracranial hypertension, prevent hemorrhage from the remnant of the tumor, and suppress the growth of any residual tumor through radiosurgery [52]. Most patients can live for an extended period with a good quality of life, even with some residual tumor, particularly for tumors with a low histopathological grade [53]. The best surgical strategy for elderly patients should be conservative and maintain a safe range.

Additionally, radiosurgical treatments such as Gamma Knife or CyberKnife can be beneficial supplementary treatments for elderly patients with meningiomas [54, 55]. For medium- and small-sized tumors, radiosurgery can provide tumor growth control equivalent to Simpson Grade I resection [56]. Although radiosurgery has disadvantages like functional status decline, edema, and neurologic deficits [57], it may be reasonable to use it in combination with maximal safe tumor debulking in elderly patients if a surgeon cannot achieve Simpson Grade I resection safely in some challenging brain regions [58].

For a long time, the accepted goal of surgery for meningiomas has been Simpson's I resection and preservation of all important neurovascular structures [9, 12, 15]. This is feasible for superficial tumors that are easy to access. However, for tumors in deep and challenging regions, or with complicated relationships with important neurovascular structures, striking the balance between total resection and the protection of function can be challenging.

For younger patients, tumor remnants and higher WHO grades [59] might have a greater impact as they would lead to recurrence of the tumor and a possible need for a second surgery [60]; however, in elderly patients, as the tumor grows slowly in most cases, pathology and subtotal resection are somewhat less likely to affect the final outcome, as the tumor growth time is more likely to exceed the patient's remaining life expectancy [61–63].

For younger patients, it is reasonable to pursue Simpson I resection with the knowledge that since regrowth of any residual is more likely given that the life expectancy is much longer. Conversely, for elderly patients, especially in the common scenario of poor general medical conditions and given their more vulnerable brain parenchyma, subtotal resection with adequate debulking in exchange for increased safety of the involved or adjacent eloquent neurovascular complexes, and with regard to their future quality of life, it is not just acceptable but indeed preferable.

Our study showed that factors related to surgical safety had a significant impact on the outcome, confirming that [15] factors associated with tumor features such as pathology and Simpson grades had less impact in terms of postoperative quality of life (QoL). Currently, the best approach to selecting any treatment for a patient is to carefully judge his general medical and neurological condition.

With radiosurgery as a supplementary therapy, the goal of elderly meningioma surgery should be more firmly focused on controlling symptoms and preserving function. A meningioma patient should be treated as a person who prioritizes a high quality of life rather than simply as a tumor that requires complete resection. In conclusion, as more intrinsic factors associated with genetic features have been revealed [64, 65], novel genetic therapies can be expected for meningiomas and additional supplemental treatments can be effective for residual tumors. However, further studies are needed to validate this strategy.

Despite differences in age, number of preexisting diseases, and ASA score, even in elderly patients, high total resection rates were achieved in 79–88.6% of elderly patients compared to 62.9–90.3% of young patients [36, 66] and were more common in elderly patients (46.5% in group II vs. 31.8% in group I). The three most common postoperative morbidities were bleeding (12.9%), cranial nerve disorder (10%), and CSF fistula (8.1%). Postoperative bleeding, palsy, speech disorder, pneumonia, and renal insufficiency are present more often in elderly patients. Patients >80 years of age had the highest bleeding rate (29.3%), and young patients had the lowest bleeding rate (9.1%). In elderly patients, males were more frequently affected by bleeding than females. In their series, Maurice-Williams and Kitchen [26] reported a bleeding rate of 20% for elderly patients and no bleeding in young patients. Cranial nerve disorder is one of the most common surgery-related complications reported in the literature [27]. Postoperative bleeding, infections, and deep vein thromboses were the three most common complications in elderly patients reported by Boviatsis et al. in a series of 108 elderly patients compared to 240 young patients [63]. Despite the higher rate of postoperative morbidities in elderly patients, many studies have shown improvement in neurological status postoperatively [28, 65]. In their series of 115 elderly patients, Zhao et al. demonstrated a significantly lower postoperative KPS (mean KPS of 79.6) than that in 413 young patients (mean KPS 88.81) [30]. Li et al. [27] showed no differences between elderly and young patients with a KPS of 90 or higher (82.9% and 81.2%, respectively) at a follow-up of 1 year after surgery. Similar results have been reported by Slot et al. [15]. The improvement in neurological conditions, even in the elderly, despite the higher rate of postoperative complications, can be explained by the fact that not all postoperative complications lead to persistent neurological disorders [26, 32].

4 Conclusions

As the population ages, the incidence of intracranial meningiomas (IMs) in the elderly is expected to increase. However, the management of meningiomas in elderly patients remains challenging due to the inconclusive surgical outcomes in this demographic. Despite this, some authors suggest that meningioma surgery can be well-tolerated in elderly patients, and that age alone should not be considered a contraindication for surgery. Instead, the characteristics of the tumor and the patient's ASA class play a more significant role in determining surgical outcomes. Although it is widely accepted that symptomatic IMs should be treated with surgery, older patients are more likely to experience postoperative complications, which can lead to a higher risk of mortality. For symptomatic IMs that are growing, radical surgical resection is typically recommended as the initial treatment; however, for this subgroup of patients, it is important to carefully consider their overall health, morbidities, and mortality rates, as well as their projected quality of life after surgery, before determining the most appropriate treatment.

In our study, we found that there is no inherent correlation between the presence of IM and factors such as biology, volume, edema, or tumor site in elderly patients. Additionally, there is no significant increase in the risk of recurrence for this demographic. Instead, elderly patients face a higher risk of reduced performance status and mortality related to comorbidities, particularly cardiovascular disease, and the inherent frailty of the aging population. Clinical and neurological evaluations, as well as operative risk assessments, play a crucial role in determining whether surgery is appropriate for each individual patient.

References

1. Eseonu C, Vivas-Buitrago T, Quiñones-Hinojosa A. Meningiomas in the elderly. Handb Clin Neurol. 2020;169:261–71.
2. Claus EB, Bondy ML, Schildkraut JM, Wiemels JL, Wrensch M, Black PM. Epidemiology of intracranial meningioma. Neurosurgery. 2005;57:1088–95.
3. Timmer M, Seibl-Leven M, Wittenstein K, Grau S, Stavrinou P, Röhn G, Krischek B, Goldbrunner R. Long-term outcome and health-related quality of life of elderly patients after meningioma surgery. World Neurosurg. 2019;125:e697–710.
4. Awad IA, Kalfas I, Hahn JF, Little JR. Intracranial meningiomas in the aged: surgical outcome in the era of computed tomography. Neurosurgery. 1989;24:557–60.
5. Chen ZY, Zheng CH, Tang Li Su XY, Lu GH, Zhang CY, Xiao SW, Tan YF. Intracranial meningioma surgery in the elderly (over 65 years): prognostic factors and outcome. Acta Neurochir. 2015;157:1549–57, discussion 1557.
6. Soyalp C, Yuzkat N, Kilic M, Akyol ME, Demir CY, Gulhas N. Operative and prognostic parameters associated with elective versus emergency surgery in a retrospective cohort of elderly patients. Aging Clin Exp Res. 2019;31:403–10.
7. Konglund A, Rogne SG, Lund-Johansen M, Scheie D, Helseth E, Meling TR. Outcome following surgery for intracranial meningiomas in the aging. Acta Neurol Scand. 2013;127:161–9.

8. Bateman BT, Pile-Spellman J, Gutin PH, Berman MF. Meningioma resection in the elderly: nationwide inpatient sample, 1998–2002. Neurosurgery. 2005;57:866–72.
9. Eksi MS, Canbolat Ç, Akbas A, Özmen BB, Akpınar E, Usseli MI, Güngör A, Güdük M, Hacıhanefioglu M, Ersen Danyeli A, et al. Elderly patients with intracranial meningioma: surgical considerations in 228 patients with a comprehensive analysis of the literature. World Neurosurg. 2019;132:e350–65.
10. World Health Organization. 10 facts on ageing and the life course. 2012. http://www.who.int/features/factfiles/ageing/ageing_facts/en/index.html. Accessed 2 Feb 2014.
11. Knuf KM, Maani CV, Cummings AK. Clinical agreement in the American Society of Anesthesiologists physical status classification. Perioper Med. 2018;7:14.
12. Bir SC, Konar S, Maiti TK, Guthikonda B, Nanda A. Surgical outcomes and predictors of recurrence in elderly patients with meningiomas. World Neurosurg. 2016;90:251–61.
13. Schul DB, Wolf S, Krammer MJ, Landscheidt JF, Tomasino A, Lumenta CB. Meningioma surgery in the elderly: outcome and validation of 2 proposed grading score systems. Neurosurgery. 2012;70:555–65.
14. Cornu P, Chatellier G, Dagreou F, Clemenceau S, Foncin J-F, Rivierez M, Philippon J. Intracranial meningiomas in elderly patients. Acta Neurochir. 1990;102:98–102.
15. Slot KM, Peters JVM, Vandertop WP, Verbaan D, Peerdeman SM. Meningioma surgery in younger and older adults: patient profile and surgical outcomes. Eur Geriatr Med. 2017;9:95–101.
16. Armocida D, Pesce A, Frati A, Santoro A, Salvati M. EGFR amplification is a real independent prognostic impact factor between young adults and adults over 45yo with wild-type glioblastoma? J Neurooncol. 2019;146:275–84.
17. Ahmeti H, Borzikowsky C, Hollander D, Röcken C, Jansen O, Synowitz M, Mehdorn MH. Risks and neurological benefits of meningioma surgery in elderly patients compared to young patients. J Neurooncol. 2021;154:335–44.
18. Arienta C, Caroli M, Balbi S. Intracranial meningiomas in patients over 70 years old. Follow-up in operated and unoperated cases. Aging (Milano). 1992;4:29–33.
19. Black P, Kathiresan S, Chung W. Meningioma surgery in the elderly: a case–control study assessing morbidity and mortality. Acta Neurochir. 1998;140:1013–7.
20. Buhl R, Hasan A, Behnke A, Mehdorn HM. Results in the operative treatment of elderly patients with intracranial meningioma. Neurosurg Rev. 2000;23:25–9.
21. Nakamura M, Roser F, Dormiani M, et al. Surgical treatment of cerebellopontine angle meningiomas in elderly patients. Acta Neurochir. 2005;147:603–10.
22. Patil C, Veeravagu A, Lad S, Boakye M. Craniotomy for resection of meningioma in the elderly: a multicentre, prospective analysis from the national surgical quality improvement program. J Neurol Neurosurg Psychiatry. 2010;81:502–5.
23. Caroli M, Locatelli F, Prada F, et al. Surgery for intracranial meningiomas in the elderly: a clinical-radiological grading system as a predictor of outcome. J Neurosurg. 2005;102:290–4.
24. Sacko O, Sesay M, Roux FE, et al. Intracranial meningioma surgery in the ninth decade of life. Neurosurgery. 2007;61:950–5.
25. Roser F, Ebner F, Ritz R, Samii M, Tatagiba M, Nakamura M. Management of skull based meningiomas in the elderly patient. J Clin Neurosci. 2007;14:224–8.
26. Maurice-Williams RS, Kitchen ND. Intracranial tumours in the elderly: the effect of age on the outcome of first time surgery for meningiomas. Br J Neurosurg. 1992;6:131–7.
27. Li Y, Lu D, Feng D, et al. Management of incidental anterior skull base large and giant meningiomas in elderly patients. J Neurooncol. 2020;148:481–8.
28. Guo S, Gerganov V, Giordano M, et al. Elderly patients with frontobasal and suprasellar meningiomas: safety and efficacy of tumor removal via frontolateral approach. World Neurosurg. 2020;135:452–8.
29. Amano T, Nakamizo A, Michiwaki Y, et al. Surgical outcome in elderly patients with intracranial meningioma. J Clin Neurosci. 2018;56:63–6.
30. Zhao X, Zhao D, Wu Y, Gao W, Cui H, Wang Y, Nakaji P, Bao Y. Meningioma in the elderly: characteristics, prognostic factors, and surgical strategy. J Clin Neurosci. 2018;56:143–9.
31. Rafiq R, Katiyar V, Garg K, Kasliwal M, Chandra PS, Kale SS. Comparison of outcomes of surgery for intracranial meningioma in elderly and young patients—a systematic review and meta-analysis. Clin Neurol Neurosurg. 2021;207:106772.
32. Poon MT-C, Fung LH-K, Pu JK-S, et al. Outcome of elderly patients undergoing intracranial meningioma resection—a systematic review and meta-analysis. Br J Neurosurg. 2014;28:303–9.
33. Poon MT-C, Fung LH-K, Pu JK-S, et al. Outcome comparison between younger and older patients undergoing intracranial meningioma resections. J Neuro Oncol. 2013;114:219–27.
34. Bartek J, Sjåvik K, Förander P, et al. Predictors of severe complications in intracranial meningioma surgery: a population-based multicenter study. World Neurosurg. 2015;83:673–8.
35. Brokinkel B, Holling M, Spille DC, et al. Surgery for meningioma in the elderly and long-term survival: comparison with an age- and sex-matched general population and with younger patients. J Neurosurg. 2017;126:1201–11.

36. Armocida D, Arcidiacono UA, Palmieri M, Pesce A, Cofano F, Picotti V, Salvati M, D'Andrea G, Garbossa D, Santoro A, Frati A. Intracranial meningioma in elderly patients. Retrospective multicentric risk and surgical factors study of morbidity and mortality. Diagnostics (Basel). 2022;12(2):351. https://doi.org/10.3390/diagnostics12020351.
37. Meling TR, Da Broi M, Scheie D, Helseth E. Skull base versus non-skull base meningioma surgery in the elderly. Neurosurg Rev. 2019;42(4):961–72. https://doi.org/10.1007/s10143-018-1005-6. Epub 2018 Jul 5.
38. Armocida D, Frati A, Salvati M, Santoro A, Pesce A. Is Ki-67 index overexpression in IDH wild type glioblastoma a predictor of shorter progression free survival? A clinical and molecular analytic investigation. Clin Neurol Neurosurg. 2020;198:106126.
39. Mandel M, Li Y, Figueiredo EG, Teixeira MJ, Steinberg GK. Presurgical planning with open-source Horos software for superficial brain arteriovenous malformations. World Neurosurg. 2021;157:3–12.
40. D'Andrea G, Roperto R, Caroli E, Crispo F, Ferrante L. Thirty-seven cases of intracranial meningiomas in the ninth decade of life: our experience and review of the literature. Neurosurgery. 2005;56:956–61.
41. Palmieri M, Armocida D, De Pietro R, Chiarello G, Rizzo F, Garbossa D, Marampon F, Santoro A, Frati A. Clinical efficacy of adjuvant radiotherapy for World Health Organization grade II intracranial meningioma. World Neurosurg. 2023;175:e1117–23. https://doi.org/10.1016/j.wneu.2023.04.075.
42. Frati A, Armocida D, Arcidiacono UA, Pesce A, D'Andrea G, Cofano F, Garbossa D, Santoro A. Peritumoral brain edema in relation to tumor size is a variable that influences the risk of recurrence in intracranial meningiomas. Tomography. 2022;8(4):1987–96. https://doi.org/10.3390/tomography8040166.
43. GBD. Italy collaborators. Italy's health performance, 1990–2017: findings from the global burden of disease study 2017. Lancet Public Health. 2017;2019(4):e645–57.
44. Petridis AK, Doukas A, Mahvash M, et al. A case of rapid-growing anaplastic meningiomas. BMJ Case Rep. 2009;2009:bcr0520091862. https://doi.org/10.1136/bcr.05.2009.1862.
45. Yamamoto J, Takahashi M, Idei M, Nakano Y, Soejima Y, Akiba D, Kitagawa T, Ueta K, Miyaoka R, Nishizawa S. Clinical features and surgical management of intracranial meningiomas in the elderly. Oncol Lett. 2017;14:909–17.
46. Czernicki T. Surgical management of intracranial meningiomas in the elderly: early and long-term outcomes. Clin Interv Aging. 2020;15:2439–51.
47. May M, Sedlak V, Pecen L, Priban V, Buchvald P, Fiedler J, Vaverka M, Lipina R, Reguli S, Malik J, Netuka D, Benes V. Role of risk factors, scoring systems, and prognostic models in predicting the functional outcome in meningioma surgery: multicentric study of 552 skull base meningiomas. Neurosurg Rev. 2023;46(1):124. https://doi.org/10.1007/s10143-023-02004-5. PMID: 37219634; PMCID: PMC10205827.
48. de la Monte SM. Quantitation of cerebral atrophy in preclinical and end-stage Alzheimer's disease. Ann Neurol. 1989;25:450–9.
49. Tomlinson BE, Blessed G, Roth M. Observations on the brains of non-demented old people. J Neurol Sci. 1968;7:331–56.
50. Atak Tel BM, Aktas G, Bilgin S, Baltaci SB, Taslamacioglu DT. Control level of type 2 diabetes mellitus in the elderly is associated with polypharmacy, accompanied comorbidities, and various increased risks according to the beers criteria. Diagnostics (Basel). 2023;13(22):3433. https://doi.org/10.3390/diagnostics13223433.
51. Lee SH, Oh CW, Han JH, et al. The effect of brain atrophy on outcome after a large cerebral infarction. J Neurol Neurosurg Psychiatry. 2010;81:1316–21.
52. Zhao L, Zhao W, Hou Y, Wen C, Wang J, Wu P, Guo Z. An overview of managements in meningiomas. Front Oncol. 2020;10:1523. https://doi.org/10.3389/fonc.2020.01523. Erratum in: Front Oncol 2020 Sep 24;10:599431. doi: https://doi.org/10.3389/fonc.2020.599431.
53. Armocida D, Marzetti F, Pesce A, Caporlingua A, D'Angelo L, Santoro A. Purely meningeal intracranial relapse of melanoma brain metastases after surgical resection and immunotherapy as a unique disease progression pattern: our experience and review of the literature. World Neurosurg. 2019;134:150–4.
54. Kaul D, Budach V, Graaf L, et al. Outcome of elderly patients with meningioma after image-guided stereotactic radiotherapy: a study of 100 cases. Biomed Res Int. 2015;2015:868401.
55. Kondziolka D, Lunsford LD, Coffey RJ, et al. Stereotactic radiosurgery of meningiomas. J Neurosurg. 1991;74:552–9.
56. Pollock BE, Stafford SL, Utter A, et al. Stereotactic radiosurgery provides equivalent tumor control to Simpson Grade 1 resection for patients with small-to medium-size meningiomas. Int J Radiat Oncol Biol Phys. 2003;55:1000–5.
57. Stafford SL, Pollock BE, Foote RL, et al. Meningioma radiosurgery: tumor control, outcomes, and complications among 190 consecutive patients. Neurosurgery. 2001;49:1029–38.
58. Chen WC, Lucas CG, Magill ST, Rogers CL, Raleigh DR. Radiotherapy and radiosurgery for meningiomas. Neurooncol Adv. 2023;5(Suppl 1):i67–83. https://doi.org/10.1093/noajnl/vdac088.
59. Wen PY, Packer RJ. The 2021 WHO classification of tumors of the central nervous system: clinical implications. Neuro Oncol. 2021;23:1215–7.

60. Louis DN, Perry A, Reifenberger G, et al. The 2016 World Health Organization Classification of tumors of the central nervous system: a summary. Acta Neuropathol. 2016;131:803–20.
61. Saklad M. Grading of patients for surgical procedures. Anesthesiology. 1941;2:281–4.
62. Karnofsky DA, Abelmann WH, Craver LF, Burchenal JH. The use of the nitrogen mustards in the palliative treatment of carcinoma. With particular reference to bronchogenic carcinoma. Cancer. 1948;1:634–56.
63. Boviatsis E, Bouras T, Kouyialis A, et al. Impact of age on complications and outcome in meningioma surgery. Surg Neurol. 2007;68:407–11.
64. Mei Y, Bi WL, Greenwald NF, et al. Genomic profile of human meningioma cell lines. PLoS One. 2017;12:e0178322.
65. Byvalsev V, Stepanov I, Belykh E, et al. Molecular biology of brain meningiomas. Patol Fiziol Eksp Ter. 2017;61:82–91.
66. Konglund A, Rogne SG, Helseth E, et al. Meningioma surgery in the very old—validating prognostic scoring systems. Acta Neurochir. 2013;155:2263–71.

Meningiomas and Hormone-Related Clinical Conditions

Marialaura Del Basso De Caro,
Rosa Maria Di Crescenzo, and Federico Russo

1 Introduction

Sex hormones seem to play a role in the development of meningiomas. About one century ago, Cushing and Eisenhardt [1] first described a case of rapid symptom progression of a meningioma during pregnancy. The hormone influence on meningiomas is suggested by many clinical and pathological data, such as higher incidence among women [2], increased growth during pregnancy and shrinkage after delivery [3–6], association with tumors of the female system and increased risk in patients with breast cancer [7, 8], presence of progesterone receptor (PR) expression in tumor histological specimens [9, 10], and "in vitro" proliferation of tumor cell lines in culture after exposure to estrogen and progesterone [11, 12].

This chapter discusses the clinical and therapeutic implications correlated with the sex hormone influence on meningiomas, including pre- and post-menopause status, pregnancy, fertility and oral contraceptive therapies, post-menopausal hormone replacement therapies, and risk of meningiomas in female hormone-related tumors.

2 Hormone Receptors in Meningiomas and Their Clinical Significance

The presence of estrogen receptor (ER) and PR in meningiomas was described more than 40 years ago [13–16] and confirmed by many other studies [9, 10, 17–25]. The PR is expressed from 72% to more than 90% of meningiomas [9, 22], whereas the rate of ER positivity ranges from 11% to 30% [26]; androgen receptors (AR) are expressed in about 45% of the histological specimens.

The rate of PR positivity is slightly higher in females versus males, in premenopausal versus postmenopausal women [27] and in meningiomas occurring during pregnancy [9, 10, 27], but with no significance; conversely, the results are less clear for ER [10]. Besides, the expression of PR in meningiomas tends to decrease with age [10, 28].

According to the pathological data, a recent literature review [10] confirms the high rate of PR expression of WHO grade I meningiomas as

M. Del Basso De Caro (✉) · R. M. Di Crescenzo
Department of Advanced Biomedical Sciences, Section of Pathology, University of Naples "Federico II", Naples, Italy
e-mail: delbasso@unina.it; marialaura.delbasso@unina.it

F. Russo
Division of Neurosurgery, Department of Neurosciences, Reproductive and Odontostomatological Sciences, University of Naples "Federico II", Naples, Italy

F. Maiuri, M. Del Basso De Caro (eds.), *Meningiomas*,
https://doi.org/10.1007/978-3-031-76680-0_3

compared to atypical and particularly the anaplastic ones; however, no significant differences of estrogen and androgen receptors are evidenced according to the WHO grade. Atypical WHO grade II meningiomas express significantly lower values of PR; besides, an inverse correlation between values of PR expression and Ki 67 MIB 1 was evidenced [9]. The differences of PR between the histological subtypes of WHO grade I meningiomas are not significant.

According to the location, most studies did not find significant differences of PR expression among the different locations (cranial versus spinal, skull base versus not skull base) [19, 20, 23]. However, two recent studies [9, 29] have shown significantly higher rates of cases with higher values of PR expression in medial skull base and spinal as compared to lateral skull base and no skull base meningiomas.

The correlation between PR expression and meningioma recurrence has been denied by some studies [30–33], whereas other studies [9, 19, 21, 34–36] have found significant inverse correlation, with high recurrence rates in meningiomas with low PR expression at initial surgery. The recurrence time seems to be also correlated [9].

Finally, meningiomas at initial diagnosis more frequently express the PR than the recurrent meningiomas.

In conclusion, the data on the PR expression are more numerous and strongest, whereas the data on ER and AR are more dubious and require further studies. Thus, the immunohistochemical evaluation of the PR expression must be included in the routine histological study of meningiomas.

3 Pregnancy

The occurrence of a meningioma during pregnancy is a well-known event, which has been described in single case reports and small case series [4, 5, 27, 37].

However, its evidence is very low; in a large population-based study [38], only 11 patients among more than 2,353,000 pregnancies had a meningioma diagnosed during the pregnancy. Neglecting the mild symptoms of most slowly growing meningiomas and avoidance of radiological examinations in this period may account for the low incidence. Besides, the meningiomas are more frequent among nulliparous women as compared to parous women [38]. Thus, there is no increased risk of occurrence of a meningioma during pregnancy or 1-year postpartum.

The onset of neurological symptoms of a known or a still undiagnosed meningioma more often occurs during the third (49%) or second trimester (30%), less frequently in the first trimester (4%), or at delivery (17%) [5]. This reflects the effect of endocrine and vascular changes of pregnancy on the meningioma growth.

Pregnancy-related meningiomas, when compared to those of the general population, are mainly supratentorial (95%); medial skull base tumors (olfactory groove, tuberculum sellae, anterior clinoid) are significantly more frequent (68%) than nonskull base and infratentorial ones [5, 39]. They are more often large (median diameter about 4 cm, up to 40% >5 cm). Vision-related symptoms (58%) at presentation are more frequent than not vision-related ones. Sudden and severe symptoms (mental changes, imminent blindness, progressive paresis) are observed in about 20% of the patients. From the histological view point, most meningiomas are WHO grade I (75–80%), with mean Ki 67 Li <4% and PR expression >80%, suggesting a more favorable clinical outcome.

The occurrence of symptoms and the often-reported rapid clinical deterioration during the second and third trimester may depend on several endocrine and vascular changes. Endocrine mechanisms include increase in progesterone, human placental lactogen, and prolactin serum levels, whereas pituitary hormones are decreased prior to childbirth [40]. Vascular mechanisms include increase in blood supply to the meningioma, peritumoral brain edema often evidenced at magnetic resonance imaging (MRI), and expression of vascular endothelial growth factor (VEGF) evidenced on the surgical specimens. It is likely that both endocrine and vascular changes play a crucial role in the clinical expression. However, this needs to be better defined in further studies. The same changes may also explain the reported cases of meningiomas shrinkage after delivery without surgery, causing spontaneous relief of symptoms [6, 41].

The best management of meningiomas occurring during pregnancy is not well defined and there are not specific guidelines, due to their rarity. The choice depends upon the trimester of pregnancy and the entity of meningioma symptoms and signs. In general, the treatment of the meningioma should be delayed up to the delivery if symptoms are slight and there are no deficits. If severe neurological deficits occur, mainly sudden visual loss, the meningioma must be resected during pregnancy [42]. Although the emergency craniotomy may be indicated regardless of the gestational age, it should be avoided during the first trimester and realized during the second trimester, when the spontaneous abortion is less likely [43]. If neurological symptoms occur during the third trimester, the preterm delivery may be first realized, followed by elective postpartum craniotomy. Although the craniotomy during pregnancy carries risks for both mater and fetus, the overall neurological, obstetrical, and neonatological outcomes are not worse if the neurosurgical operation is limited to selected cases.

4 Sex Hormone Therapies

Sex hormone therapies are currently used in many clinical conditions and for many indications, including contraception, fertility, hormone replacement in postmenopausal women, and management of sex-related tumors [44]. The potential risk of occurrence of a meningioma or progression of a known meningioma with hormonal therapies is largely discussed, but the results are still controversial.

4.1 Progesterone Agonists

Progesterone agonists are implicated in the meningioma growth. The cyproterone acetate (CPA) is often used for contraception, hirsutism, transgender male-to-female patients, and palliative treatment of prostatic cancer. Its use has been found to be correlated with the behavior of meningiomas, with tumor growth during the treatment and stabilization or regression on treatment withdrawal [45–53]. This effect is mainly evident for high doses of CPA and long-term exposure (>1 year) [54]. CPA-associated meningiomas, when compared with those of the general population, are diagnosed at younger age, are more frequently located in anterior and medial skull base, are more often multiple, and show strong PR expression [22, 51, 52, 55, 56].

Data on other progesterone agonists, such as chlormadinone acetate and nomegestrol acetate, and meningioma growth are scarce and less conclusive [51, 57–59]. However, the risk remains lower than that seen with CPA.

In conclusion, the use of progesterone agonists should be avoided in patients with a known meningioma.

4.2 Oral Contraceptives

The use of oral contraceptives is widely diffuse in young women. The risk of meningioma correlated with contraceptive use has been discussed in several studies over the past 20 years [12, 44, 48, 60–68] and in two literature reviews [44, 69]. These provide controversial results, depending on several different factors, such as type and dose of contraceptive drugs, current or past use, and duration of the treatment.

Studies including patients using progesterone-only contraceptives [63, 64, 68] have shown an increased risk of meningioma in those taking therapy for more than 5 years [67] and in those with PR positive meningiomas [64] and increase in recurrence and decrease in progression-free survival [68]. Studies including patients who used estrogen-only or estrogen–progesterone contraceptives report increased risk of meningioma occurrence only after use of more than 5 years [61, 62, 67].

The relationship between oral contraceptives and hormone status of meningiomas is controversial, with studies reporting no statistically significant differences [27, 68] and another showing high rate of meningiomas with low PR expression during contraceptive therapy [61].

In conclusion, it is advisable to avoid contraceptive therapy, particularly with progesterone, at high doses and for long-term administration, in patients with a known meningioma.

4.3 Hormone Fertility Therapies

The fertility treatments include a variety of methods: pharmacological ovarian stimulation, intrauterine insemination, "in vitro" fertilization, and injection of human chorionic gonadotropin.

The relationship between fertility treatments and meningiomas has scarcely been discussed. The study of Korhonen et al. [30] did not show increased risk of meningiomas after fertility treatments. In another review [70], the group of female patients with meningiomas and history of fertility treatments were found to have a significantly younger age and a high rate of multiple non-skull base meningiomas as compared to the group with no treatment. Single case reports also suggest the influence of these treatments on the meningioma growth [71–73]. However, this relationship must be confirmed by further studies.

4.4 Hormone Replacement Therapies

The association between the use of hormone replacement therapies and the risk of meningioma remains a debated topic, with the available data from reported studies still being inconclusive.

The risk of meningioma occurrence in postmenopausal women who are current users of hormone replacement therapies was first suggested by Jhawar et al. [60]. Other studies have confirmed this risk in both current and former users [63, 74, 75] and in women treated with estrogen as compared to those treated by combined estrogen–progesterone therapies [76, 77]. A large case control study [78] and some meta-analyses [67, 79, 80] report increased risk of meningiomas for prior use of hormone-related therapies, multiple therapies, estrogen therapies, and longer exposure to drugs. Conversely, other studies failed to show an increase in risk in postmenopausal women on hormone replacement therapies [12, 48, 61, 62, 66, 81].

Dresser et al. [82] have also demonstrated that hormone replacement therapies with estrogen do not influence the growth rate of meningiomas.

5 Hormone-Related Extraneural Tumors

Meningiomas sometimes occur in patients with personal history of hormone-related conditions and tumors, including breast cancer, ovarian cancer, endometrial and cervical cancers, uterine fibromyoma, endometriosis, and prostatic cancer. However, the risk of meningioma development in these patients is different and not still well defined.

5.1 Breast Cancer

The association between meningioma and breast cancer is the most frequent and more extensively studied since the first observations [83, 84]. More recent studies conducted using cancer registries have reported divergent results, including an increased risk of meningioma after breast cancer diagnosis [8], no higher risk [85], and a similar incidence compared to the normal population [86]. A recent systematic review with meta-analysis [87] including 18 studies found that female patients with meningioma had a tenfold higher incidence of breast cancer compared to the baseline prevalence of breast cancer in the general population. Conversely, in female patients with breast cancer, the risk of developing a meningioma is 1.42 folds higher compared to the general population, with no statistical significance. This may also depend on the shorter life expectancy and survival rates of patients with breast cancer.

The association between meningioma and breast cancer is supported by several findings, including hormone receptor expression in both diseases [9, 88], the role of hormones in their genesis and progression [44, 55], and an increased risk of both tumors after hormone therapies, mainly synthetic progestogens [55, 89, 90]. In case of occurrence of meningiomas and brain metastases from breast cancer (0.52) [91], an overactivation of the MYC oncogene has been suggested [92, 93].

5.2 Tumors of Female Reproductive System and Endometriosis

The uterine myoma is the most common benign tumor in females, mainly occurring during the reproductive years; hormone factors are implicated in the pathogenesis development and growth.

The association between meningioma and uterine myoma has been discussed in several studies [66, 94, 95]. In a cooperative study, Claus et al. [94] found that meningioma patients were more likely to develop an uterine myoma than control subjects. In the report by Johnson et al. [66], a history of uterine myoma was positively associated with meningioma risk in a multivariate analysis. In the study by Yen et al. [95], the risk of developing meningioma was significantly higher among women with uterine myoma than in the control group without uterine myoma; this difference was more significant in the middle-age range. The risk did not show significant change after myomectomy. These data reflect the influence of sex hormones in both uterine myoma and meningioma.

Female patients with meningioma seem to have a history of endometriosis significantly more frequently than female patients without meningioma [94]. Conversely, the risk of ovarian cancer and endometrial and cervical cancers is still not defined.

6 Conclusions

The risk of developing meningioma is higher in female patients undergoing sex hormone therapies, but it does not influence medical conduct. More interestingly, female patients with incidental, known, or recurrent meningioma should be informed that pregnancy, oral contraceptives, hormonal fertility treatments, and hormone replacement therapies may promote more rapid growth of the meningiomas. In these patients, long-term, high dose hormonal therapies should be avoided.

References

1. Cushing H, Eisenhardt L. Meningiomas arising from the tuberculum sellae. Arch Ophthalmol. 1929;1(2):168–206.
2. Wiemels J, Wrensch M, Claus EB. Epidemiology and etiology of meningioma. J Neurooncol. 2010;99(3):307–14. https://doi.org/10.1007/s11060-010-0386-3.
3. Wahab M, Al-Azzawi F. Meningioma and hormonal influences. Climacteric. 2003;6(4):285–92.
4. Lusis EA, Scheithauer BW, Yachnis AT, Fischer BR, Chicoine MR, Paulus W, et al. Meningiomas in pregnancy: a clinicopathologic study of 17 cases. Neurosurgery. 2012;71(5):951–61. https://doi.org/10.1227/NEU.0b013e31826adf65.
5. Laviv Y, Bayoumi A, Mahadevan A, Young B, Boone M, Kasper EM. Meningiomas in pregnancy: timing of surgery and clinical outcomes as observed in 104 cases and establishment of a best management strategy. Acta Neurochir. 2018;160(8):1521–9. https://doi.org/10.1007/s00701-017-3146-8.
6. Kerschbaumer J, Freyschlag CF, Stockhammer G, Taucher S, Maier H, Thomé C, et al. Hormone-dependent shrinkage of a sphenoid wing meningioma after pregnancy: case report. J Neurosurg. 2016;124(1):137–40. https://doi.org/10.3171/2014.12.JNS142112.
7. Custer BS, Koepsell TD, Mueller BA. The association between breast carcinoma and meningioma in women. Cancer. 2002;94(6):1626–35. https://doi.org/10.1002/cncr.10410.
8. Lopez-Rivera V, Zhu P, Dono A, Lee S, Chen PR, Ballester LY, et al. Increased risk of subsequent meningioma among women with malignant breast cancer. World Neurosurg. 2020;139:e271–85. https://doi.org/10.1016/j.wneu.2020.03.203.
9. Maiuri F, Mariniello G, de Divitiis O, Esposito F, Guadagno E, Teodonno G, et al. Progesterone receptor expression in meningiomas: pathological and prognostic implications. Front Oncol. 2021;11:611218. https://doi.org/10.3389/fonc.2021.611218.
10. Agopiantz M, Carnot M, Denis C, Martin E, Gauchotte G. Hormone receptor expression in meningiomas: a systematic review. Cancers (Basel). 2023;15(3):980. https://doi.org/10.3390/cancers15030980.
11. Maiuri F, Montagnani S, Gallicchio B, Carandente M, Lanza GG, D'Andrea F. Oestrogen and progesterone

sensitivity in cultured meningioma cells. Neurol Res. 1989;11(1):9–13. https://doi.org/10.1080/01616412.1989.11739853.

12. Claus EB, Calvocoressi L, Bondy ML, Wrensch M, Wiemels JL, Schildkraut JM. Exogenous hormone use, reproductive factors, and risk of intracranial meningioma in females. J Neurosurg. 2013;118(3):649–56. https://doi.org/10.3171/2012.9.JNS12811.
13. Donnell MS, Meyer GA, Donegan WL. Estrogen-receptor protein in intracranial meningiomas. J Neurosurg. 1979;50(4):499–502. https://doi.org/10.3171/jns.1979.50.4.0499.
14. Poisson M, Magdelenat H, Foncin JF, Bleibel JM, Philippon J, Pertuiset B, et al. [Estrogen and progestin receptors in meningiomas: a study in 22 cases (author's transl)]. Rev Neurol (Paris). 1980;136(3):193–203.
15. Martuza RL, MacLaughlin DT, Ojemann RG. Specific estradiol binding in schwannomas, meningiomas, and neurofibromas. Neurosurgery. 1981;9(6):665–71. https://doi.org/10.1227/00006123-198112000-00009.
16. Magdelenat H, Pertuiset BF, Poisson M, Martin PM, Philippon J, Pertuiset B, et al. Progestin and oestrogen receptors in meningiomas. Biochemical characterization, clinical and pathological correlations in 42 cases. Acta Neurochir. 1982;64(3–4):199–213. https://doi.org/10.1007/BF01406053.
17. Carroll RS, Zhang J, Dashner K, Black PM. Progesterone and glucocorticoid receptor activation in meningiomas. Neurosurgery. 1995;37(1):92–7. https://doi.org/10.1227/00006123-199507000-00014.
18. Hsu DW, Efird JT, Hedley-Whyte ET. Progesterone and estrogen receptors in meningiomas: prognostic considerations. J Neurosurg. 1997;86(1):113–20. https://doi.org/10.3171/jns.1997.86.1.0113.
19. Roser F, Nakamura M, Bellinzona M, Rosahl SK, Ostertag H, Samii M. The prognostic value of progesterone receptor status in meningiomas. J Clin Pathol. 2004;57(10):1033–7. https://doi.org/10.1136/jcp.2004.018333.
20. Wolfsberger S, Doostkam S, Boecher-Schwarz HG, Roessler K, van Trotsenburg M, Hainfellner JA, et al. Progesterone-receptor index in meningiomas: correlation with clinico-pathological parameters and review of the literature. Neurosurg Rev. 2004;27(4):238–45. https://doi.org/10.1007/s10143-004-0340-y.
21. Pravdenkova S, Al-Mefty O, Sawyer J, Husain M. Progesterone and estrogen receptors: opposing prognostic indicators in meningiomas. J Neurosurg. 2006;105(2):163–73. https://doi.org/10.3171/jns.2006.105.2.163.
22. Portet S, Banor T, Bousquet J, Simonneau A, Flores M, Ingrand P, et al. New insights into expression of hormonal receptors by meningiomas. World Neurosurg. 2020;140:e87–96. https://doi.org/10.1016/j.wneu.2020.04.168.
23. Telugu RB, Chowhan AK, Rukmangadha N, Patnayak R, Phaneendra BV, Mowliswara Prasad BC, et al. Estrogen and progesterone receptor in meningiomas: an immunohistochemical analysis. J Cancer Res Ther. 2020;16(6):1482–7. https://doi.org/10.4103/jcrt.JCRT_1075_16.
24. Mnango L, Mwakimonga A, Ngaiza AI, Yahaya JJ, Vuhahula E, Mwakigonja AR. Expression of progesterone receptor and its association with clinicopathological characteristics in meningiomas: a cross-sectional study. World Neurosurg X. 2021;12:100111. https://doi.org/10.1016/j.wnsx.2021.100111.
25. Maiuri F, Montagnani S, Gallicchio B. Estrogen and progesterone receptors in meningiomas. Surg Neurol. 1986;26(5):435–40. https://doi.org/10.1016/0090-3019(86)90255-7.
26. Hua L, Zhu H, Li J, Tang H, Kuang D, Wang Y, et al. Prognostic value of estrogen receptor in WHO Grade III meningioma: a long-term follow-up study from a single institution. J Neurosurg. 2018;128(6):1698–706. https://doi.org/10.3171/2017.2.JNS162566.
27. Maiuri F, Mariniello G, Somma T, Guadagno E, Corvino S, Pagano S, et al. Meningiomas in premenopausal women: role of the hormone related conditions. Front Oncol. 2020;10:556701. https://doi.org/10.3389/fonc.2020.556701.
28. Maiuri F, Corvino S, Lorenzetti M, Franca RA, Esposito F, Del Basso De Caro M. Intracranial meningiomas in patients aged ≥80 years: pathological features and surgical problems. World Neurosurg. 2023;173:e498–508. https://doi.org/10.1016/j.wneu.2023.02.086.
29. Maiuri F, Mariniello G, Guadagno E, Barbato M, Corvino S, Del Basso De Caro M. WHO grade, proliferation index, and progesterone receptor expression are different according to the location of meningioma. Acta Neurochir. 2019;161(12):2553–61. https://doi.org/10.1007/s00701-019-04084-z.
30. Korhonen K, Salminen T, Raitanen J, Auvinen A, Isola J, Haapasalo H. Female predominance in meningiomas cannot be explained by differences in progesterone, estrogen, or androgen receptor expression. J Neurooncol. 2006;80(1):1–7. https://doi.org/10.1007/s11060-006-9146-9.
31. Guevara P, Escobar-Arriaga E, Saavedra-Perez D, Martinez-Rumayor A, Flores-Estrada D, Rembao D, et al. Angiogenesis and expression of estrogen and progesterone receptors as predictive factors for recurrence of meningioma. J Neurooncol. 2010;98(3):379–84. https://doi.org/10.1007/s11060-009-0086-z.
32. Iplikcioglu AC, Hatiboglu MA, Ozek E, Ozcan D. Is progesteron receptor status really a prognostic factor for intracranial meningiomas? Clin Neurol Neurosurg. 2014;124:119–22. https://doi.org/10.1016/j.clineuro.2014.06.015.
33. de Carvalho GTC, da Silva-Martins WC, de Magalhães KCSF, Nunes CB, Soares AN, Tafuri LSA, et al. Recurrence/regrowth in grade I meningioma: how to predict? Front Oncol. 2020;10:1144. https://doi.org/10.3389/fonc.2020.01144.
34. Strik HM, Strobelt I, Pietsch-Breitfeld B, Iglesias-Rozas JR, Will B, Meyermann R. The impact of

progesterone receptor expression on relapse in the long-term clinical course of 93 benign meningiomas. In Vivo. 2002;16(4):265–70.
35. Metellus P, Nanni I, Dussert C, Trinkhaus M, Fuentes S, Chinot O, et al. [Prognostic implications of biologic markers in intracranial meningiomas: 120 cases]. Neurochirurgie. 2008;54(6):750–6. https://doi.org/10.1016/j.neuchi.2008.07.005.
36. Franca RA, Della Monica R, Corvino S, Chiariotti L, Del Basso De Caro M. WHO grade and pathological markers of meningiomas: clinical and prognostic role. Pathol Res Pract. 2023;243:154340. https://doi.org/10.1016/j.prp.2023.154340.
37. Carbone L, Somma T, Iorio GG, Vitulli F, Conforti A, Raffone A, et al. Meningioma during pregnancy: what can influence the management? A case series and review of the literature. J Matern Fetal Neonatal Med. 2021;35:8767. https://doi.org/10.1080/14767058.2021.2004585.
38. Pettersson-Segerlind J, Mathiesen T, Elmi-Terander A, Edström E, Talbäck M, Feychting M, et al. The risk of developing a meningioma during and after pregnancy. Sci Rep. 2021;11(1):9153. https://doi.org/10.1038/s41598-021-88742-2.
39. Chakravarthy V, Kaplan B, Gospodarev V, Myers H, De Los Reyes K, Achiriloaie A. Houdini tumor: case report and literature review of pregnancy-associated meningioma. World Neurosurg. 2018;114:e1261–5. https://doi.org/10.1016/j.wneu.2018.03.187.
40. Hortobágyi T, Bencze J, Murnyák B, Kouhsari MC, Bognár L, Marko-Varga G. Pathophysiology of meningioma growth in pregnancy. Open Med (Wars). 2017;12:195–200. https://doi.org/10.1515/med-2017-0029.
41. Moscovici S, Fraifeld S, Cohen JE, Dotan S, Elchalal U, Shoshan Y, et al. Parasellar meningiomas in pregnancy: surgical results and visual outcomes. World Neurosurg. 2014;82(3–4):e503–12. https://doi.org/10.1016/j.wneu.2013.06.019.
42. Cohen-Gadol AA, Friedman JA, Friedman JD, Tubbs RS, Munis JR, Meyer FB. Neurosurgical management of intracranial lesions in the pregnant patient: a 36-year institutional experience and review of the literature. J Neurosurg. 2009;111(6):1150–7. https://doi.org/10.3171/2009.3.JNS081160.
43. ACOG Committee Opinion No. 474: nonobstetric surgery during pregnancy. Obstet Gynecol. 2011;117(2 Pt 1):420–1. https://doi.org/10.1097/AOG.0b013e31820eede9.
44. Hage M, Plesa O, Lemaire I, Raffin Sanson ML. Estrogen and progesterone therapy and meningiomas. Endocrinology. 2022;163(2):bqab259. https://doi.org/10.1210/endocr/bqab259.
45. Gazzeri R, Galarza M, Gazzeri G. Growth of a meningioma in a transsexual patient after estrogen-progestin therapy. N Engl J Med. 2007;357(23):2411–2. https://doi.org/10.1056/NEJMc071938.
46. Cebula H, Pham TQ, Boyer P, Froelich S. Regression of meningiomas after discontinuation of cyproterone acetate in a transsexual patient. Acta Neurochir. 2010;152(11):1955–6. https://doi.org/10.1007/s00701-010-0787-2.
47. Gil M, Oliva B, Timoner J, Maciá MA, Bryant V, de Abajo FJ. Risk of meningioma among users of high doses of cyproterone acetate as compared with the general population: evidence from a population-based cohort study. Br J Clin Pharmacol. 2011;72(6):965–8. https://doi.org/10.1111/j.1365-2125.2011.04031.x.
48. Cea-Soriano L, Blenk T, Wallander MA, Rodríguez LA. Hormonal therapies and meningioma: is there a link? Cancer Epidemiol. 2012;36(2):198–205. https://doi.org/10.1016/j.canep.2011.08.003.
49. Bernat AL, Oyama K, Hamdi S, Mandonnet E, Vexiau D, Pocard M, et al. Growth stabilization and regression of meningiomas after discontinuation of cyproterone acetate: a case series of 12 patients. Acta Neurochir. 2015;157(10):1741–6. https://doi.org/10.1007/s00701-015-2532-3.
50. Kalamarides M, Peyre M. Dramatic shrinkage with reduced vascularization of large meningiomas after cessation of progestin treatment. World Neurosurg. 2017;101:814.e7–10. https://doi.org/10.1016/j.wneu.2017.03.013.
51. Malaize H, Samoyeau T, Zanello M, Roux A, Benzakoun J, Peeters S, et al. Evolution of the neurosurgical management of progestin-associated meningiomas: a 23-year single-center experience. J Neurooncol. 2021;152(2):279–88. https://doi.org/10.1007/s11060-021-03696-9.
52. Samarut E, Lugat A, Amelot A, Scharbarg E, Hadjadj S, Primot C, et al. Meningiomas and cyproterone acetate: a retrospective, monocentric cohort of 388 patients treated by surgery or radiotherapy for intracranial meningioma. J Neurooncol. 2021;152(1):115–23. https://doi.org/10.1007/s11060-020-03683-6.
53. Weill A, Nguyen P, Labidi M, Cadier B, Passeri T, Duranteau L, et al. Use of high dose cyproterone acetate and risk of intracranial meningioma in women: cohort study. BMJ. 2021;372:n37. https://doi.org/10.1136/bmj.n37.
54. Lee KS, Zhang JJY, Kirollos R, Santarius T, Nga VDW, Yeo TT. A systematic review and meta-analysis of the association between cyproterone acetate and intracranial meningiomas. Sci Rep. 2022;12(1):1942. https://doi.org/10.1038/s41598-022-05773-z.
55. Champeaux-Depond C, Weller J, Froelich S, Sartor A. Cyproterone acetate and meningioma: a nationwide-wide population based study. J Neurooncol. 2021;151(2):331–8. https://doi.org/10.1007/s11060-020-03672-9.
56. Graillon T, Boissonneau S, Appay R, Boucekine M, Peyrière H, Meyer M, et al. Meningiomas in patients with long-term exposition to progestins: characteristics and outcome. Neurochirurgie. 2021;67(6):556–63. https://doi.org/10.1016/j.neuchi.2021.04.018.
57. Passeri T, Champagne PO, Bernat AL, Hanakita S, Salle H, Mandonnet E, et al. Spontaneous regression of meningiomas after interruption of nomegestrol acetate: a series of three patients. Acta Neurochir.

2019;161(4):761–5. https://doi.org/10.1007/s00701-019-03848-x.
58. Nguyen P, Hoisnard L, Neumann A, et al. Use prolonged use of nomegestrol acetate and risk of meningioma intracranial: a cohort study based on data of the SNDS. EPI-PHARE, 2021. https://www.epi-phare.fr/app/uploads/2021/04/epi-phare_rap-port_acetate_nomegetrol_avril-2021.pdf2021.
59. Nguyen P, Hoisnard L, Neumann A, et al. Use prolonged use of chlormadinone acetate and risk of meningioma intracranial. Eur J Contracept Reprod Health Care. 2022;27(5):355–6. https://doi.org/10.1080/13625187.2022.2114792. cohort study based on data of the SND. EPI-PHARE, 2021. https://www.epi-phare.fr/app/uploads/2021/04/epiphare_rap-port_acetate_chlormadinone_avril-2021-1.pdf2022.
60. Jhawar BS, Fuchs CS, Colditz GA, Stampfer MJ. Sex steroid hormone exposures and risk for meningioma. J Neurosurg. 2003;99(5):848–53. https://doi.org/10.3171/jns.2003.99.5.0848.
61. Custer B, Longstreth WT, Phillips LE, Koepsell TD, Van Belle G. Hormonal exposures and the risk of intracranial meningioma in women: a population-based case-control study. BMC Cancer. 2006;6:152. https://doi.org/10.1186/1471-2407-6-152.
62. Lee E, Grutsch J, Persky V, Glick R, Mendes J, Davis F. Association of meningioma with reproductive factors. Int J Cancer. 2006;119(5):1152–7. https://doi.org/10.1002/ijc.21950.
63. Wigertz A, Lönn S, Mathiesen T, Ahlbom A, Hall P, Feychting M, et al. Risk of brain tumors associated with exposure to exogenous female sex hormones. Am J Epidemiol. 2006;164(7):629–36. https://doi.org/10.1093/aje/kwj254.
64. Korhonen K, Raitanen J, Isola J, Haapasalo H, Salminen T, Auvinen A. Exogenous sex hormone use and risk of meningioma: a population-based case-control study in Finland. Cancer Causes Control. 2010;21(12):2149–56. https://doi.org/10.1007/s10552-010-9634-2.
65. Michaud DS, Gallo V, Schlehofer B, Tjønneland A, Olsen A, Overvad K, et al. Reproductive factors and exogenous hormone use in relation to risk of glioma and meningioma in a large European cohort study. Cancer Epidemiol Biomarkers Prev. 2010;19(10):2562–9. https://doi.org/10.1158/1055-9965.EPI-10-0447.
66. Johnson DR, Olson JE, Vierkant RA, Hammack JE, Wang AH, Folsom AR, et al. Risk factors for meningioma in postmenopausal women: results from the Iowa Women's Health Study. Neuro oncol. 2011;13(9):1011–9. https://doi.org/10.1093/neuonc/nor081.
67. Qi ZY, Shao C, Huang YL, Hui GZ, Zhou YX, Wang Z. Reproductive and exogenous hormone factors in relation to risk of meningioma in women: a meta-analysis. PLoS One. 2013;8(12):e83261. https://doi.org/10.1371/journal.pone.0083261.
68. Harland TA, Freeman JL, Davern M, McCracken DJ, Celano EC, Lillehei K, et al. Progesterone-only contraception is associated with a shorter progression-free survival in premenopausal women with WHO Grade I meningioma. J Neurooncol. 2018;136(2):327–33. https://doi.org/10.1007/s11060-017-2656-9.
69. Yang X, Liu F, Zheng J, Cheng W, Zhao C, Di J. Relationship between oral contraceptives and the risk of gliomas and meningiomas: a dose-response meta-analysis and systematic review. World Neurosurg. 2021;147:e148–62. https://doi.org/10.1016/j.wneu.2020.11.175.
70. Shahin MN, Magill ST, Dalle Ore CL, Viner JA, Peters PN, Solomon DA, et al. Fertility treatment is associated with multiple meningiomas and younger age at diagnosis. J Neurooncol. 2019;143(1):137–44. https://doi.org/10.1007/s11060-019-03147-6.
71. Frassanito P, De Bonis P, Mattogno PP, Novello M, Anile C. Hormonal therapy for fertility and huge meningioma: a purely random association? Acta Neurol Belg. 2012;112(3):299–301. https://doi.org/10.1007/s13760-012-0046-9.
72. Patterson A, Elashaal A. Fast-growing meningioma in a woman undergoing fertility treatments. Case Rep Neurol Med. 2016;2016:3287381. https://doi.org/10.1155/2016/3287381.
73. Ter Wengel PV, Martin E, Gooren L, Den Heijer M, Peerdeman SM. Meningiomas in three male-to-female transgender subjects using oestrogens/progestogens and review of the literature. Andrologia. 2016;48(10):1130–7. https://doi.org/10.1111/and.12550.
74. Blitshteyn S, Crook JE, Jaeckle KA. Is there an association between meningioma and hormone replacement therapy? J Clin Oncol. 2008;26(2):279–82. https://doi.org/10.1200/JCO.2007.14.2133.
75. Andersen L, Friis S, Hallas J, Ravn P, Schrøder HD, Gaist D. Hormone replacement therapy increases the risk of cranial meningioma. Eur J Cancer. 2013;49(15):3303–10. https://doi.org/10.1016/j.ejca.2013.05.026.
76. Benson VS, Pirie K, Green J, Bull D, Casabonne D, Reeves GK, et al. Hormone replacement therapy and incidence of central nervous system tumours in the million women study. Int J Cancer. 2010;127(7):1692–8. https://doi.org/10.1002/ijc.25184.
77. Korhonen K, Auvinen A, Lyytinen H, Ylikorkala O, Pukkala E. A nationwide cohort study on the incidence of meningioma in women using postmenopausal hormone therapy in Finland. Am J Epidemiol. 2012;175(4):309–14. https://doi.org/10.1093/aje/kwr335.
78. Shu X, Jiang Y, Wen T, Lu S, Yao L, Meng F. Association of hormone replacement therapy with increased risk of meningioma in women: a hospital-based multicenter study with propensity score matching. Asia Pac J Clin Oncol. 2019;15(5):e147–53. https://doi.org/10.1111/ajco.13138.
79. Fan ZX, Shen J, Wu YY, Yu H, Zhu Y, Zhan RY. Hormone replacement therapy and risk of meningioma in women: a meta-analysis. Cancer

Causes Control. 2013;24(8):1517–25. https://doi.org/10.1007/s10552-013-0228-7.
80. Benson VS, Kirichek O, Beral V, Green J. Menopausal hormone therapy and central nervous system tumor risk: large UK prospective study and meta-analysis. Int J Cancer. 2015;136(10):2369–77. https://doi.org/10.1002/ijc.29274.
81. Hatch EE, Linet MS, Zhang J, Fine HA, Shapiro WR, Selker RG, et al. Reproductive and hormonal factors and risk of brain tumors in adult females. Int J Cancer. 2005;114(5):797–805. https://doi.org/10.1002/ijc.20776.
82. Dresser L, Yuen CA, Wilmington A, Walker M, Vogel TJ, Merrell RT, et al. Estrogen hormone replacement therapy in incidental intracranial meningioma: a growth-rate analysis. Sci Rep. 2020;10(1):17960. https://doi.org/10.1038/s41598-020-74344-x.
83. Lapresle J, Netsky MG, Zimmerman HM. [The pathology of meningiomas; a study of 121 cases]. Am J Pathol. 1952;28(5):757–91.
84. Schoenberg BS, Christine BW, Whisnant JP. Nervous system neoplasms and primary malignancies of other sites. The unique association between meningiomas and breast cancer. Neurology. 1975;25(8):705–12. https://doi.org/10.1212/wnl.25.8.705.
85. Criscitiello C, Disalvatore D, Santangelo M, Rotmensz N, Bazolli B, Maisonneuve P, et al. No link between breast cancer and meningioma: results from a large monoinstitutional retrospective analysis. Cancer Epidemiol Biomarkers Prev. 2014;23(1):215–7. https://doi.org/10.1158/1055-9965.EPI-13-1041.
86. Ji J, Sundquist J, Sundquist K. Association of tamoxifen with meningioma: a population-based study in Sweden. Eur J Cancer Prev. 2016;25(1):29–33. https://doi.org/10.1097/CEJ.0000000000000133.
87. Degeneffe A, De Maertelaer V, De Witte O, Lefranc F. The association between meningioma and breast cancer: a systematic review and meta-analysis. JAMA Netw Open. 2023;6(6):e2318620. https://doi.org/10.1001/jamanetworkopen.2023.18620.
88. Grann VR, Troxel AB, Zojwalla NJ, Jacobson JS, Hershman D, Neugut AI. Hormone receptor status and survival in a population-based cohort of patients with breast carcinoma. Cancer. 2005;103(11):2241–51. https://doi.org/10.1002/cncr.21030.
89. Abenhaim HA, Suissa S, Azoulay L, Spence AR, Czuzoj-Shulman N, Tulandi T. Menopausal hormone therapy formulation and breast cancer risk. Obstet Gynecol. 2022;139(6):1103–10. https://doi.org/10.1097/AOG.0000000000004723.
90. Hoisnard L, Laanani M, Passeri T, Duranteau L, Coste J, Zureik M, et al. Risk of intracranial meningioma with three potent progestogens: a population-based case-control study. Eur J Neurol. 2022;29(9):2801–9. https://doi.org/10.1111/ene.15423.
91. Maiuri F, Cappabianca P, Iaconetta G, D'Acunzi G. Meningiomas associated with brain metastases. Zentralbl Neurochir. 2002;63(3):111–5. https://doi.org/10.1055/s-2002-35823.
92. Elmaci L, Ekinci G, Kurtkaya O, Sav A, Pamir MN. Tumor in tumor: metastasis of breast carcinoma to intracranial meningioma. Tumori. 2001;87(6):423–7. https://doi.org/10.1177/030089160108700613.
93. Caroli E, Salvati M, Giangaspero F, Ferrante L, Santoro A. Intrameningioma metastasis as first clinical manifestation of occult primary breast carcinoma. Neurosurg Rev. 2006;29(1):49–54. https://doi.org/10.1007/s10143-005-0395-4.
94. Claus EB, Calvocoressi L, Bondy ML, Schildkraut JM, Wiemels JL, Wrensch M. Family and personal medical history and risk of meningioma. J Neurosurg. 2011;115(6):1072–7. https://doi.org/10.3171/2011.6.JNS11129.
95. Yen YS, Sun LM, Lin CL, Chang SN, Sung FC, Kao CH. Higher risk for meningioma in women with uterine myoma: a nationwide population-based retrospective cohort study. J Neurosurg. 2014;120(3):655–61. https://doi.org/10.3171/2013.10.JNS131357.

Meningiomas in Genetic and Familial Syndromes: Biomolecular Correlations

Marica Eoli

1 Introduction

Meningiomas, the most common primary intracranial neoplasms, account for over a third of primary CNS tumors, inferring that around 1 in 200 people will be diagnosed with meningioma in their lifetime. The percentage rises to 1 in 110 people counting also who have asymptomatic meningiomas on MRI at an average age of 63 years [1]. Meningiomas can be sporadic, hereditary, or radiation-induced. In fact, there are many tumor predisposition syndromes that place a patient at increased risk for meningioma development (see Table 1). However, because the most common genetic syndrome causing meningiomas, *NF2* related Schwannomatosis, has a birth incidence of between 1/25,000 to 1/33,000 and about 50% of cases develop meningiomas [2], we can infer that less than 5% of meningiomas are syndromic. Radiation-induced meningiomas are the most common radiation-induced neoplasm [3]. However, Flint-Richter and Sadetzki provide strong evidence that a genetic susceptibility is present also in patients

Table 1 Tumor predisposition syndromes

Gene	Genetic tumor syndrome(s)	Other tumor(s) involved
NF2	Schwannomatosis *NF2* related	Vestibular schwannoma, other cranial nerve and peripheral nerves schwannomas, ependymoma
LZTR1	Schwannomatosis LZTR1 related	Schwannoma
SMARCB1	Schwannomatosis SMARCB1 related	Schwannoma
SMARCE1	*SMARCE1* syndrome	No
SUFU	Gorlin syndrome/NBCCS	Pediatric medulloblastoma, basal cell carcinomas ovarian fibroma, cardiac fibroma
PTCH1	Gorlin syndrome/NBCCS	Pediatric medulloblastoma, basal cell carcinomas (BCCs), meningioma, ovarian fibroma, cardiac fibroma
BAP1	*BAP1* tumor predisposition syndrome	Uveal melanoma, mesothelioma, cutaneous melanoma and renal carcinoma, basal cell carcinoma, cholangiocarcinoma

M. Eoli (✉)
Unit of Experimental Neuro-Oncology, Department of Neuro-Oncology, Fondazione IRCCS Istituto Neurologico Carlo Besta, Milan, Italy
e-mail: marica.eoli@istituto-besta.it

F. Maiuri, M. Del Basso De Caro (eds.), *Meningiomas*,
https://doi.org/10.1007/978-3-031-76680-0_4

that developed meningiomas after exposure to ionizing radiation [4].

A genetic susceptibility tumor syndrome should be suspected when there is an unusual early-onset of the tumor, as far as meningioma diagnosis is concerned before age 40 years, multiple primary independent tumors are observed in a single individual, a familial clustering of the tumors is present, a typical tumor spectrum (in the case of meningioma for example the association of meningioma and schwannoma), a rare tumor such as optic sheath meningioma.

2 Meningiomatosis

Meningiomatosis or multiple meningiomas is a condition characterized by the presence of ≥2 spatially separated meningiomas in a single patient [3]. It occurs in 1–10% of patients affected by meningiomas [5]. The lesions can be synchronous when diagnosed all at the same time or metachronous if observed in different phases of the follow-up. The incidence of multiple meningiomas is higher in women and increases with age, with a median age at diagnosis ranging from 53.7 to 71 years. It is rare in childhood and adolescence.

It represents a heterogeneous group of conditions with different etiologies. Familial meningiomatosis can be attributed to numerous inherited cancer syndromes, in some of them, such as familial meningiomatosis *SMARCE1* related meningiomas are the only tumors observed [6].

3 NF2-Related Schwannomatosis

NF2 related Schwannomatosis, previously called Neurofibromatosis type 2 (NF2-SCHW MIM # 101000) is a rare autosomal dominant familial cancer syndrome caused by pathogenic variants in the *NF2* gene. Its diagnostic criteria have been recently updated (see Figs. 1 and 2) [7]. The presence of two or more meningiomas is considered a

A diagnosis of *NF2*-related schwannomatosis (previously termed neurofibromatosis 2,NF2) can be made when an individual has one of the following:

1. Bilateral vestibular schwannomas (VS)
2. An identical *NF2* pathogenic variant in at least 2 anatomically distinct NF2-related tumors (schwannoma, meningioma, and/or ependymoma). (Note: if the variant allele fraction (VAF) in unaffected tissues such as blood is clearly <50%, the diagnosis is mosaic *NF2*-related schwannomatosis)
3. Either 2 major or 1 major and 2 minor criteria as described in the following:

Major criteria:

- Unilateral VS
- First-degree relative other than sibling with *NF2*-related schwannomatosis
- 2 or more meningiomas (Note: single meningioma qualifies as minor criteria).
- *NF2* pathogenic variant[a] in an unaffected tissue such as blood (Note: if the VAF is clearly <50%, the diagnosis is mosaic *NF2*-related schwannomatosis)

Minor criteria:

Can count >1 of a type (eg, 2 distinct schwannomas would count as 2 minor criteria)

- Ependymoma, meningioma (Note: multiple meningiomas qualify as a major criteria), schwannoma (Note: if the major criterion is unilateral VS, at least 1 schwannoma must be dermal in location)

Can count only once (eg, bilateral cortical cataracts count as a single minor criterion)

- Juvenile subcapsular or cortical cataract, retinal hamartoma, epiretinal membrane in a person aged <40 years, meningioma

Pattern of genetic changes in unaffected and tumor tissue in *NF2*-related schwannomatosis

Gene locus	Unaffected Tissue[b]	Tumor 1	Tumor 2	Comment
NF2				
Allele 1	PV1[c]	PV1	PV1	Shared *NF2* pathogenic variant
Allele 2	WT	LOH or *NF2* PV2	LOH or *NF2* PV3	Tumor-specific partial loss of 22q in trans position or a different *NF2* somatic second PV in every anatomically unrelated tumor

LOH, loss of heterozygosity; *NF2*, neurofibromatosis type 2; *PV*, pathogenic variant; *WT*, wildtype.

[a]If a likely pathogenic variant is identified, tumor analysis may aid upward classification to pathogenic variant.

[b]Tissues unaffected by tumors such as blood or skin.

[c]If the variant allele fraction is clearly <50%, the diagnosis is mosaic *NF2*-related schwannomatosis.

Fig. 1 Diagnostic criteria for NF2-related schwannomatosis

A diagnosis of *SMARCB1*- or *LZTR1*-related schwannomatosis can be made when an individual meets 1 of the following criteria:

- At least 1 pathologically confirmed schwannoma or hybrid nerve sheath tumor and a *SMARCB1* (or *LZTR1*) pathogenic variant in an unaffected tissue such as blood[a]
- A shared *SMARCB1* or *LZTR1* pathogenic variant in 2 schwannomas or hybrid nerve sheath tumors.

Pattern of genetic changes in unaffected and tumor tissue in *SMARCB1*- and *LZTR1*-related schwannomatosis[b]

Gene locus	Unaffected Tissue[c]	Tumor 1	Tumor 2	Comment
SMARCB1/LZTR1				
Allele 1	PV1[d]	PV1	PV1	Shared *SMARCB1* or *LZTR1* pathogenic variant
Allele 2	WT	LOH	LOH	Tumor-specific partial loss of 22q in trans position, LOH typically entails deletion of 22q region encompassing *LZTR1/SMARCB1/NF2*
NF2				
Allele 1	WT	PV2	PV3	Tumor-specific pathogenic *NF2* variant in cis to pathogenic *SMARCB1* variant
Allele 2	WT	LOH	LOH	Tumor-specific partial loss of 22q in trans position, LOH typically entails deletion of 22q region encompassing *LZTR1/SMARCB1/NF2*

LOH, loss of heterozygosity; *PV*, pathogenic variant; *WT*, wildtype.

[a]If a likely pathogenic variant is identified, tumor analysis may aid upward classification to pathogenic variant.

[b]See also Supplemental Figure 1A and B.

[c]Tissues unaffected by tumors such as blood or skin.

[d]If the variant allele fraction is clearly <50%, then the diagnosis is mosaic *SMARCB1*- or *LZTR1*-related schwannomatosis.

Fig. 2 Diagnostic criteria for SMARCB1- and LZTR1-related schwannomatosis

major criteria, while a single meningioma qualifies as minor criteria.

The hallmark feature is the presence of bilateral vestibular schwannoma, observed in more than 90% of cases, but *NF2*-SCHW patients can also develop other tumors, among which are schwannomas of the cranial, spinal, peripheral, and cutaneous nerves, but also meningiomas and ependymomas [8]. Other clinical manifestations include ophthalmic signs, along with dermatological features, counting subcutaneous nodular tumors, and café-au-lait spots. Meningiomas are the most common extra-axial tumors in the brain and are present in 45–58% of adult patients. Around 80% are asymptomatic, discovered through cranial and spinal magnetic resonance imaging (MRI) as part of the diagnostic work-up or follow-up [9] .

NF2-SCHW patients develop meningiomas at an earlier age than sporadic meningiomas often multiple. If meningiomas are seen in the pediatric age group, then a diagnosis of *NF2*-SCHW should be considered; they are often multiple with a mean number of 5 meningiomas per patient in a study of 119 patients [10]. Spinal meningiomas are detected in 20–70% in cervicomedullary junction or cervical spine.

Meningiomas have typical features on the MRI. They are dural-based lesions with the isointense signal on T1-weighted images and iso to the hyperintense signal on T2-weighted images with intense enhancement and an enhancing dural tail [11].

The presence of meningiomas is associated with a 2.5-fold rise in relative risk of mortality, also because in this disorder a greater rate of atypical and anaplastic grade than in sporadic meningiomas has been observed [8].

Furthermore, *NF2*-SCHW has a very high rate a mosaicism ranging between 25% and 50% of the de novo cases [12]. In this condition, only a proportion of cells harbors pathogenic variant of the gene resulting in a mosaic phenotype. In those individuals, the disease onset can be later than the others and with milder phenotype [13]. *NF2*-SCHW is an autosomal dominant disorder caused by a pathological sequence variant of the *NF2* gene a tumor suppressor gene [14]

NF2 gene encodes Moesin-Ezrin-Radixin-Like (Merlin), a 595 amino acid protein that belongs to the ERM (ezrin, radixin, moesin) protein family, known to mediate the interaction of cytoskeletal components with cell membrane proteins, involved in the control of cytoskeletal dynamics and ion transport [15]. Merlin represents a critical modulator of contact-dependent inhibition of cell proliferation, regulating the Hippo/SWH (Sav/Wts/Hpo) signaling pathway, which plays a pivotal role in tumor suppression by restricting proliferation and promoting apoptosis [16]. Merlin also exerts a negative control toward other proliferative signaling pathways including RAS-MAPK, FAK/Src, PI3K/AKT/mTOR, Rac, and Wnt/β-catenin pathways [17]. Furthermore, several somatic *NF2* mutations have been identified in neoplasm in particular sporadic meningiomas and schwannomas.

The development of *NF2*-SCHW associated tumors follows Knudson's two-hit hypothesis: the germinal mutation of *NF2* is followed by the somatic mutation of *NF2* in the wild type allele acting as the initiating event for tumor formation [8].

Two others schwannomatosis *SMARCB1*-related schwannomatosis (*SMARCB1*-SCHW MIM # 162091) and *LZTR1*-related schwannomatosis (*LZTR1*-SCHW MIM # 615670) predispose to the development of meningiomas. Those schwannomatosis are even more rare with an incidence estimated at around half of that of *NF2*-SCHW (1:60,000) [18]. They can be sporadic, but approximately 20% of cases appears to be familial [19] with an autosomal dominant mode of transmission with incomplete penetrance. Their diagnostic criteria has been updated and does not include meningioma as criteria (see Fig. 1) [7].

LZTR1/*SMARCB1*-SCHW are characterized by multiple schwannomas located in cranial, spinal, cutaneous nerves and less commonly meningiomas. Differently with *NF2*-SCHW, in the majority of cases, they don't present bilateral vestibular schwannoma, which are the pathognomonic sign of *NF2*-SCHW [8]. However, sometimes, *LZTR1*/SMARCB1-schwannomatosis patients develop unilateral vestibular schwannomas [18, 20, 21]. Individuals most commonly present symptoms and signs between the second and fourth decade of life. The most common presenting symptoms are localized or diffuse pain or an asymptomatic mass. Schwannomas most often affect peripheral nerves and spinal nerves [22].

Meningiomas have mostly been reported in about 5% of individuals with *SMARCB1*-SCHW [23, 24]. However, recently they have been described also in *LZTR1*-SCHW [25]. There are often multiple tumors with falx cerebri being a preferential location for cranial meningiomas.

SMARCB1-SCHW is caused by a pathogenic variant of *SMARCB1* gene [26]. *SMARCB1*, located on 22q, is classified as a tumor suppressor gene. *SMARCB1* encodes for protein that is a highly conserved subunit of the SWI/SNF complex, which plays a key role for the transcriptional activation/repression of numerous genes through an ATP-dependent chromatin remodeling process [27]. *LZTR1*-SCHW is due to a pathogenic variant of *LZTR1* gene [28]. *LZTR1*-SCHW are more frequent than *SMARCB1*-SCHW accounting for around 50% of familial cases and 30% of sporadic ones [21]. *LZTR1* is located on 22q, centromeric to SMARCB1 and NF2, and it encodes the leucine zipper-like transcription regulator 1, located in the Golgi apparatus. LZTR1 is a tumor suppressor gene [28].

Recently, a new hereditary tumor predisposition syndrome has been reported, arising in an increased risk for spinal and intracranial clear cell meningiomas (CCMs) in young patients. In 2013, Smith described three cases with multiple spinal CCMs in a single family, who on sequencing showed *SMARCE1* loss of function and no germline pathogenic variant in *NF2* and S*MARCB1* genes [29]. Following reports confirmed a role for *SMARCE1* loss-of-function germline mutations giving rise to an autosomal dominant inheritance pattern, in familial with spinal and intracranial CCM meningiomas [29, 30].

Clear cell meningioma is a rare histopathological type grade 2 highly aggressive meningioma, affects predominantly younger individuals, and only represents 0.2–1% of all meningiomas and mostly affects the lumbosacral spine area [6].

The *SMARCE1* gene is located on chromosome 17q21. This gene encodes a component of the SWI/SNF (BAF57) protein complex which per se is a modulator for chromatin remodeling and gene transcription. In addition, sporadic meningiomas frequently exhibit loss of the tumor suppressor *NF2*; however, *SMARCE1* mutations occur in a mutually exclusive manner, signifying a separate and different histotype of meningiomas [31].

Germline pathogenic variant in *SUFU* gene has been associated with increased risk of inherited multiple meningiomas based on report on a single large Finnish family with five affects [32].

4 Gorlin Syndrome

Meningiomas are rarely observed in Gorlin syndrome (also known as nevoid basal cell carcinoma syndrome (NBCCS) MIM 109400), an autosomal dominant cancer predisposition syndrome characterized by an increased risk of several tumors including pediatric medulloblastoma, basal cell carcinomas (BCCs), meningioma, ovarian fibroma, cardiac fibroma, and jaw cysts and a vast spectrum of bone abnormalities [33–35].

Notably, *SUFU* pathogenic variants account for only approximately 5% of NBCCS cases whereas approximately 70% of NBCCS cases are caused by heterozygous germline variants in the *PTCH1* gene [33].

Lee reviewed the phenotype and tumors associated of all the cases with *SUFU* mutations published until October 21 [36]. There were 176 patients identified with a confirmed germline *SUFU* variant and 34 different neoplasms (benign and malignant) documented among patients; the most common were medulloblastoma (59 patients), BCC (21 patients), and meningioma (19 patients). There were a total of ten patients with meningioma who did not have a history of prior radiation and seven patients who did and data were unavailable for two patients.

The review clarified that in at least a subset of patients, the risk of meningioma was due to a genetic predisposition and not to the secondary effects of radiation treatment.

There is also evidence that meningiomas may be more frequent among patients with NBCCS associated with *SUFU* variant compared to *PTCH1* variant where the incidence of meningiomas occurred in 4/9 patients with a *SUFU* mutation and in 2/126 patients with PTCH1 mutation [33, 35, 36] .

5 BAP1 Tumor Predisposition Syndrome

Meningioma is one of the rare tumors that can occur in *BAP1* tumor predisposition syndrome (MIM# 614327). This syndrome predisposes to uveal melanoma, mesothelioma, cutaneous melanoma and renal carcinoma and less frequently meningioma, basal cell carcinoma and cholangiocarcinoma [37]. So far more than 180 families have been described; the carrier frequency of germiline pathogenic variants is about 1 in 50,000 people, the frequency of meningiomas associated is about 2% [37] and they display an aggressive behavior and frequently have rhabdoid and papillary morphology [38, 39]. *BAP1* syndrome is an autosomal dominant disorder caused by pathogenic germline variants in *BAP1* gene. The gene encodes a deubiquitinating enzyme that interacts with *BRCA1* and functions as a tumor suppressor protein through its deubiquitinase activity that regulates target genes in cell cycle control, cellular differentiation, and DNA damage repair [40].

6 Other Tumor Predisposition Syndromes

Meningiomas have also been reported in patients or families with other tumor predisposition syndromes such as Cowden syndrome [41], Li-Fraumeni syndrome [42], with a very low frequency and the incomplete clinical description of some cases. It remains to be determined whether these are true components of the syndrome or they developed simply by chance.

References

1. Wilson TA, Huang L, Ramanathan D, Lopez-Gonzalez M, Pillai P, De Los Reyes K, et al. Review of atypical and anaplastic meningiomas: classification, molecular biology, and management. Front Oncol. 2020;10:565582.
2. Evans DG, Huson SM, Donnai D, Neary W, Blair V, Teare D, et al. A genetic study of type 2 neurofibromatosis in the United Kingdom. I. Prevalence, mutation rate, fitness, and confirmation of maternal transmission effect on severity. J Med Genet. 1992;29(12):841–6.
3. Fahlström A, Dwivedi S, Drummond K. Multiple meningiomas: epidemiology, management, and outcomes. Neurooncol Adv. 2023;5(Suppl 1):i35–48.
4. Flint-Richter P, Sadetzki S. Genetic predisposition for the development of radiation-associated meningioma: an epidemiological study. Lancet Oncol. 2007;8(5):403–10.

5. Ramos-Fresnedo A, Domingo RA, Vivas-Buitrago T, Lundy L, Trifiletti DM, Jentoft ME, et al. Multiple meningiomas: does quantity matter? A population-based survival analysis with underlined age and sex differences. J Neurooncol. 2020;149(3):413–20.
6. Evans LT, Van Hoff J, Hickey WF, Smith MJ, Evans DG, Newman WG, et al. SMARCE1 mutations in pediatric clear cell meningioma: case report. J Neurosurg Pediatr. 2015;16(3):296–300.
7. Plotkin SR, Messiaen L, Legius E, Pancza P, Avery RA, Blakeley JO, et al. Updated diagnostic criteria and nomenclature for neurofibromatosis type 2 and schwannomatosis: an international consensus recommendation. Genet Med. 2022;24(9):1967–77.
8. Halliday D, Parry A, Evans DG. Neurofibromatosis type 2 and related disorders. Curr Opin Oncol. 2019;31(6):562–7.
9. Riemenschneider MJ, Perry A, Reifenberger G. Histological classification and molecular genetics of meningiomas. Lancet Neurol. 2006;5(12):1045–54.
10. Goutagny S, Bah AB, Henin D, Parfait B, Grayeli AB, Sterkers O, et al. Long-term follow-up of 287 meningiomas in neurofibromatosis type 2 patients: clinical, radiological, and molecular features. Neuro Oncol. 2012;14(8):1090–6.
11. Gregory GE, Islim AI, Hannan CJ, Jones AP, Hammerbeck-Ward C, Rutherford SA, et al. The clinical, genetic, and immune landscape of meningioma in patients with NF2-schwannomatosis. Neurooncol Adv. 2023;5(Suppl 1):i94–104.
12. Evans DG, Hartley CL, Smith PT, King AT, Bowers NL, Tobi S, et al. Incidence of mosaicism in 1055 de novo NF2 cases: much higher than previous estimates with high utility of next-generation sequencing. Genet Med. 2020;22(1):53–9.
13. Halliday D, Emmanouil B, Vassallo G, Lascelles K, Nicholson J, Chandratre S, et al. Trends in phenotype in the English paediatric neurofibromatosis type 2 cohort stratified by genetic severity. Clin Genet. 2019;96(2):151–62.
14. Gusella JF, Ramesh V, MacCollin M, Jacoby LB. Neurofibromatosis 2: loss of merlin's protective spell. Curr Opin Genet Dev. 1996;6(1):87–92.
15. Golovnina K, Blinov A, Akhmametyeva EM, Omelyanchuk LV, Chang L-S. Evolution and origin of merlin, the product of the neurofibromatosis type 2 (NF2) tumor-suppressor gene. BMC Evol Biol. 2005;5:69.
16. Curto M, McClatchey AI. Nf2/Merlin: a coordinator of receptor signalling and intercellular contact. Br J Cancer. 2008;98(2):256–62.
17. Grusche FA, Richardson HE, Harvey KF. Upstream regulation of the hippo size control pathway. Curr Biol. 2010;20(13):R574–82.
18. Evans DG, Bowers NL, Tobi S, Hartley C, Wallace AJ, King AT, et al. Schwannomatosis: a genetic and epidemiological study. J Neurol Neurosurg Psychiatry. 2018;89(11):1215–9.
19. Antinheimo J, Sankila R, Carpén O, Pukkala E, Sainio M, Jääskeläinen J. Population-based analysis of sporadic and type 2 neurofibromatosis-associated meningiomas and schwannomas. Neurology. 2000;54(1):71–6.
20. Smith MJ, Kulkarni A, Rustad C, Bowers NL, Wallace AJ, Holder SE, et al. Vestibular schwannomas occur in schwannomatosis and should not be considered an exclusion criterion for clinical diagnosis. Am J Med Genet A. 2012;158A(1):215–9.
21. Smith MJ, Bowers NL, Bulman M, Gokhale C, Wallace AJ, King AT, et al. Revisiting neurofibromatosis type 2 diagnostic criteria to exclude LZTR1-related schwannomatosis. Neurology. 2017;88(1):87–92.
22. Merker VL, Esparza S, Smith MJ, Stemmer-Rachamimov A, Plotkin SR. Clinical features of schwannomatosis: a retrospective analysis of 87 patients. Oncologist. 2012;17(10):1317–22.
23. Bacci C, Sestini R, Provenzano A, Paganini I, Mancini I, Porfirio B, et al. Schwannomatosis associated with multiple meningiomas due to a familial SMARCB1 mutation. Neurogenetics. 2010;11(1):73–80.
24. Christiaans I, Kenter SB, Brink HC, van Os TAM, Baas F, van den Munckhof P, et al. Germline SMARCB1 mutation and somatic NF2 mutations in familial multiple meningiomas. J Med Genet. 2011;48(2):93–7.
25. Neupane S, Kashyap A, Paudel S, Bhattarai G, Kharel SK, Adhikari A, et al. A rare case of schwannomatosis with meningioma: a case report. Ann Med Surg. 2024;86(3):1724–8.
26. Hulsebos TJM, Kenter SB, Jakobs ME, Baas F, Chong B, Delatycki MB. SMARCB1/INI1 maternal germ line mosaicism in schwannomatosis. Clin Genet. 2010;77(1):86–91.
27. Kalimuthu SN, Chetty R. Gene of the month: SMARCB1. J Clin Pathol. 2016;69(6):484–9.
28. Piotrowski A, Xie J, Liu YF, Poplawski AB, Gomes AR, Madanecki P, et al. Germline loss-of-function mutations in LZTR1 predispose to an inherited disorder of multiple schwannomas. Nat Genet. 2014;46(2):182–7.
29. Smith MJ, O'Sullivan J, Bhaskar SS, Hadfield KD, Poke G, Caird J, et al. Loss-of-function mutations in SMARCE1 cause an inherited disorder of multiple spinal meningiomas. Nat Genet. 2013;45(3):295–8.
30. Gerkes EH, Fock JM, den Dunnen WFA, van Belzen MJ, van der Lans CA, Hoving EW, et al. A heritable form of SMARCE1-related meningiomas with important implications for follow-up and family screening. Neurogenetics. 2016;17(2):83–9.
31. St Pierre R, Collings CK, Samé Guerra DD, Widmer CJ, Bolonduro O, Mashtalir N, et al. SMARCE1 deficiency generates a targetable mSWI/SNF dependency in clear cell meningioma. Nat Genet. 2022;54(6):861–73.
32. Aavikko M, Li S-P, Saarinen S, Alhopuro P, Kaasinen E, Morgunova E, et al. Loss of SUFU function in familial multiple meningioma. Am J Hum Genet. 2012;91(3):520–6.
33. Smith MJ, Beetz C, Williams SG, Bhaskar SS, O'Sullivan J, Anderson B, et al. Germline mutations in SUFU cause Gorlin syndrome-associated

childhood medulloblastoma and redefine the risk associated with PTCH1 mutations. J Clin Oncol. 2014;32(36):4155–61.
34. Evans DG, Oudit D, Smith MJ, Rutkowski D, Allan E, Newman WG, et al. First evidence of genotype-phenotype correlations in Gorlin syndrome. J Med Genet. 2017;54(8):530–6.
35. Guerrini-Rousseau L, Smith MJ, Kratz CP, Doergeloh B, Hirsch S, Hopman SMJ, et al. Current recommendations for cancer surveillance in Gorlin syndrome: a report from the SIOPE host genome working group (SIOPE HGWG). Fam Cancer. 2021;20(4):317–25.
36. Lee SG, Evans G, Stephen M, Goren R, Bondy M, Goodman S. Medulloblastoma and other neoplasms in patients with heterozygous germline SUFU variants: a scoping review. Am J Med Genet A. 2024;194(6):e63496.
37. Masoomian B, Shields JA, Shields CL. Overview of BAP1 cancer predisposition syndrome and the relationship to uveal melanoma. J Curr Ophthalmol. 2018;30(2):102–9.
38. Shankar GM, Abedalthagafi M, Vaubel RA, Merrill PH, Nayyar N, Gill CM, et al. Germline and somatic BAP1 mutations in high-grade rhabdoid meningiomas. Neuro Oncol. 2017;19(4):535–45.
39. Shankar GM, Santagata S. BAP1 mutations in high-grade meningioma: implications for patient care. Neuro Oncol. 2017;19(11):1447–56.
40. Walpole S, Pritchard AL, Cebulla CM, Pilarski R, Stautberg M, Davidorf FH, et al. Comprehensive study of the clinical phenotype of germline BAP1 variant-carrying families worldwide. J Natl Cancer Inst. 2018;110(12):1328–41.
41. Yakubov E, Ghoochani A, Buslei R, Buchfelder M, Eyüpoglu IY, Savaskan N. Hidden association of Cowden syndrome, PTEN mutation and meningioma frequency. Oncoscience. 2016;3(5–6):149–55.
42. Kim B, Tabori U, Hawkins C. An update on the CNS manifestations of brain tumor polyposis syndromes. Acta Neuropathol. 2020;139(4):703–15.

Part II

Meningiomas: Pathological–Clinical Correlations

Meningioma Development: Meningeal Embryology and Tumorigenesis

Vratko Himic, Sanjeeva Jeyaretna, and Daniel M. Fountain

1 Introduction

This chapter describes the spatial phenotype of meningiomas and its relationship with the putative embryonic origin of the meninges from neural crest and mesodermal lineages. After presenting a descriptive outline of the genomic characteristics of spatially distinct meningiomas, we explore the evidence base in human and animal model studies. We highlight where gaps in the evidence remain and consider the outstanding questions in the field and highlight the key genetic pathways implicated to describe how meningeal embryology and development influences future tumorigenesis.

V. Himic
St Bartholomew's Hospital, Barts Health NHS Trust, London, UK

S. Jeyaretna · D. M. Fountain (✉)
Department of Neurosurgery, John Radcliffe Hospital, Oxford University Hospitals NHS Foundation Trust, Oxford, UK
e-mail: daniel.fountain@nhs.net

D. M. Fountain (✉)
Department of Neurosurgery, John Radcliffe Hospital, Oxford University Hospitals NHS Foundation Trust, Oxford, UK

MRC Weatherall Institute of Molecular Medicine, University of Oxford, Oxford, UK
e-mail: daniel.fountain@nhs.net

2 The Spatial Phenotype of Genotypically Distinct Meningioma

Recent molecular classifications have associated an underlying genotype with a propensity for meningiomas to develop in a particular location [1–9]. These classifications form the base of much of our understanding on the distribution of meningiomas and, at a molecular level, help us make inferences about the embryological and pre-tumorigenic traits that those cells may have harboured. This section divides groups of pathogenic variants into those that involve:

- Neurofibromatosis type 2 (*NF2*) and associated variants.
- *TRAF7* and the PI3K-AKT-mTOR pathway.
- *SMO* and the Hedgehog signalling pathway.
- Other variants with individual associations.

We explore the evidence for a correlation between the location of a meningioma and its underlying genomic profile.

2.1 NF2 and Associated Variants

Neurofibromatosis type 2 (*NF2*), the gene most associated with meningioma development, gives rise to meningiomas particularly at the falx and the convexity [10]. *NF2* encodes the protein

F. Maiuri, M. Del Basso De Caro (eds.), *Meningiomas*,
https://doi.org/10.1007/978-3-031-76680-0_5

Merlin and is responsible for the eponymous neurofibromatosis type 2 syndrome (where patients develop meningiomas alongside vestibular schwannomas and other tumours). It is present in about half of non-syndromic (i.e., sporadic) meningiomas [8]. Biallelic inactivation of *NF2* through the loss of the long arm of chromosome 22 (specifically 22q12.2), in association with a loss-of-function mutation, culminate in a loss of tumour suppressor gene activity. The adjacent neighbour of *NF2*, the gene coding for the SWI/SNF-related matrix-associated actin-dependent regulator of chromatin subfamily B member 1 protein (*SMARCB1*), is particularly implicated in anterior falcine meningiomas [11, 12]. There is also an interesting demarcation at the level of the sphenoid bone at the boundary of embryonic intramembranous (convexity) and endochondral ossification (skull base) where meninges from both neural crest and mesoderm meet (Fig. 1) [13].

Moreover, spinal meningiomas, comprising 7.7% of meningiomas (as reported by the cancer registry in England [14] and 4.1% in the United States [15]) were found to be associated with pathogenic *NF2* variants [13] as well as loss-of-function mutations in *SMARCE1*; associated with multiple spinal meningiomas and a clear cell histological subtype [16, 17].

The advent of whole genome and exome sequencing has brought about the discovery of non-*NF2* meningioma variants [2, 18]. These genes are involved in varied and often intercepting signalling pathways, by which we classify them below.

2.2 *TRAF7*, *KLF4*, and the PI3K-AKT-mTOR Pathway

TRAF7 is an E3 ubiquitin ligase that is central to cell survival and proliferation. It is the most common mutation found in meningiomas after *NF2* [13] and is present in about 25% of meningiomas [19], and in 5.5% where it is the sole pathogenic variant [1]. They classically develop in the anterior and middle skull base [1]. Several mutations have been reported at a specific intronic locus on *TRAF7*; this intron contributes to the WD40 repeats on the TRAF7 protein, serving as a scaffold for the NF-κB and Jun/AP1 pathways [19, 20]. Work in frogs and zebrafish revealed a pleiotropic role for *Traf7* in congenital heart disease as well as meningiomas of the skull base. Mutated *Traf7* was able to interfere

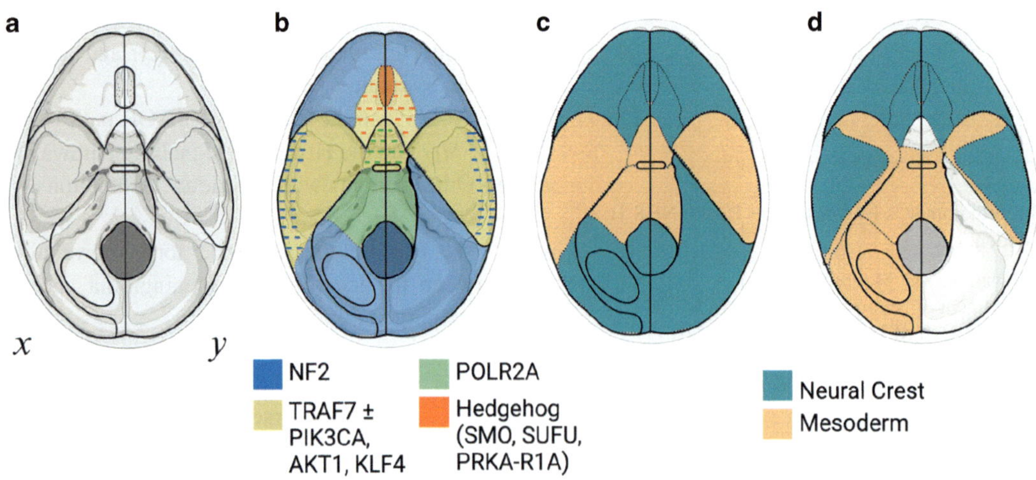

Fig. 1 The spatial phenotype of meningiomas. (**a**) The skull base. The left-hand side of the image (*x*) represents the skull base including the posterior cranial fossa, whilst the right-hand side (*y*) represent the skull base with the tentorium cerebelli. This convention is maintained throughout the rest of the figure. (**b**) Distribution of pathogenic variants of meningioma along the skull base. (**c**) Meningeal tissue origin at the skull base, with the tentorium present on the right-hand side posteriorly and posterior fossa on the left. (**d**) Bone tissue origin at the skull base. (Adapted with permission from Fountain et al. Oncogene 2021)

with normal wild-type TRAF7 protein function, disrupting downstream signalling with MEKK3 and JNK; it did this through heterodimerisation, having implications on cardiac as well as craniofacial development [21].

Pathogenic variants in *TRAF7* can occur in combination with variants in critical genes of the PI3K-AKT-mTOR pathway [22, 23], which is crucial in regulating various stages of the cell cycle. Here, genes including *AKT1/3, PIK3CA* and *PIK3R1* have pathogenic variants associated with *TRAF7*-mutated meningiomas [2, 3, 6, 24]. *TRAF7* has also been shown to interact with the RAS-RAF-ERK pathway, including through actin cytoskeleton regulation via RAS-Like GTPase signalling [25, 26]. Mutations in *KLF4* commonly co-occur with those in *TRAF7*, and they even cooperate to further promote tumorigenesis in meningeal cells; potentially through interfering with the feedback loop that is involved in regulating GTPase activity [25]. Specific *KLF4* mutations have, for example, been associated with increasing hypoxia-related signalling in tissue taken from skull base meningiomas [27]. The pattern of *TRAF7* variants, as described in Clarke's seminal paper, show three combinations: either mutated alone, in combination with *KLF4* or in concert with *AKT1*. For example, the K409Q mutation in *KLF4* almost always co-occurred with *TRAF7*, and the E17K mutation in *AKT1* occurred with *TRAF7* two thirds of the time [2].

2.3 *SMO* and the Hedgehog Signalling Pathway

This pathway is highly evolutionarily conserved and plays a role in embryonic development across various species [28]. There are various effectors in this pathway, but the following have been reported to be particularly implicated in meningiomas, notably *SMO* [4] but also *SUFU* [29] and *PRKA-R1A* [18]. In contrast to those meningiomas harbouring the 22q deletions, meningiomas with aberrant Hedgehog signalling do not display genomic instability [1, 6] and are more likely to be found at the anterior fossa midline [1, 2, 4]. The relevance of *SUFU* was reported in a family with multiple meningiomas through wild-type allele loss as per the classic two-hit model [29]. The association of *PRKA-R1A* and *SMO* variants has also been reported [4, 18]; indeed, changes to functioning of PKA go on to affect the presence of SMO on the cell surface as part of the signalling pathway [30]. And whilst we present these pathways separately to the aforementioned PI3K-AKT-mTOR pathway, there is crosstalk between them [31]. In fact, meningiomas harbouring this spectrum of mutations exhibit some overlap in spatial phenotype, particularly at the anterior and middle medial skull base [2, 4, 32].

2.4 Other Variants

Pathogenic variants in certain genes show a clear, mutually exclusive pattern of meningioma distribution. Meningiomas containing variants in *POLR2A* are stable, and localise to the tuberculum sellae [18]. And despite overlapping pathways, whilst *KLF4* is often seen in meningiomas of the lateral skull base (including the lateral sphenoid wing) [1], those tumours with *TRAF7* variants shows a propensity for the medial and anterior skull base [2, 3, 18]. Lastly, some pathogenic variants, such as *BAP1*, do not yet show evidence convincing enough to be able to assign a tentative spatial phenotype.

2.5 The Relationship of Histopathology with Genotype

Meningiomas are graded according to a WHO grade and have a relatively large variety of histological subtypes compared to other types of tumour [33]. Much like the association of pathogenic variants with meningioma location, there are also associations between the variants and the histological subtypes (see Table 1). There are some pathogenic variants that show exclusive association with histopathological subtype. For example, alterations in *NF2* and *SMARCB1* exhibit a relationship with fibrous meningiomas,

Table 1 A summary of the relationship between major meningioma genotypes with known spatial phenotypes [1–6, 10, 11, 18, 24, 34, 35]

Category	Genotype(s)	Location	Histology
NF2-associated	NF2	Convexity Spinal	Fibrous Transitional Atypical Anaplastic
	SMARCB1	Falx cerebri	Atypical
TRAF7-associated	TRAF7-alone	Medial anterior cranial fossa Middle cranial fossa	Meningothelial
	PIK3CA[a]		
	AKT1[a]		
	KLF4[a]		Secretory
Classical hedgehog	SMO	Medial anterior cranial fossa	Meningothelial
	SUFU		
	PRKA-R1A		
Other	POLR2A	Medial middle and posterior cranial fossae	Meningothelial
	SMARCE1	Spinal	Clear cell
	BAP1	Convexity	Rhabdoid
	Hyperdiploid	Convexity	Angiomatous Transitional

[a] Pathogenic variants in KLF4, AKT1, and PIK3CA occurring independently or in combination with TRAF7

SMARCE1 with clear cell, combined *KLF4* and *TRAF7* with secretory meningiomas, and *BAP1* with the rhabdoid subtype [1–6, 10, 11, 18, 24, 34, 35]. Secretory meningiomas are a subtype of meningioma that are particularly notable for causing significant cytotoxic disproportionate oedema that can lead to complications in the peri-operative phase, although they otherwise have a similar prognosis to other benign subtypes [36].

3 Development of the Meninges in Humans

In a landmark study in 1986, O'Rahilly and Müller investigated serially sectioned embryonic meninges from postovulatory day 18 to postovulatory week 8 (from Carnegie stage 8 to 23), revealing much of what we understand about *human* meningeal development (see Fig. 2) [37]. Between days 22 and 24, migrating neural crest cells envelop the neural axis, after which another thicker and looser mesenchymal layer also migrates through days 24–28, starting from the hindbrain to cover the neural tube [38]. At around the fifth week of gestation (Carnegie stage 15), this mesenchyme envelops the brain and becomes the primary meninx; this tissue does not only give rise to a future mature meninges, but also to the overlying bones of the skull, periosteum, the future vasculature and the dermal layers of the scalp [37, 39]. At Carnegie stage 16 the notochord's cellular sheath contributes to the medial aspect of the future tentorium cerebelli; this mesenchymal tissue will later be replaced by the dura after displacement away from the midline [37].

By day 42 (Carnegie stage 17), 3 separate layers exist: first the dermal layer, then the calvarial layer (future bone and its sutures) and then the meninges [40]. At this stage the dural limiting layer is forming in the basal regions alongside mesenchymal condensation. At Carnegie stage 18 (day 44), the primary meninx cavitates. Then in Carnegie stage 19 (day 48), the veins that will become the future sinuses appear, as does the choroid plexus of the fourth ventricle. Over the next week of development, the outer ecto/pachymeninx (future dura) forms over the endo/leptomeninx (future arachnoid and pia) establishing the intermediate stage between the early primary meninx and the differentiated meninges. By day 55 the dura, arachnoid and the pia mater have begun their assembly, developing in a basolateral to apical direction (a pattern discovered in *Foxc1* mutant mice) [41]. By Carnegie stage 23 there are three clearly distinct meningeal layers, and a

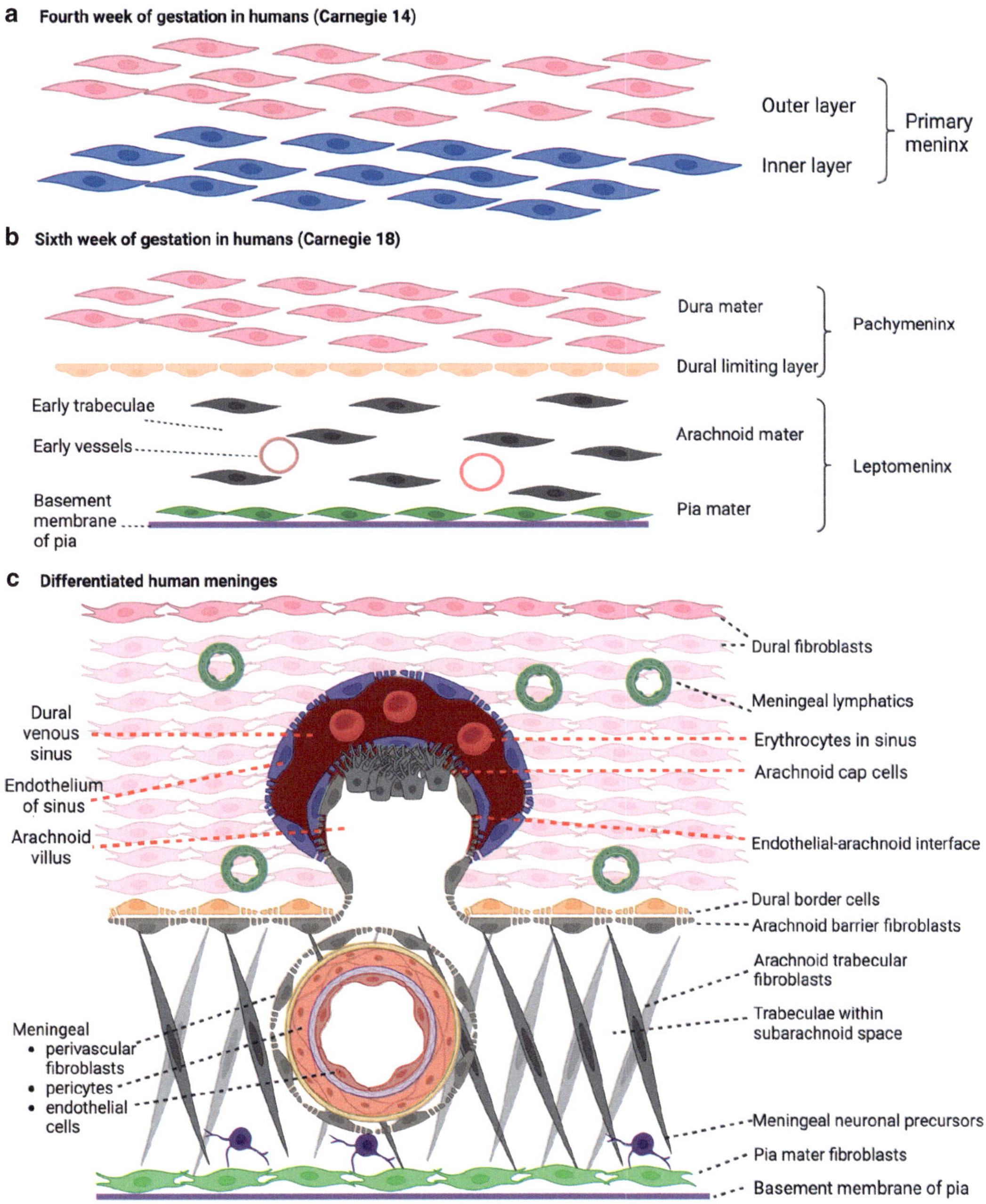

Fig. 2 Meningeal development. (**a**) The establishment of the primary meninx with an inner and outer layer. At Carnegie stage 14. (**b**) The progress to the pachymeninx and leptomeninx. At Carnegie stage 18. (**c**) The meninges and its differentiated constituent cell types. The subarachnoid space contains vasculature with an endothelium, pericytes and perivascular fibroblasts. The arachnoid villi invaginate through the dura into the dural venous sinuses where the arachnoid and endothelial layers form a direct interface. At the top of these villi sit the arachnoid cap cells. (Created with BioRender.com)

closed system is formed as the dural limiting layer continues to develop and set its margins. Most of the future adult cisternae are present, and the sigmoid and transverse sinuses now sit within the dura. The demarcation between developing brain parenchyma and meninges is made by the pial basement membrane which begins to be synthesised by extracellular matrix proteins.

After the establishment of these three layers and the subarachnoid space through a cavitation process, distinct cell populations arise [38, 39]. These include: dural border cells located at the dura-arachnoid interface with inter-junctional cellular connections [42]; arachnoid barrier cells with tight fibroblast packing [43]; intercalating arachnoid trabeculae filled with CSF that bridge across the subarachnoid space [44], and lastly arachnoid cap cells which, in their deep position on the apices of the arachnoid villi, have access to the venous sinuses [38].

At the spinal meninges, Sensenig pioneered work in the 1950s with 57 staged embryos showing that the neural crest and mesoderm have different adult derivates after development, those being the pia and arachnoid respectively [45, 46].

4 Developmental Origin of Meninges in Mammals from Animal Models

Advances in tracking the cell changes and migration in genetically modified animal models have been a major contributor in our understanding of the developmental origin of the meninges [38, 39]. In the 1990s pioneering work from Nicole Le Douarin used quail-chick chimeras created using microsurgical, isotropic and isochronic substitution (transplantation) methods to track the migration of neural crest cells to the adult forebrain meninges and separately the movement of mesodermal tissue to form the meninges of the midbrain and hindbrain [47]. The biological differences between these early cells in the stage-matched quail and chick model (such as the structural differences in the interphase nuclei between the two species) facilitated this [48]. Later, the neural crest marker *HNK1* was used to show that the spinal meninges in avian species is derived from the neural crest, and a *Wnt1-Cre* mouse model showed the same for the truncal meninges [49]. Regarding endothelial cells, these are generally agreed to all be derived from mesoderm [50].

The advent of conditional knockout models specific to developmental stages clarified the relationships between precursor and daughter cell lineages, as well as their migration during embryogenesis, compared to the previous snapshot views of early pioneering studies [51]. This heterogeneity in early cell populations is notable in the context of the histopathological variety which can be seen in meningioma subtypes. It also raises the question of what the common cellular origin of meningiomas is. Whilst the arachnoid cap cell is often described as this origin, the meningeal precursor cells that are positive for the prostaglandin D2 synthase (PTGDS) protein have been shown to potentially play a role [38]. *Ptgds* is expressed in arachnoid barrier cells in the adult mouse [52], and meningiomas in humans have shown that PTGDS is a good histological marker of arachnoid cells [53]. Precursor cells expressing *Ptgds* can differentiate into both arachnoid barrier or dural border cells, and *NF2* inactivation is all that is required for these cells to develop into meningothelial or fibroblastic meningiomas, respectively; we can now trace meningioma subtypes not only to the cells they derive from, but even further, to their initial precursors [38, 52].

The meninges also influences cranial development; these two developmental processes do not occur in isolation. Mouse models have shed light on the development of the viscerocranium (which houses the facial skeleton) and the neurocranium (which encases the brain). Murine and avian models have shown that the viscerocranium is derived exclusively from the neural crest whilst the vault develops from both neural crest and mesoderm [54–57]. Using C57BL/6J mouse embryos, McBratney-Owen and colleagues showed that the cranial base forms early in development, first as undifferentiated mesenchymal tissue that forms a solidifying chondrocranium between embryological days 11–16. The parachordal cartilage appears first near the notochord at embryological day 11, after which the structures chondrify and then ossify, starting in the posterior chondrocranium [58]. Furthermore, X-Gal staining of *Wnt1-Cre/R26R* and *Mesp1-Cre/R26R* transgenic mice has revealed that the cartilages that give rise to the ethmoid, presphenoid and basisphenoid arise from neural crest (apart from the hypochiasmatic cartilages) whilst the basioccipital and non-squamous temporal bones are mesodermal in origin [58].

The cranial base and the calvarium develop in concert, and this interplay will influence the covariation structure and the resulting shape in the adult form [59]. Knockout mice models demonstrated that the coronal and sagittal sutures form along the boundaries of a neural crest-mesodermal interface, similar to meningeal spatial phenotypes [55, 56]. This interplay between meningeal and calvarial development is also seen in *Foxc1* mutated mice which have defects in both structures [60, 61]. Further evidence comes from mice with *Tgfbr2* conditional gene ablation at the neural crest; these models developed skull defects such as calvarial agenesis and cleft palate, mediated by a disruption of TGFβ signalling [62]. In later development the role of the dura in cranial development was revealed through the study of postnatal rat and rabbit models; the dura demonstrated the ability to maintain the patency of the cranial sutures and also re-ossified the calvarium when overlying bone was surgically removed [63–65] This is thought to be due to the secretion of bone re-modelling regulators, such as BMP, FGF and TGFβ [66, 67].

The meninges also has a long list of contributory effects on the brain. Indeed, in the clinical setting, defects in the meninges are associated with cortical manifestations such as cobblestone lissencephaly [68] and Dandy-Walker malformations [69]. The meningeal functions include being a stem cell niche, angiogenesis, neurogenesis (and subsequent neuronal migration and alignment), and facilitating cellular survival [39]. Many of these traits are achieved by the secretion of various regulatory molecules. As a stem cell niche, evidence from rats has shown that glia-like cells in the leptomeninges house an extracellular matrix that can provide an initial scaffold for cortical neurons [70]. Several of the meningeal functions during development are due to retinoic acid (RA) signalling. For example, when this chemokinetic signal is removed, the cortical layering, specifically the multipolar-to-bipolar change in the intermediate zone, is disrupted [71]. RA signalling can also rescue cortical development in *Foxc1* mutant mice with an absent meninges; whether it is absolutely indispensable to this role remains to be seen [61, 71]. RA signalling is also important for regulating angiogenesis [72]. Lastly, besides RA, meningeal cells can release both chemoattractant [73] and chemorepulsant [74] molecules to affect neuronal and glial translocation.

More recently, single-cell transcriptomic approaches have uncovered different markers in the dura, arachnoid and pia, through the distinct epigenomic profile seen in fibroblast diversity [75]. Arachnoid cells in the developing telencephalon have lower *Ptgds* expression in comparison to those at the skull base. The deployment of more novel techniques that exploit the epigenomic profile of tumour samples is already being used at the cutting edge to understand biological drivers of meningiomas [6, 76]. In addition, conditional activation mouse models have shown that there are specific developmental periods where the activation of certain genes and pathways results in the development of meningiomas. For example, when *Smo* was conditionally activated in the mesodermal *Ptgds*-expressing layer of the mouse skull base, meningiomas developed in this region [77]. The proliferation of *SmoM2* arachnoidal cells was inhibited by Sonidegib in vitro, showing early evidence that specific inhibitors of these developmental implicated genetic pathways could perhaps target meningiomas in the future [77].

5 Hypotheses for Meningioma Tumourigenesis

To what extent does our understanding of meningeal embryology help us establish hypotheses for meningioma tumorigenesis? Are there driver mutations and epigenomic profiles that make certain meningeal cell lineages more sensitive to tumorigenesis than others? Why is this relevant? Recent studies show that meningiomas have different epigenomic, genomic and malignant profiles in the skull base relative to those not at the skull base [24, 78, 79]. Moreover, why does the neural crest-derived meninges have a propensity to develop tumours following biallelic loss of *NF2*? Or why does the skull base have a propensity for more common markers that are seen in tumours in other body systems?

5.1 The Merlin, the Meninges and the Meningioma

Merlin, the protein resulting from *NF2* transcription, is a key central regulatory protein in embryogenesis [80]. Many of the embryological mechanisms it is involved in are mediated through contact-dependent growth inhibition, a key conduit for organising developing tissue in a regulated and restricted way. These mechanisms include establishing junctional polarity complexes, linking the plasma membrane to the cortical cytoskeleton and playing in role in tight/adherens-junctions [81, 82] It also plays a role in the development of neural crest-derived meninges and Schwann cells. Merlin, being part of the membrane-cytoskeleton linker family, acts as a tumour suppressor gene and inhibits mitogenic signalling [83, 84]. We know this through some key murine evidence: when *Nf2* is mutated in embryonic stem cells, the (homozygous) mice embryos fail by embryonic day 7 as the extraembryonic ectoderm does not organise and no gastrulation takes place [85]; when *Nf2* is deleted in a homozygous manner in arachnoid or Schwann cells, the tissue exhibits hyperplasia and tumorigenesis [86, 87]. We also have localising evidence for *Nf2* activity; when tracking its promoter (β-Gal) in mice models, its RNA expression can be first detected by embryonic day 5.5 and even more so by day 6.6 where it is prominent at the ectoderm with high mitotic activity. Additionally, the activity of this promoter shows transient changes during neural crest cell migration, as well as differences in the intensity of expression in different sites. It also demonstrates activity in various tissues that are affected in NF2, such as the acoustic, trigeminal and dorsal root ganglia [88].

The pathological loss of effective *NF2* transcription in the NF2 syndrome drives meningioma development through Merlin inactivation [38]. Whilst this mechanism is not yet fully understood, it is known to affect Hippo pathway signalling. One of the potential mechanisms involves the lipid binding to the plasma membrane in response to osmotic stress (with mediators including ARF6 and PIP proteins) [89]. In doing so the *NF2*-Hippo pathway interaction can regulate key cell cycle stages (including programmed cell death) [90] as well as cellular spread through its membrane-cytoskeleton linker function [91]. Furthermore, both in vitro [92] and knockout mouse [93] models have shown that the activity of the Yap oncoprotein can drive early neural crest migration. The closely related *SMARCB1* gene and its pathogenic variants are also highly relevant in our discussion of tumorigenesis mechanisms; not only is it important for neural induction and pluripotency regulation in embryonic stem cells via the SWI/SNF complex [94], the stage at which it is lost affects the type of tumour that develops (earlier, solitary loss causes rhabdoid tumours, whereas later loss jointly with *Nf2* inactivation predisposes to schwannomas) [95].

Moreover, why do cells derived from the neural crest often demonstrate this loss of 22q? In the three step/four-hit model (as hypothesised for neurofibromatosis type 2), an initial early somatic or germline variant in *SMARCB1* or *LZTR1* (both found on 22q) is followed by a loss of heterozygosity in 22q resulting in monosomy with subsequent expression of only the pathogenic variants [96]. Other signalling genes that distribute along 22q, such as *SOX10* and *PDGFB* (which both contribute to specification of the neural crest lineage), further compound the consequences of monosomy of 22q [97].

5.2 Hedgehog, mTOR, and Wnt

The Hedgehog pathway's central role in most embryogenesis cascades, being particularly critical to organisation of the CNS and the establishment of polarity during vertebral limb and digit development. In the CNS, it has wide-ranging roles including acting as a morphogen to pattern the cortex, pituitary, and spinal cord, as well as acting as a guide for developing axons and its role in the developing notochord and floor plate [98]. This makes it a likely contender for being involved in tumorigenesis in meningiomas. Its activation in meningiomas was shown at the mRNA level [99] and mutations in its component genes, *SMO* and *SUFU*, are seen in up to 5% of meningiomas [1, 100], which some authors classify as its own group of "hedgehog-activated" tumours [38]. Dysregulation of this pathway is also linked to disorders of craniofacial development, like Gorlin syndrome [101], as well as evidence in zebrafish [102].

The mTOR pathway is upregulated in both *NF2* mutations with loss of Merlin, as well as in about 10% of meningiomas with *PIK3CA* and *AKT1* pathogenic variants [38, 103], and there is an association of the mTOR pathway with Proteus syndrome. Some authors argue that both Proteus syndrome and sporadic meningiomas are on a spectrum [38]; they both implicate *AKT1* mutations and lead to meningiomas that are hyperostotic, meningothelial, and at the anterior and middle skull base [104, 105].

The Wnt pathway is another central player in meningeal, and indeed global, development [106]. Through regulatory networks involving *TWIST1* [72], *FOXC1* [41] *and ZIC* [107] genes, it is involved in meningeal embryogenesis, including differentiation and progress of associated cerebral vasculature. Evidence from transcriptomic profiles of almost 300 meningiomas shows that many of the aforementioned mechanisms converge on the *FOXM1*/Wnt pathway, and that meningiomas with this trait have worse clinical outcomes and show mitotic gene expression [108].

5.3 Bringing It Together. Outstanding Questions

Questions remain, and much of it has to do with the models that are available at present. For example, cell lines consist of immortalised malignant cells that may not necessarily represent the natural pathophysiology that underlies most benign meningiomas [13].

Further uncertainties remain regarding:

- the true precursor(s) to the developing meningioma,
- the extent to which the tumours are heterogenous,
- their microenvironment,
- immune cell milieu surrounding meningiomas,
- international consensus on reporting and defining meningioma locations.

Additionally, the underlying genetic changes in the tumour may not yet explain the whole story; in a study of almost 2000 meningiomas, a quarter of them did not possess an underlying identifiable mutation [1]. This could be due to unidentified pathogenic variants. In this way, the barrier to discovering these new mutations may simply be the need for more time and perseverance with current approaches, or alternatively could be explained by the limitations of whole exome sequencing and the need for a yet-to-be deployed new assay or technology.

Furthermore, whilst the evidence suggests a relatively clear demarcation between the meninges derived from the neural crest and those derived from the mesoderm, how certain is it to say that they contribute to embryogenesis (and hence possibly tumorigenesis) in a mutually exclusive way? What happens along the borders/fault lines between the different meningeal spatial regions?

The pathways involved in meningeal and cranial bone development are intertwined, and we know that Merlin interacts with several developmental signalling pathways, including Hippo, WNT/β-catenin, TGF-β, Notch and Hedgehog [80]. This is particularly relevant to the Hedgehog signalling pathways which have been shown to control polar patterning of the extremities, height and, crucially, the central nervous system, including regulating neural progenitors and neural stem cells. Many of its effects are due to intra- and intercellular gradients that are set up in these developing tissues; much of this information has been gained from *Drosophila*, zebrafish, chick and other animal models [109]. However, the Hedgehog signalling pathway may follow patterns that are specific to each species; therefore, it is important to caveat the success of animal model findings and try look for further evidence in humans. Evidence using next generation sequencing, multi-omic transcriptional and epigenomic studies can play a key role going forward. Identification of dynamic markers at the prenatal stage in human embryonic studies could pave the way for investigating the migration and differentiation of meningioma precursor cells.

Furthermore, meningiomas are, in general, slow growing tumours. They therefore usually depend on biallelic disruption as per Knudson's two-hit hypothesis. Whilst many meningiomas are amenable to resection with good outcomes, more aggressively behaving and recurrent meningiomas require further treatment and manifest more aggressive molecular subtypes including ultimately WHO grade progression. The tumours that progress in this way can have lower pre-operative Karnofsky Performance Scale (KPS) scores and are more irregular in their morphology [110]. Whilst we know that aggressive meningiomas (even after total resection) exhibit loss of the DREAM complex function and subsequent activation of the cell cycle [24], it would be useful to compare the profile of these meningiomas before and after they underwent progression. A recent cohort study found that, in general, those meningiomas that progress through to higher WHO grades may have clues in their original (pre-progression) biopsy, through somatic mutation and copy number alteration analysis [111].

6 Conclusion

Ultimately, efforts to elucidate the molecular markers of tumours are to guide future clinical therapies. Discerning meningeal development on a spatiotemporal level will be a key in further expanding on our understanding of meningioma development, including its proclivity for specific locations, and perhaps explain why some subtypes develop more than others, why some are aggressive, and some are less so, alongside the central questions of developing effective targeted therapies tailored to the spatial and genomic profile of each patient's meningioma. Furthermore, what about the approximately quarter of meningiomas that do not have an identified underlying mutation? Or indeed the interaction of early tumorigenic cells in the complex immune and signalling milieu of the meninges? The fact that the expression of genes crucial in development, tumour suppressor genes and oncogenes all converge on key cellular proliferation, apoptotic, communication, and survival pathways make parsing out the traits that are exclusive or specific to a given meningioma that much harder. Many of these questions likely await the development of techniques with even more spatial and temporal resolution for tracking intra- and intercellular signalling at the (epi)genetic level.

References

1. Youngblood MW, Duran D, Montejo JD, Li C, Omay SB, Özduman K, et al. Correlations between genomic subgroup and clinical features in a cohort of more than 3000 meningiomas. J Neurosurg. 2019;133(5):1345–54.
2. Clark VE, Erson-Omay EZ, Serin A, Yin J, Cotney J, Ozduman K, et al. Genomic analysis of non-NF2 meningiomas reveals mutations in TRAF7, KLF4, AKT1, and SMO. Science. 2013;339(6123):1077–80.
3. Abedalthagafi M, Bi WL, Aizer AA, Merrill PH, Brewster R, Agarwalla PK, et al. Oncogenic PI3K mutations are as common as AKT1 and SMO mutations in meningioma. Neuro Oncol. 2016;18(5):649–55.
4. Brastianos PK, Horowitz PM, Santagata S, Jones RT, McKenna A, Getz G, et al. Genomic sequencing of meningiomas identifies oncogenic SMO and AKT1 mutations. Nat Genet. 2013;45(3):285–9.
5. Mawrin C, Perry A. Pathological classification and molecular genetics of meningiomas. J Neurooncol. 2010;99(3):379–91.
6. Sahm F, Schrimpf D, Stichel D, Jones DTW, Hielscher T, Schefzyk S, et al. DNA methylation-based classification and grading system for meningioma: a multicentre, retrospective analysis. Lancet Oncol. 2017;18(5):682–94.

7. Nassiri F, Liu J, Patil V, Mamatjan Y, Wang JZ, Hugh-White R, et al. A clinically applicable integrative molecular classification of meningiomas. Nature. 2021;597(7874):119–25.
8. Riemenschneider MJ, Perry A, Reifenberger G. Histological classification and molecular genetics of meningiomas. Lancet Neurol. 2006;5(12):1045–54.
9. Maas SLN, Stichel D, Hielscher T, Sievers P, Berghoff AS, Schrimpf D, et al. Integrated molecular-morphologic meningioma classification: a multicenter retrospective analysis. Retrospectively and Prospectively Validated JCO. 2021;39(34):3839–52.
10. Kros J, de Greve K, van Tilborg A, Hop W, Pieterman H, Avezaat C, et al. NF2 status of meningiomas is associated with tumour localization and histology. J Pathol. 2001;194(3):367–72.
11. Schmitz U, Mueller W, Weber M, Sévenet N, Delattre O, von Deimling A. INI1 mutations in meningiomas at a potential hotspot in exon 9. Br J Cancer. 2001;84(2):199–201.
12. van den Munckhof P, Christiaans I, Kenter SB, Baas F, Hulsebos TJM. Germline SMARCB1 mutation predisposes to multiple meningiomas and schwannomas with preferential location of cranial meningiomas at the falx cerebri. Neurogenetics. 2012;13(1):1–7.
13. Fountain DM, Smith MJ, O'Leary C, Pathmanaban ON, Roncaroli F, Bobola N, et al. The spatial phenotype of genotypically distinct meningiomas demonstrate potential implications of the embryology of the meninges. Oncogene. 2021;40(5):875–84.
14. Brodbelt AR, Barclay ME, Greenberg D, Williams M, Jenkinson MD, Karabatsou K. The outcome of patients with surgically treated meningioma in England: 1999–2013. A cancer registry data analysis. Br J Neurosurg. 2019;33(6):641–7.
15. Ostrom QT, Price M, Neff C, Cioffi G, Waite KA, Kruchko C, et al. CBTRUS statistical report: primary brain and other central nervous system tumors diagnosed in the United States in 2016–2020. Neuro Oncol. 2023;25(Suppl 4):iv1–99.
16. Smith MJ, O'Sullivan J, Bhaskar SS, Hadfield KD, Poke G, Caird J, et al. Loss-of-function mutations in SMARCE1 cause an inherited disorder of multiple spinal meningiomas. Nat Genet. 2013;45(3):295–8.
17. Smith MJ, Wallace AJ, Bennett C, Hasselblatt M, Elert-Dobkowska E, Evans LT, et al. Germline SMARCE1 mutations predispose to both spinal and cranial clear cell meningiomas. J Pathol. 2014;234(4):436–40.
18. Clark VE, Harmancı AS, Bai H, Youngblood MW, Lee TI, Baranoski JF, et al. Recurrent somatic mutations in POLR2A define a distinct subset of meningiomas. Nat Genet. 2016;48(10):1253–9.
19. Zotti T, Scudiero I, Vito P, Stilo R. The emerging role of TRAF7 in tumor development. J Cell Physiol. 2017;232(6):1233–8.
20. Bouwmeester T, Bauch A, Ruffner H, Angrand PO, Bergamini G, Croughton K, et al. A physical and functional map of the human TNF-alpha/NF-kappa B signal transduction pathway. Nat Cell Biol. 2004;6(2):97–105.
21. Mishra-Gorur K, Barak T, Kaulen LD, Henegariu O, Jin SC, Aguilera SM, et al. Pleiotropic role of TRAF7 in skull-base meningiomas and congenital heart disease. Proc Natl Acad Sci. 2023;120(16):e2214997120.
22. Porta C, Paglino C, Mosca A. Targeting PI3K/Akt/mTOR signaling in cancer. Front Oncol. 2014;4:64. https://www.frontiersin.org/journals/oncology/articles/10.3389/fonc.2014.00064/full.
23. Reuss DE, Piro RM, Jones DTW, Simon M, Ketter R, Kool M, et al. Secretory meningiomas are defined by combined KLF4 K409Q and TRAF7 mutations. Acta Neuropathol. 2013;125(3):351–8.
24. Patel AJ, Wan YW, Al-Ouran R, Revelli JP, Cardenas MF, Oneissi M, et al. Molecular profiling predicts meningioma recurrence and reveals loss of DREAM complex repression in aggressive tumors. Proc Natl Acad Sci U S A. 2019;116(43):21715–26.
25. Najm P, Zhao P, Steklov M, Sewduth RN, Baietti MF, Pandolfi S, et al. Loss-of-function mutations in TRAF7 and KLF4 cooperatively activate RAS-like GTPase signaling and promote meningioma development. Cancer Res. 2021;81(16):4218–29.
26. Szulzewsky F, Thirimanne HN, Holland EC. Meningioma: current updates on genetics, classification, and mouse modeling. Ups J Med Sci. 2024;129:e10579. https://doi.org/10.48101/ujms.v129.10579.
27. von Spreckelsen N, Waldt N, Poetschke R, Kesseler C, Dohmen H, Jiao HK, et al. KLF4K409Q-mutated meningiomas show enhanced hypoxia signaling and respond to mTORC1 inhibitor treatment. Acta Neuropathol Commun. 2020;8(1):41.
28. Skoda AM, Simovic D, Karin V, Kardum V, Vranic S, Serman L. The role of the hedgehog signaling pathway in cancer: a comprehensive review. Bosn J Basic Med Sci. 2018;18(1):8–20.
29. Aavikko M, Li SP, Saarinen S, Alhopuro P, Kaasinen E, Morgunova E, et al. Loss of SUFU function in familial multiple meningioma. Am J Hum Genet. 2012;91(3):520–6.
30. Jia J, Tong C, Wang B, Luo L, Jiang J. Hedgehog signalling activity of smoothened requires phosphorylation by protein kinase A and casein kinase I. Nature. 2004;432(7020):1045–50.
31. Montagnani V, Stecca B. Role of protein kinases in hedgehog pathway control and implications for cancer therapy. Cancers. 2019;11(4):449.
32. Sahm F, Bissel J, Koelsche C, Schweizer L, Capper D, Reuss D, et al. AKT1E17K mutations cluster with meningothelial and transitional meningiomas and can be detected by SFRP1 immunohistochemistry. Acta Neuropathol. 2013;126(5):757–62.

33. Louis DN, Perry A, Wesseling P, Brat DJ, Cree IA, Figarella-Branger D, et al. The 2021 WHO classification of tumors of the central nervous system: a summary. Neuro Oncol. 2021;23(8):1231–51.
34. Shankar GM, Abedalthagafi M, Vaubel RA, Merrill PH, Nayyar N, Gill CM, et al. Germline and somatic BAP1 mutations in high-grade rhabdoid meningiomas. Neuro Oncol. 2017;19(4):535–45.
35. Wellenreuther R, Kraus JA, Lenartz D, Menon AG, Schramm J, Louis DN, et al. Analysis of the neurofibromatosis 2 gene reveals molecular variants of meningioma. Am J Pathol. 1995;146(4):827–32.
36. Regelsberger J, Hagel C, Emami P, Ries T, Heese O, Westphal M. Secretory meningiomas: a benign subgroup causing life-threatening complications. Neuro Oncol. 2009;11(6):819–24.
37. O'Rahilly R, Müller F. The meninges in human development. J Neuropathol Exp Neurol. 1986;45(5):588–608.
38. Boetto J, Peyre M, Kalamarides M. Meningiomas from a developmental perspective: exploring the crossroads between meningeal embryology and tumorigenesis. Acta Neurochir. 2021;163(1):57–66.
39. Dasgupta K, Jeong J. Developmental biology of the meninges. Genesis. 2019;57(5):e23288.
40. Siegenthaler JA, Pleasure SJ. We have got you 'covered': how the meninges control brain development. Curr Opin Genet Dev. 2011;21(3):249–55.
41. Vivatbutsiri P, Ichinose S, Hytönen M, Sainio K, Eto K, Iseki S. Impaired meningeal development in association with apical expansion of calvarial bone osteogenesis in the Foxc1 mutant. J Anat. 2008;212(5):603–11.
42. Adeeb N, Mortazavi MM, Tubbs RS, Cohen-Gadol AA. The cranial dura mater: a review of its history, embryology, and anatomy. Childs Nerv Syst. 2012;28(6):827–37.
43. Haines DE, Harkey HL, al-Mefty O. The 'subdural' space: a new look at an outdated concept. Neurosurgery. 1993;32(1):111–20.
44. Barshes N, Demopoulos A, Engelhard HH. Anatomy and physiology of the leptomeninges and CSF space. Cancer Treat Res. 2005;125:1–16.
45. Sensenig EC. The early development of the meninges of the spinal cord in human embryos. Contrib Embryol. 1951;228:145–57.
46. Sensenig EC. The development of the occipital and cervical segments and their associated structures in human embryos. Contrib Embryol Carnegie Inst. 1957;36:141–51.
47. Couly GF, Coltey PM, Le Douarin NM. The developmental fate of the cephalic mesoderm in quail-chick chimeras. Development. 1992;114(1):1–15.
48. Le Lièvre CS, Le Douarin NM. Mesenchymal derivatives of the neural crest: analysis of chimaeric quail and chick embryos. J Embryol Exp Morphol. 1975;34(1):125–54.
49. Batarfi M, Valasek P, Krejci E, Huang R, Patel K. The development and origins of vertebrate meninges. Biol Commun. 2017;62(2):73–81.
50. Douarin NL, Kalcheim C. The neural crest. Cambridge: Cambridge University Press; 1999. 494 p
51. Kalamarides M, Peyre M, Giovannini M. Meningioma mouse models. J Neurooncol. 2010;99(3):325–31.
52. Kalamarides M, Stemmer-Rachamimov AO, Niwa-Kawakita M, Chareyre F, Taranchon E, Han ZY, et al. Identification of a progenitor cell of origin capable of generating diverse meningioma histological subtypes. Oncogene. 2011;30(20):2333–44.
53. Yamashima T, Sakuda K, Tohma Y, Yamashita J, Oda H, Irikura D, et al. Prostaglandin D synthase (beta-trace) in human arachnoid and meningioma cells: roles as a cell marker or in cerebrospinal fluid absorption, tumorigenesis, and calcification process. J Neurosci. 1997;17(7):2376–82.
54. Gross JB, Hanken J. Review of fate-mapping studies of osteogenic cranial neural crest in vertebrates. Dev Biol. 2008;317(2):389–400.
55. Yoshida T, Vivatbutsiri P, Morriss-Kay G, Saga Y, Iseki S. Cell lineage in mammalian craniofacial mesenchyme. Mech Dev. 2008;125(9–10):797–808.
56. Jiang X, Iseki S, Maxson RE, Sucov HM, Morriss-Kay GM. Tissue origins and interactions in the mammalian skull vault. Dev Biol. 2002;241(1):106–16.
57. Evans DJR, Noden DM. Spatial relations between avian craniofacial neural crest and paraxial mesoderm cells. Dev Dyn. 2006;235(5):1310–25.
58. McBratney-Owen B, Iseki S, Bamforth SD, Olsen BR, Morriss-Kay GM. Development and tissue origins of the mammalian cranial base. Dev Biol. 2008;322(1):121–32.
59. Lieberman DE, Hallgrímsson B, Liu W, Parsons TE, Jamniczky HA. Spatial packing, cranial base angulation, and craniofacial shape variation in the mammalian skull: testing a new model using mice. J Anat. 2008;212(6):720–35.
60. Kume T, Deng KY, Winfrey V, Gould DB, Walter MA, Hogan BL. The forkhead/winged helix gene Mf1 is disrupted in the pleiotropic mouse mutation congenital hydrocephalus. Cell. 1998;93(6):985–96.
61. Siegenthaler JA, Ashique AM, Zarbalis K, Patterson KP, Hecht JH, Kane MA, et al. Retinoic acid from the meninges regulates cortical neuron generation. Cell. 2009;139(3):597–609.
62. Ito Y, Yeo JY, Chytil A, Han J, Bringas P, Nakajima A, et al. Conditional inactivation of Tgfbr2 in cranial neural crest causes cleft palate and calvaria defects. Development. 2003;130(21):5269–80.
63. Levine JP, Bradley JP, Roth DA, McCarthy JG, Longaker MT. Studies in cranial suture biology: regional dura mater determines overlying suture biology. Plast Reconstr Surg. 1998;101(6):1441–7.
64. Mabbutt LW, Kokich VG. Calvarial and sutural redevelopment following craniectomy in the neonatal rabbit. J Anat. 1979;129(Pt 2):413–22.
65. Grova M, Lo DD, Montoro D, Hyun JS, Chung MT, Wan DC, et al. Models of cranial suture biology. J Craniofac Surg. 2012;23(7 Suppl 1):1954–8.

66. Warren SM, Brunet LJ, Harland RM, Economides AN, Longaker MT. The BMP antagonist noggin regulates cranial suture fusion. Nature. 2003;422(6932):625–9.
67. Spector JA, Greenwald JA, Warren SM, Bouletreau PJ, Detch RC, Fagenholz PJ, et al. Dura mater biology: autocrine and paracrine effects of fibroblast growth factor 2. Plast Reconstr Surg. 2002;109(2):645–54.
68. Devisme L, Bouchet C, Gonzalès M, Alanio E, Bazin A, Bessières B, et al. Cobblestone lissencephaly: neuropathological subtypes and correlations with genes of dystroglycanopathies. Brain. 2012;135(Pt 2):469–82.
69. Aldinger KA, Lehmann OJ, Hudgins L, Chizhikov VV, Bassuk AG, Ades LC, et al. FOXC1 is required for normal cerebellar development and is a major contributor to chromosome 6p25.3 Dandy-Walker malformation. Nat Genet. 2009;41(9):1037–42.
70. Bifari F, Decimo I, Pino A, Llorens-Bobadilla E, Zhao S, Lange C, et al. Neurogenic radial glia-like cells in meninges migrate and differentiate into functionally integrated neurons in the neonatal cortex. Cell Stem Cell. 2017;20(3):360–373.e7.
71. Haushalter C, Schuhbaur B, Dollé P, Rhinn M. Meningeal retinoic acid contributes to neocortical lamination and radial migration during mouse brain development. Biol Open. 2017;6(2):148–60.
72. Bonney S, Harrison-Uy S, Mishra S, MacPherson AM, Choe Y, Li D, et al. Diverse functions of retinoic acid in brain vascular development. J Neurosci. 2016;36(29):7786–801.
73. Zarbalis K, Choe Y, Siegenthaler JA, Orosco LA, Pleasure SJ. Meningeal defects alter the tangential migration of cortical interneurons in Foxc1hith/hith mice. Neural Dev. 2012;7:2.
74. Choe Y, Huynh T, Pleasure SJ. Migration of oligodendrocyte progenitor cells is controlled by transforming growth factor β family proteins during corticogenesis. J Neurosci. 2014;34(45):14973–83.
75. DeSisto J, O'Rourke R, Jones HE, Pawlikowski B, Malek AD, Bonney S, et al. Single-cell transcriptomic analyses of the developing meninges reveal meningeal fibroblast diversity and function. Dev Cell. 2020;54(1):43–59.e4.
76. Choudhury A, Magill ST, Eaton CD, Prager BC, Chen WC, Cady MA, et al. Meningioma DNA methylation groups identify biological drivers and therapeutic vulnerabilities. Nat Genet. 2022;54(5):649–59.
77. Boetto J, Apra C, Bielle F, Peyre M, Kalamarides M. Selective vulnerability of the primitive meningeal layer to prenatal Smo activation for skull base meningothelial meningioma formation. Oncogene. 2018;37(36):4955–63.
78. Cornelius JF, Slotty PJ, Steiger HJ, Hänggi D, Polivka M, George B. Malignant potential of skull base versus non-skull base meningiomas: clinical series of 1,663 cases. Acta Neurochir. 2013;155(3):407–13.
79. Himič V, Burman RJ, Fountain DM, Hofer M, Livermore LJ, Jeyaretna DS. Metastatic meningioma: a case series and systematic review. Acta Neurochir. 2023;165(10):2873–83.
80. Mota M, Shevde LA. Merlin regulates signaling events at the nexus of development and cancer. Cell Commun Signal. 2020;18(1):63.
81. Gladden AB, Hebert AM, Schneeberger EE, McClatchey AI. The NF2 tumor suppressor, Merlin, regulates epidermal development through the establishment of a junctional polarity complex. Dev Cell. 2010;19(5):727–39.
82. Yi C, Troutman S, Fera D, Stemmer-Rachamimov A, Avila JL, Christian N, et al. A tight junction-associated Merlin-angiomotin complex mediates Merlin's regulation of mitogenic signaling and tumor suppressive functions. Cancer Cell. 2011;19(4):527–40.
83. Hamaratoglu F, Willecke M, Kango-Singh M, Nolo R, Hyun E, Tao C, et al. The tumour-suppressor genes NF2/Merlin and expanded act through hippo signalling to regulate cell proliferation and apoptosis. Nat Cell Biol. 2006;8(1):27–36.
84. Li W, You L, Cooper J, Schiavon G, Pepe-Caprio A, Zhou L, et al. Merlin/NF2 suppresses tumorigenesis by inhibiting the E3 ubiquitin ligase CRL4DCAF1 in the nucleus. Cell. 2010;140(4):477–90.
85. McClatchey AI, Saotome I, Ramesh V, Gusella JF, Jacks T. The Nf2 tumor suppressor gene product is essential for extraembryonic development immediately prior to gastrulation. Genes Dev. 1997;11(10):1253–65.
86. Giovannini M, Robanus-Maandag E, van der Valk M, Niwa-Kawakita M, Abramowski V, Goutebroze L, et al. Conditional biallelic Nf2 mutation in the mouse promotes manifestations of human neurofibromatosis type 2. Genes Dev. 2000;14(13):1617–30.
87. Kalamarides M, Niwa-Kawakita M, Leblois H, Abramowski V, Perricaudet M, Janin A, et al. Nf2 gene inactivation in arachnoidal cells is rate-limiting for meningioma development in the mouse. Genes Dev. 2002;16(9):1060–5.
88. Akhmametyeva EM, Mihaylova MM, Luo H, Kharzai S, Welling DB, Chang LS. Regulation of the neurofibromatosis 2 gene promoter expression during embryonic development. Dev Dyn. 2006;235(10):2771–85.
89. Hong AW, Meng Z, Plouffe SW, Lin Z, Zhang M, Guan KL. Critical roles of phosphoinositides and NF2 in hippo pathway regulation. Genes Dev. 2020;34(7–8):511–25.
90. Gutmann DH, Giovannini M. Mouse models of neurofibromatosis 1 and 2. Neoplasia. 2002;4(4):279–90.
91. Laulajainen M, Muranen T, Carpén O, Grönholm M. Protein kinase A-mediated phosphorylation of the NF2 tumor suppressor protein merlin at serine 10 affects the actin cytoskeleton. Oncogene. 2008;27(23):3233–43.

92. Hindley CJ, Condurat AL, Menon V, Thomas R, Azmitia LM, Davis JA, et al. The hippo pathway member YAP enhances human neural crest cell fate and migration. Sci Rep. 2016;6:23208.
93. Zhang N, Bai H, David KK, Dong J, Zheng Y, Cai J, et al. The Merlin/NF2 tumor suppressor functions through the YAP oncoprotein to regulate tissue homeostasis in mammals. Dev Cell. 2010;19(1):27–38.
94. Langer LF, Ward JM, Archer TK. Tumor suppressor SMARCB1 suppresses super-enhancers to govern hESC lineage determination. elife. 2019;8:e45672.
95. Vitte J, Gao F, Coppola G, Judkins AR, Giovannini M. Timing of Smarcb1 and Nf2 inactivation determines schwannoma versus rhabdoid tumor development. Nat Commun. 2017;8(1):300.
96. Fountain DM, Sauka-Spengler T. The SWI/SNF complex in neural crest cell development and disease. Annu Rev Genomics Hum Genet. 2023;24:203–23.
97. Smith CL, Tallquist MD. PDGF function in diverse neural crest cell populations. Cell Adhes Migr. 2010;4(4):561–6.
98. Patten I, Placzek M. The role of sonic hedgehog in neural tube patterning. Cell Mol Life Sci. 2000;57(12):1695–708.
99. Laurendeau I, Ferrer M, Garrido D, D'Haene N, Ciavarelli P, Basso A, et al. Gene expression profiling of the hedgehog signaling pathway in human meningiomas. Mol Med. 2010;16(7–8):262–70.
100. Boetto J, Bielle F, Sanson M, Peyre M, Kalamarides M. SMO mutation status defines a distinct and frequent molecular subgroup in olfactory groove meningiomas. Neuro Oncol. 2017;19(3):345–51.
101. Kimonis VE, Goldstein AM, Pastakia B, Yang ML, Kase R, DiGiovanna JJ, et al. Clinical manifestations in 105 persons with nevoid basal cell carcinoma syndrome. Am J Med Genet. 1997;69(3):299–308.
102. Wada N, Javidan Y, Nelson S, Carney TJ, Kelsh RN, Schilling TF. Hedgehog signaling is required for cranial neural crest morphogenesis and chondrogenesis at the midline in the zebrafish skull. Development. 2005;132(17):3977–88.
103. Saxton RA, Sabatini DM. mTOR signaling in growth, metabolism, and disease. Cell. 2017;168(6):960–76.
104. Gilbert-Barness E, Cohen MM, Opitz JM. Multiple meningiomas, craniofacial hyperostosis and retinal abnormalities in Proteus syndrome. Am J Med Genet. 2000;93(3):234–40.
105. Keppler-Noreuil KM, Baker EH, Sapp JC, Lindhurst MJ, Biesecker LG. Somatic AKT1 mutations cause meningiomas colocalizing with a characteristic pattern of cranial hyperostosis. Am J Med Genet A. 2016;170(10):2605–10.
106. Dasgupta K, Chung JU, Asam K, Jeong J. Molecular patterning of the embryonic cranial mesenchyme revealed by genome-wide transcriptional profiling. Dev Biol. 2019;455(2):434–48.
107. Inoue T, Ogawa M, Mikoshiba K, Aruga J. Zic deficiency in the cortical marginal zone and meninges results in cortical lamination defects resembling those in type II lissencephaly. J Neurosci. 2008;28(18):4712–25.
108. Vasudevan HN, Braunstein SE, Phillips JJ, Pekmezci M, Tomlin BA, Wu A, et al. Comprehensive molecular profiling identifies FOXM1 as a key transcription factor for meningioma proliferation. Cell Rep. 2018;22(13):3672–83.
109. Jiang J, Hui C, chung. Hedgehog signaling in development and cancer. Dev Cell. 2008;15(6):801–12.
110. Zhang Q, Wen Z, Ni M, Li D, Wang K, Jia GJ, et al. Malignant progression contributes to the failure of combination therapy for atypical meningiomas. Front Oncol. 2020;10:608175.
111. Cain SA, Pope B, Mangiola S, Mantamadiotis T, Drummond KJ. Somatic mutation landscape in a cohort of meningiomas that have undergone grade progression. BMC Cancer. 2023;23(1):216.

Molecular Biology and Genetics of Meningiomas: Classification and Diagnostic Methods

Luca Bertero, Felix Sahm, and Rosa Maria Di Crescenzo

1 Introduction

For diagnostic purposes, meningiomas are currently classified according to the latest WHO Classification of Central Nervous System Tumors published in 2021 [1]. Similarly to other tumor types, their comprehensive diagnostic assessment relies on both histopathological findings and molecular hallmarks. Histopathology still plays a critical role in the diagnostic work up of these tumors since the morphological spectrum of meningiomas is very wide and differential diagnosis can prove challenging. Moreover, correctly distinguishing the specific histological types is still necessary for tumor grading and, more recently, molecular markers also became diagnostically relevant according to the latest WHO classification grading criteria.

L. Bertero (✉)
Pathology Unit, Department of Medical Sciences, University of Turin, Turin, Italy
e-mail: luca.bertero@unito.it

F. Sahm
Department of Neuropathology, Institute of Pathology, University Hospital Heidelberg, Heidelberg, Germany

Clinical Cooperation Unit Neuropathology, German Consortium for Translational Cancer Research (DKTK), German Cancer Research Center (DKFZ), Heidelberg, Germany

R. M. Di Crescenzo
Department of Advanced Biomedical Sciences, Section of Pathology, University "Federico II" of Naples, Naples, Italy

2 Morphological Subtypes and Molecular Features

Multiple (up to 15) meningioma subtypes have long been recognized based on their histopathology and meningothelial, fibrous and transitional are the most frequent subtypes. Frequency of molecular alterations, including SNV and CNV, vary according to the morphology but, in most of the cases, they cannot be considered diagnostic surrogates due to their limited specificity. Mutations in *AKT1*, *TRAF7*, *SMO*, *POLR2A*, and *PIK3CA* are usually observed in meningothelial meningiomas, while chromosome 22q deletion associated with a *NF2* mutation is a common signature of fibrous, transitional and psammomatous meningiomas [2]. These alterations and meningioma histological types are also associated with clinical variables like tumor site (e.g. meningothelial meningiomas usually arise within the skull base, while fibrous and transitional meningiomas are associated with a convexity location).

Mutations, usually combined, of *TRAF7* and *KLF4* are a molecular hallmark of secretory meningioma, while gain of chromosome 5 is a common feature of angiomatous, microcystic and metaplastic meningiomas, and clear cell

F. Maiuri, M. Del Basso De Caro (eds.), *Meningiomas*,
https://doi.org/10.1007/978-3-031-76680-0_6

meningioma is characterized by somatic or germline *SMARCE1* mutations. Papillary and rhabdoid meningiomas show a morphological and molecular overlap since they share mutations in *PBRM1*, which is more common in papillary meningioma, and *BAP1*, which is more common in rhabdoid meningiomas, although the mutations can also coexist.

In the current diagnostic practice, the assessment of these alterations is usually reserved for selected cases or enrollment into clinical trials, since the conventional histopathological assessment is usually sufficient to achieve the correct diagnosis in most of the cases. Nevertheless, molecular profiling can prove useful for differential diagnosis in challenging cases or for research purposes.

DNA methylation profiling is an alternative, comprehensive, and effective diagnostic tool which is now frequently used for diagnosing CNS tumors, including meningiomas. Concerning the diagnostic assessment, meningiomas harbor a specific epigenetic signature which can be reliably identified and distinguished across the whole spectrum of CNS tumors and other histological mimics (e.g., solitary fibrous tumor, schwannoma, fibrous dysplasia) [3]. By analyzing a large series of samples, methylation profiling allowed to identify 6 epigenetic classes (defined as benign 1–3, intermediate 1–2, and malignant) associated with meningioma subtypes, although without a strict match [4]. For example, fibrous, psammomatous and transitional meningiomas preferentially clustered within the benign 1 class, while meningothelial and secretory meningiomas were associated with the benign 2 class. The overlap in terms of epigenetic profiles can be due to the acquired oncogenic molecular alterations but can also suggest differences in terms of cells of origin and biological background.

3 Grading and Prognostic Makers

Meningioma grading is mainly based on morphological features. A grade 2 is assigned either due to the increased mitotic count (≥4/10HPF), presence of brain invasion or presence of at least three of the following five findings: hypercellularity, prominent nucleoli, patternless growth, small cell appearance, and necrosis. Grade 3 is established based on the presence of a higher mitotic count (≥20/10HPF) or of an overtly malignant morphology. Moreover, clear cell meningioma and chordoid meningioma are considered grade 2 irrespectively of the other morphological characteristics due to their higher recurrence rate and potential aggressive behavior. Conversely, papillary and rhabdoid subtypes are no longer considered grade 3 neoplasms if other aggressive features are lacking. Nevertheless, these features are frequently observed in these subtypes, especially if papillary/rhabdoid morphology is widespread.

Among immunohistochemical markers, loss of H3 K27 trimethylation (i.e., H3 K27me3) has been demonstrated to be associated with increased risk of recurrence, although this marker is not a grading criterion according to the latest WHO classification [5, 6].

The main, diagnostically relevant, molecular hallmarks associated with meningioma grading are *TERT* promoter mutations and *CDKN2A/B* homozygous deletion. The TERT gene codes for telomerase reverse transcriptase whose activity is critical to avoid replicative senescence by telomers preservation; *CDKN2A/B* codes for two proteins, p16 and p14arf, which act as tumor suppressors by cell cycle regulation. Presence of one of these alterations has been found to be associated with early recurrence and poorer prognosis irrespectively of other morphology-based grading features [7–9]. Due to this data, the 2021 WHO classification provides that if either one of these alterations is detected, a grade 3 should be assigned. More recently, it has been reported that even a heterozygous deletion of *CDNK2A/B* is prognostically relevant and associated with poorer outcome [10].

Focusing on the mutational landscape of higher grade meningiomas, distinct groups have been observed based on NF2 status: (1) a first, more frequent in male patients, characterized by *NF2* mutations associated with *CDKN2A/B*, *ARID1A* and *KDM6A* alterations; (2) a second, *NF2*-agnostic group, harboring a *TERT* promoter or *TP53* mutations; (3) a third, *NF2*-intact group, including further subgroups characterized by *BAP1/PBRM1* mutations, *AKT1/PIK3CA/SMO* mutations or heterogenous, non-recurrent alterations [11]. This data further supports the molecular complexity and heterogeneity of SNV in meningiomas, hindering their use for prognostic classification purposes in daily diagnostics.

As previously mentioned, DNA methylation profiling is a further option for meningioma molecular characterization. The identified six methylation classes proven to be prognostically relevant and included three benign, two intermediate, and one malignant groups. A limited overlap between WHO grade (defined according to the previous 2016 WHO classification) and these classes was observed. In particular, while grade 1 and grade 3 meningiomas were enriched in benign and malignant methylation classes, respectively, grade 2 neoplasms were dispersed among the different classes. Prognostic analysis showed a stricter correlation between methylation classes and progression free survival compared with histology-based grading. Concerning the other prognostically relevant molecular markers, *CDKN2A/B* occurred with a significant rate in the malignant class.

A further study explored the epigenetic landscape of meningiomas identifying three main subgroups: a first one with NF2-intact status associated with favorable prognosis, an immune-enriched group with intermediate prognosis, and a third hypermitotic group with poor prognosis [12].

Overall, this data show that comprehensive molecular profiling can enable a more reliable prognostic stratification, but the required resources and the complexity of these analyses pose some challenges, especially considered the relatively high incidence of this tumor. Performing a comprehensive molecular profiling in all meningioma cases would be unfeasible in most settings, while limiting the analysis to aggressive/recurrent cases would limit the informative value of the analyses. For these reasons, data-driven diagnostic workflows are warranted to streamline the analyses, reduce the costs/workload, and maximize the clinical significance of the gathered information. To achieve this aim, an integrated approach based on histological examination followed, if deemed necessary, by molecular analyses has been proposed [13]. In particular, in this algorithm, the possibility to employ different types of molecular assays has been considered to adjust for the available analytical platforms in each specific laboratory.

4 Diagnostic Methods

Multiple assays can be used to address meningioma heterogeneity and contribute to their diagnostic and prognostic classification.

Limited-scope assays play a significant role during daily diagnostics and mainly include immunohistochemistry and FISH. Immunohistochemical stainings can be used to confirm meningioma diagnosis and even investigate specific molecular alterations. For example, the canonical *SMARCE1* mutations observed in clear cell meningioma result in a loss of protein expression which can be detected by immunohistochemistry [14]. Similarly, BAP1 expression is lost in case of *BAP1* mutations [15]. Immunohistochemistry, despite the narrow scope, has several advantages in terms of the wide availability, low costs, fast turnaround time, and the limited material which is required. FISH analysis can also be used to explore single markers like *CDKN2A/B* status, but this technique is time consuming and thus it cannot be used for screening purposes. For this reason, the efficacy of immunohistochemical surrogates has been explored. For example, the correlation between *CDKN2A/B* status and p16 protein expression has been tested with varying results [16, 17]. Focused molecular assays, like Sanger sequencing, can also be used to detect SNV like *TERT* promoter mutations. Conversely, comprehensive molecular profiling can be performed through different analytical platforms including DNA/RNA NGS, DNA methylation profiling, and array-based CNV assays. Some important caveats should be kept in mind for all these assays which are based on bulky analysis of nucleic acids. First, the submitted sample should be representative of the lesion and the rate of tumor versus non-neoplastic tissue should be enriched, if feasible. Second, an adequate amount of DNA/RNA should be submitted according to assay requirements. If the available sample is limited, a careful triage of the material should be performed.

DNA NGS is mainly used to investigate SNV and small insertions/deletions (InDels), but, based on assay characteristics and targeted regions, can also provide CNV information. DNA NGS assays can vary significantly, both in terms of technical characteristics and of covered regions, extending from small, targeted panels evaluating a limited number of genes to whole exome/genome sequencing. For routine diagnostics, targeted panels are usually adopted, and a wide range of commercial assays is available, usually focused on the most altered genes in solid tumors. For this reason, some of the previously reported meningioma-associated genes can be missing. Custom panels can thus be considered based on laboratory characteristics: meningioma caseload, available facilities and expertise, clinical requirements (e.g. available therapeutic trials). These targeted assays have proven effective and cost-efficient for meningioma molecular profiling [18]. Whole exome/genome sequencing, are currently mostly limited to the research setting due to the higher costs and technical complexity, especially in terms of bioinformatic analysis. Of note, using this kind of tools it is possible to detect potential germline variants: these findings should be managed carefully according to relevant guidelines and national regulations.

RNA NGS is another relevant tool for diagnostic neuropathology and is mainly used to detect gene fusions. Based on currently available data, these alterations are infrequent occurrences in meningiomas [19], thus this assay type is usually not included in the diagnostic workflow. RNA NGS has also been evaluated in neuropathology for tumor stratification based on transcriptomic analysis, for example, for medulloblastoma subgrouping. Prognostically relevant signatures have been identified also in meningioma [20], but further validation is needed.

DNA methylation profiling can be performed through multiple assay types like NGS, but array-based assays are usually used. Array analysis allows to evaluate the methylation status of several hundred thousand genome loci, gathering the comprehensive epigenetic profile of the analyzed sample. This profile is then compared with other relevant samples using a classifier like the Heidelberg classifier (molecularneuropathology.org/mnp/) to search for a potential match with one of the already established diagnostic classes (e.g., the most recent classifier versions require a $\geq$0.9 classifier score to define a match). An extensive knowledge about this tool is warranted to maximize the gathered information and avoid potential pitfalls [21]. CNV can also be inferred by array data providing additional hints for diagnostic and prognostic classification. According to the available data, the probability to correctly classify a tumor sample relies on multiple variables including tumor type and sample characteristics like age, DNA quantity and quality, tumor content. Novel classifier versions are being developed including new entities which are progressively being discovered/defined using this approach. Further strategies are also being explored to improve classification rates, for instance, by removing the confounding effect of intermixed inflammatory elements [22]. It is also important to note that these tools are not certified for diagnostic use by themselves. Their use should thus be validated in each laboratory according to applicable regulations.

Finally, array-based CNV analysis is another relevant for CNS tumor diagnostics, enabling the detection of gene/chromosome copy number events like amplifications, gains and deletions. In meningiomas, this analysis can be used to detect, for example, chromosome arms 1p and 22q losses, chromosome 5 gain and *CDKN2A/B* deletion. Even though, as previously mentioned, CNV data can be inferred from array-based DNA methylation profiling results, a dedicated CNV analysis provides a higher resolution.

References

1. WHO Classification of Tumours. Editorial Board. Central nervous system tumours. Lyon: International Agency for Research on Cancer; 2021.
2. Birzu C, Peyre M, Sahm F. Molecular alterations in meningioma: prognostic and therapeutic perspectives. Curr Opin Oncol. 2020;32(6):613–22.

3. Capper D, Jones DTW, Sill M, Hovestadt V, Schrimpf D, Sturm D, et al. DNA methylation-based classification of central nervous system tumours. Nature. 2018;555(7697):469–74.
4. Sahm F, Schrimpf D, Stichel D, Jones DTW, Hielscher T, Schefzyk S, et al. DNA methylation-based classification and grading system for meningioma: a multicentre, retrospective analysis. Lancet Oncol. 2017;18(5):682–94.
5. Katz LM, Hielscher T, Liechty B, Silverman J, Zagzag D, Sen R, et al. Loss of histone H3K27me3 identifies a subset of meningiomas with increased risk of recurrence. Acta Neuropathol. 2018;135(6):955–63.
6. Behling F, Fodi C, Gepfner-Tuma I, Kaltenbach K, Renovanz M, Paulsen F, et al. H3K27me3 loss indicates an increased risk of recurrence in the Tubingen meningioma cohort. Neuro Oncol. 2021;23(8):1273–81.
7. Sahm F, Schrimpf D, Olar A, Koelsche C, Reuss D, Bissel J, et al. TERT promoter mutations and risk of recurrence in meningioma. J Natl Cancer Inst. 2016;108(5):djv377.
8. Mirian C, Duun-Henriksen AK, Juratli T, Sahm F, Spiegl-Kreinecker S, Peyre M, et al. Poor prognosis associated with TERT gene alterations in meningioma is independent of the WHO classification: an individual patient data meta-analysis. J Neurol Neurosurg Psychiatry. 2020;91(4):378–87.
9. Sievers P, Hielscher T, Schrimpf D, Stichel D, Reuss DE, Berghoff AS, et al. CDKN2A/B homozygous deletion is associated with early recurrence in meningiomas. Acta Neuropathol. 2020;140(3):409–13.
10. Khan AB, English CW, Chen WC, Athukuri P, Bayley JC, Brandt VL, et al. Even heterozygous loss of CDKN2A/B greatly accelerates recurrence in aggressive meningioma. Acta Neuropathol. 2023;145(4):501–3.
11. Williams EA, Santagata S, Wakimoto H, Shankar GM, Barker FG II, Sharaf R, et al. Distinct genomic subclasses of high-grade/progressive meningiomas: NF2-associated, NF2-exclusive, and NF2-agnostic. Acta Neuropathol Commun. 2020;8(1):171.
12. Choudhury A, Magill ST, Eaton CD, Prager BC, Chen WC, Cady MA, et al. Meningioma DNA methylation groups identify biological drivers and therapeutic vulnerabilities. Nat Genet. 2022;54(5):649–59.
13. Maas SLN, Stichel D, Hielscher T, Sievers P, Berghoff AS, Schrimpf D, et al. Integrated molecular-morphologic meningioma classification: a multicenter retrospective analysis, retrospectively and prospectively validated. J Clin Oncol. 2021;39(34):3839–52.
14. Tauziede-Espariat A, Parfait B, Besnard A, Lacombe J, Pallud J, Tazi S, et al. Loss of SMARCE1 expression is a specific diagnostic marker of clear cell meningioma: a comprehensive immunophenotypical and molecular analysis. Brain Pathol. 2018;28(4):466–74.
15. Shankar GM, Santagata S. BAP1 mutations in high-grade meningioma: implications for patient care. Neuro Oncol. 2017;19(11):1447–56.
16. Guyot A, Duchesne M, Robert S, Lia AS, Derouault P, Scaon E, et al. Analysis of CDKN2A gene alterations in recurrent and non-recurrent meningioma. J Neuro Oncol. 2019;145(3):449–59.
17. Tang V, Lu R, Mirchia K, Van Ziffle J, Devine P, Lee J, et al. Loss of p16 expression is a sensitive marker of CDKN2A homozygous deletion in malignant meningiomas. Acta Neuropathol. 2023;145(4):497–500.
18. Mawrin C, Koch R, Waldt N, Sandalcioglu IE, Braunsdorf WEK, Warnke JP, et al. A new amplicon-based gene panel for next generation sequencing characterization of meningiomas. Brain Pathol. 2022;32(2):e13046.
19. Khan AB, Gadot R, Shetty A, Bayley JC, Hadley CC, Cardenas MF, et al. Identification of novel fusion transcripts in meningioma. J Neuro Oncol. 2020;149(2):219–30.
20. Viaene AN, Zhang B, Martinez-Lage M, Xiang C, Tosi U, Thawani JP, et al. Transcriptome signatures associated with meningioma progression. Acta Neuropathol Commun. 2019;7(1):67.
21. Capper D, Stichel D, Sahm F, Jones DTW, Schrimpf D, Sill M, et al. Practical implementation of DNA methylation and copy-number-based CNS tumor diagnostics: the Heidelberg experience. Acta Neuropathol. 2018;136(2):181–210.
22. Wu Z, Abdullaev Z, Pratt D, Chung HJ, Skarshaug S, Zgonc V, et al. Impact of the methylation classifier and ancillary methods on CNS tumor diagnostics. Neuro Oncol. 2022;24(4):571–81.

Comprehensive Overview of Genetic Alterations in Meningioma

Sybren L. N. Maas, Felix Sahm, and Tareq A. Juratli

1 Introduction

In adult patients, meningiomas are the most common intracranial tumors [1]. Although the precise etiology of meningioma formation remains to be fully elucidated, the cytological similarities with meningothelial cells of the arachnoid, strongly suggests that meningiomas originate from these cells. This phenomenon also explains the anatomical localization of meningiomas, which closely surround and, at times, invade the brain parenchyma. Historically, the WHO grading system for predicting the risk of meningioma relied solely on histological and cytomorphological criteria [2]. Efforts have been made toward identifying histological markers associated with an unfavorable prognosis, such as increased mitotic activity, distinct histological subtypes, and the presence of brain invasion [2]. One of the pivotal challenges remains the accurate identification of patients who may benefit from more aggressive initial therapy. Inspired by the genomic subclassification seen in other neoplasms, there is a growing interest in detecting genomic signatures that indicate an adverse prognosis in meningiomas. In recent years, comprehensive molecular studies have contributed to an improved understanding of the molecular and genetic alterations underpinning meningioma development and disease progression [3–5]. Some genetic alterations are observed across different WHO grades, while others are specific to either low risk WHO grade 1 cases, or are linked to an increased risk, thus influencing the assigned WHO grade when detected [6, 7].

S. L. N. Maas
Department of Pathology, Leiden University Medical Center, Leiden, Netherlands

Department of Pathology, Erasmus MC Cancer Institute, University Medical Center Rotterdam, Rotterdam, The Netherlands

F. Sahm
Department of Neuropathology, University Hospital Heidelberg and CCU Neuropathology, German Consortium for Translational Cancer Research (DKTK), German Cancer Research Center (DKFZ), Heidelberg, Germany

T. A. Juratli (✉)
Department of Neurosurgery, Faculty of Medicine and University Hospital Carl Gustav Carus, Technische Universität Dresden, Dresden, Germany

National Center for Tumor Diseases (NCT) Dresden, University Hospital Carl Gustav Carus, Dresden, Germany
e-mail: Tareq.juratli@ukdd.de

2 The Molecular Landscape of WHO Grade 1 Meningiomas

The majority (approximately 70%) of meningiomas are grade 1 tumors, indicating that their biological course is relatively benign. Histologically, grade 1 means that the tumor shows little to no

F. Maiuri, M. Del Basso De Caro (eds.), *Meningiomas*,
https://doi.org/10.1007/978-3-031-76680-0_7

mitotic activity, does not invade the brain tissue, lacks a combination of morphological features associated with increased risk, and is not of a specific histological type associated with increased risk for recurrence (i.e., chordoid or clear cell meningioma) [7]. With the introduction of the 2021 fifth edition of the WHO classification of CNS tumors, it also implies that there are no molecular alterations that are associated with an increased risk for progression [7]. This introduction of molecular features into meningioma grading represents the culmination of various efforts to uncover the molecular characteristics of meningiomas. Using diverse methodologies, specific mutations, copy number variations (CNV), and complete epigenetic-, proteomic and transcriptomic profiles were unveiled that present at different levels of aggressiveness of meningiomas [6]. Some molecular alterations are almost exclusive to WHO grade 1 meningiomas and could potentially serve as markers for predicting a benign course in a meningioma patient.

In addition to the *NF2* gene, several other genes have emerged as recurrently altered driver genes in meningiomas. *NF2*-wildtype meningiomas are predominantly located at the skull base and often harbor alterations that are mutually exclusive with *NF2* alterations [8–10]. These tumors are frequently associated with the meningiothelial and secretory histological subtype and significantly influence the biological behavior of the tumor [11]. Some of these genetic markers are exclusively detected in WHO grade 1 cases (Table 1).

Table 1 Overview of genomic alterations in meningioma

WHO grade	Genomic alteration	Comment
1	*NF2*	Mainly in convexity, parasagittal, and falx meningiomas Mutation frequency correlates with the WHO tumor grade, showing higher frequencies in tumors with higher WHO grades
	TRAF7	Most frequently detected alteration in non-*NF2* altered meningioma. Occurs predominantly in skull base meningiomas
	KLF4	Together with *TRAF7* often detected in secretory skull base meningiomas
	AKT1	Almost exclusively in meningothelial skull base meningioma Commonly co-occur with *TRAF7* mutations in skull base meningiomas but are found independently in spinal meningiomas
	PIK3CA	Almost exclusively in meningothelial meningioma at the skull base Often occurring concurrently with *TRAF7* mutations
	SMO	Found almost exclusively in meningothelial meningiomas of the olfactory groove and rarely co-occur with *TRAF7* mutations
2	*SMARCB1*	Occur alongside *NF2* mutations, both germline and somatic
	SMARCE1	Tied to the clear cell subtype Predominantly found in children and young adults
	Chromosome 1p loss	Up to 82% in WHO grade 2 and 3 meningiomas
	Loss of H3K27me3	Denotes a subset of WHO grade 2 meningiomas with increased recurrence risk
	ARID1A mutations	Present in about 5.5% of WHO grades 2 and 3 meningiomas
3	*TERT*-promotor alterations	Found in 2.5% of histologic WHO grade 1, 6.5% of histologic WHO grade 2, and 15% in WHO grade 3
	CDKN2A/B deletions	In 15% of all high-grade meningiomas
	BAP1 inactivation	Associated with rhabdoid histo-morphology
	PBRM1 inactivation	A rare altertion in meningiomas, associated with papillary histo-morphology Overlaps with *BAP1* mutations
	DMD deletion/ inactivation	X-linked and thus more common in male patients In 20% of high-grade/progressive meningiomas

2.1 Key Genetic Alterations in Grade 1 Meningioma

- *NF2* mutations: A significant portion of meningiomas, with a prevalence that increases with the WHO grade, display alterations affecting the *NF2* gene (Table 1) [3, 12]. *NF2,* located on chromosome 22q, encodes the Merlin protein, also referred to as Neurofibromin 2 [13]. Merlin is a tumor suppressor protein that regulates multiple proliferative signaling pathways. *NF2* alterations meningiomas are primarily of the "loss of function" type, resulting in a large array of changes detected in meningiomas, including intragenic deletions, frameshift-, nonsense mutations or other alterations leading to Merlin inactivation [3, 13]. While complete loss of Merlin function can occur when both copies of chromosome 22q are affected, most meningiomas typically have an inactivating alteration on one arm and loss of the other arm [3, 12]. The frequency of *NF2* mutations correlates with the WHO tumor grade, with approximately 30% of grade 1 tumors, increasing to 60% in grade 2, and reaching 70% in grade 3 meningiomas [3, 12, 14]. *NF2* mutations are mainly present in convexity, parasagittal, and falx meningiomas.
- *TRAF7* mutations*:* Mutations in *TRAF7*, a member of the Tumor necrosis factor (TNF) receptor-associated factor (TRAF) family, particularly in the WD40 domain, can result in the activation of apoptosis inhibition through targeted P53 degradation [15]. Mutations in *TRAF7* are detected in 20–25% of meningiomas, making *TRAF7* the most frequently altered gene in meningiomas after *NF2* [3, 8]. *TRAF7*-alterated meningiomas are predominantly found in the medial and sphenoid wing regions of the skull base [8].
- *KLF4* mutations*: KLF4* p.K409Q mutations trigger the activation of hypoxia inducible factor 1-alpha (HIF-1α), subsequently resulting in increased angiogenesis, proliferation, and cellular differentiation [16]. *KLF4* mutations are associated with a secretory histological subtype of meningioma, which is commonly found at the skull base and often exhibits peritumoral edema [11]. Secretory meningioma feature PAS-positive eosinophilic secretions and, in the majority of cases, harbor a combination of *TRAF7* and *KLF4* mutations and are also located in the medial and sphenoid wing regions of the skull base [8, 11]. Secretory meningioma are almost exclusively WHO grade 1, suggesting that the identification of *TRAF7* and *KLF4* mutations in skull base meningiomas may serve as proxy markers for low risk (grade 1) meningioma [3].
- *AKT1 mutations: AKT1* is a member of the phosphatidylinositol-3 kinase (PI3K) signaling pathway, is involved in different cellular programs, including proliferation, differentiation, motility, survival, and apoptosis inhibition [17]. Mutations in *AKT1*, particularly p.E17K mutations, lead to PI3K-independent activation of this signaling pathway [17]. *AKT1* mutations occur frequently in the combination with *TRAF7* in skull base meningiomas, while they are found independently in spinal meningiomas [3, 10, 18]. Approximately, 10% of meningiomas harbor *AKT1* mutations, and these tumors are most often located in the anterior medial section of the skull base, including the anterior rim of foramen magnum [8, 19, 20].
- *SMO mutations:* Smoothened, the protein encoded by *SMO*, is a member of the hedgehog signaling pathway involved in cellular differentiation and proliferation [21]. *SMO*-altered meningiomas are predominantly located in the olfactory groove region [8, 22]. SMO-mutant olfactory groove meningiomas were found to be significantly larger than those with other driver mutations [23].
- *PIK3CA mutations:* Activating mutations in *PIK3CA* lead to constitutive activation of *AKT1, resulting in* growth-stimulating effects similar to *AKT1* hotspot mutations [24] Anatomically, *PIK3CA*-altered meningioma are located in the sphenoid wing, clival and anterior tuberculum sellae regions of the skull base [24]. Recently, *PIK3CA* mutations in meningiomas have attracted interest for their significant role in the tumorigenesis of hormone-associated meningiomas [25].
- *POLR2A mutations:* DNA-directed RNA polymerase II (EC 2.7.7.6) transcribes all nuclear

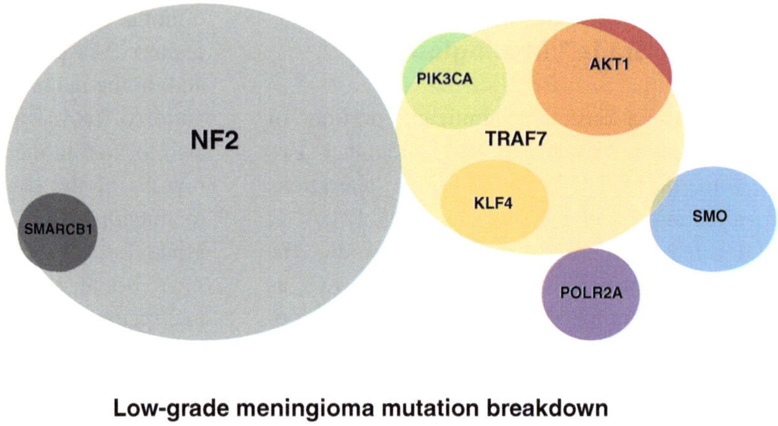

Fig. 1 Venn diagram indicating the overlap between mutation co-occurrence in different cohorts of low-grade meningioma

mRNA and consists of 12 subunits. The largest subunit A is encoded by the *POLR2A* gene and includes essential domains for catalytic function and transcriptional regulation. In meningioma samples lacking known driver genes, recurrent *POLR2A* mutations—specifically p. Gln403Lys and p.Leu438_His439del—were identified [26]. *POLR2A* mutations are infrequent in meningiomas (6%), which limits our understanding of their role; however, recent studies indicate that they may act as a poor prognostic marker, especially in cerebellopontine angle meningiomas [26, 27].

- *SMARCB1*: Similar to *NF2*, the *SMARCB1* gene is also located on chromosome 22q [28]. Somatic mutations in *SMARCB1* are rare in meningiomas and enriched in WHO grade 1 and 2 meningiomas, occurring concomitantly with *NF2* mutations [28, 29]. Of note, sporadic meningiomas occurring as single tumors in the absence of any other findings of schwannomatosis may contain a somatic variant in *SMARCB1* that is not present in the germline.

AKT1, *PIK3CA*, *POLR2A*, and *SMO* mutations are mutually exclusive, can be detected together with *TRAF7* mutations and are in general WHO grade 1 meningioma (Fig. 1). Therefore, the combination of anatomical location, meningothelial subtype and an activating mutation in *AKT1*, *PIK3CA* or *SMO* can help determine low risk for progression in meningioma. Noteworthy, *AKT1* mutations, together with *NF2* mutations are the main driver genes in WHO grade 1 spinal meningiomas [18]. Both mutations are mutually exclusive and are associated with distinct patient characteristics and tumor features. *AKT1*-mutant meningiomas originate in the cervical spine ventrally to the spinal cord and are almost exclusively associated with meningothelial histology [18]. In contrast, *NF2*-mutant meningiomas show strong female gender predominance, arise with a wider anatomic distribution, most frequently in the thoracic spine dorsally to the spinal cord [18].

3 The Molecular Landscape of WHO Grade 2 and 3 Meningiomas

Approximately, 30% of all meningiomas are histologically classified as WHO grade 2 and 3. A substantial fraction of these meningiomas recur and display remarkable resistance to conventional therapeutic interventions. While WHO grade 1 meningiomas have a recognized female predilection, aggressive tumor phenotypes tend to be more prevalent in males, although the underlying molecular mechanisms that dictate this gender influence remain elusive. The post-operative and post-radiation therapeutic challenges encountered in these tumors frequently end in disease progression, increasing the associated patient morbidity. Given the limitations of traditional chemotherapy, the need for new prognostic biomarkers and therapeutic methods is pressing.

3.1 Key Genetic Alterations in High-Grade Meningiomas and Their Implications on the WHO 2021 Grading System (Table 1)

- *NF2 mutations: NF2*, located on chromosome 22q, encodes the merlin protein, also known as neurofibromin 2, and is a key gene that has been discussed in the section above in relation to meningioma alterations.
- *The telomerase reverse transcriptase (TERT) alterations: TERT* remains transcriptionally dormant in most non-cancerous cells. However, its reactivation has implications for cellular immortality. An estimated 90% of malignant tumors demonstrate elevated telomerase levels, and around 73% harbor *TERT* gene alterations, including promoter mutations (namely C228T and C250T), gene translocations, and amplifications [30]. The presence of *TERT* promoter mutations spans across histological meningioma grades: 2.5% in grade 1, 6.5% in grade 2, and 15% in grade 3 [31]. In addition, high-grade meningiomas infrequently display genomic rearrangements leading to gene fusions like *LPCAT1::TERT* and enhanced copy numbers. Notably, *TERT* alterations are late events in tumor evolution, with significant temporal and spatial heterogeneity [32, 33]. They correlate with upregulated TERT expression, augmented telomerase activity, and clinically, are associated with increased recurrence rates and worse survival outcomes [32–34]. Most importantly, independent of the WHO grade, patients with TERT-altered meningiomas (mutation or rearrangement) did consistently worse compared with *TERT* wild-type patients. Even histological WHO 1 and WHO 2 meningiomas with *TERT* alteration demonstrated a poorer recurrence-free and overall survival compared with *TERT* wild-type WHO grade 3 meningioma patients [35, 36]. Subsequently, *TERT* promoter mutations have been identified as markers for grade 3 meningiomas, regardless of the histological grade, in the 2021 WHO classification for brain tumors [7].
- *CDK Pathway Regulators:* The tumor suppressor genes *CDKN2A* and *CDKN2B* are central to cell cycle regulation, exhibit loss of function primarily due to homozygous deletions or singular nucleotide variations, predominantly in meningiomas histologically classified as WHO grade 2 or 3 [37, 38]. In a recent large-scale study, about 15% of WHO grade 2 and 3 meningiomas harbored a *CDKN2A/B* deletion or mutation [31]. The prognostic importance of the *CDKN2A/B* status has been validated by multiple comprehensive studies. Noteworthy, two recent analyses of large set of meningiomas showed that even heterozygous deletions of *CDKN2A/B* was associated with a shorter time to recurrence [12, 38]. Biallelic *CDKN2A* inactivation is meanwhile an established biomarker for the identification of meningioma patients with a high risk of early recurrence [7]. Consequently, and similar to *TERT*-promotor mutations, homozygous deletion of *CDKN2A/B*, is a grading criterion for CNS WHO grade 3 tumors in the 2021 WHO classification for brain tumors.
- Other less common CDK pathway alterations include *CDKN2C* mutations and *CDK4/6* amplifications (with a rate of 2% for each) [31].
- *Breast cancer-associated protein 1 (BAP1) mutations:* The inactivation of this tumor suppressor gene has been linked to aggressive meningiomas with rhabdoid histo-morphology and is associated with a shorter progression-free survival [39]. *BAP1* alterations (mutations and deletions) has been observed in 2.7% of high-grade meningiomas [31]. Negative nuclear staining for BAP1 by immunohistochemistry reveals absent BAP1 expression and can be used for screening samples, but so far has not been including in WHO grading.
- *Alterations in the SWI/SNF chromatin remodeling complex:* Different somatic mutations related to the SWI/SNF chromatin remodeling complex have been identified in higher grade meningiomas. Examples are mutations in *SMARCB1* (discussed above), *SMARCE1*,

ARID1A, and *PBRM1*, which lead to dysregulation of chromatin remodeling.

1. *SMARCE1 mutations*: Mutations in *SMARCE1* are closely tied to the clear cell histological subtype of meningiomas [40]. Interestingly, while this mutation is prevalent in clear cell meningiomas, it appears to be unique to this subtype [41, 42]. A comprehensive analysis revealed no *SMARCE1* alterations in a sample of 705 other meningiomas. Furthermore, it's worth highlighting that *SMARCE1* mutations are predominantly found in children and young adults. In terms of clinical implications, patients with these mutations exhibited disease progression or recurrence patterns akin to those diagnosed with meningioma WHO grade 2 [42]. This observation has solidified the classification of these tumors under WHO grade 2.
2. *ARID1A mutations*: The *ARID1A* gene is a tumor suppressor gene that is linked to the progression of various cancer types. *ARID1A* mutations has shown to alter gene activity and promote oncogenesis. *ARID1A* usually occupies the *TERT* promoter to suppress its expression [43, 44]. In patients with high-grade neuroblastoma, lower *ARID1A* levels were associated with elevated *TERT* levels [43]. It seems that alterations in *ARID1A* lead to disinhibition of *TERT*'s promoter site, reactivating its expression, and promoting tumor progression. In a recent work, mutations in *ARID1A* in primary atypical meningiomas, were associated with a worse patients' prognosis, when compared with their *ARID1A* wild-type counterparts [45]. About 5.5% of WHO grades 2 and 3 meningiomas exhibit a somatic *ARID1A* mutation [31].
3. *PBRM1 mutations*: Biallelic inactivation of *PBRM1* has been described in papillary meningiomas [46]. *PBRM1* is a 37-exon gene residing on chromosome 3p21, adjacent to *BAP1* [47]. *PBRM1* encodes the BAF180 protein, the chromatin targeting subunit of the PBAF chromatin remodeling complex [47]. *PBRM1* is a tumor suppressor gene, mutated in 40% of clear cell renal cell carcinoma (RCC), as well as a subset of papillary RCC and bladder carcinoma [47, 48]. Mutations in *PBRM1* are most often truncations and result in loss of protein expression. It overlaps with *BAP1* mutations, and its prognostic role is yet to be established.

- Further rare X-linked gene alterations: It is recognized that males have a tendency for more biologically aggressive tumors. The mechanisms by which sex can affect meningioma grade and growth are not fully understood yet. Recently, genomic alterations on the sex chromosomes that could partly explain these phenomena has been described. The X-linked gene *DMD*, commonly associated with Duchenne muscular dystrophy, encodes dystrophin. This protein is expressed in the meninges and stabilizes cells by connecting the cytoskeleton to the cellular membrane. Beyond its known myogenic role, *DMD* is emerging as a potential cancer gene, with intragenic deletions observed in tumors of neural crest and mesodermal origin. *DMD* inactivation, found in 20% of high-grade/progressive meningiomas, is linked to tumor progression, metastasis, and unfavorable outcomes, irrespective of *TERT* mutational status [49, 50]. However, the detection of *DMD* deletions is limited in clinical routine due to technical challenges in analyzing sex chromosomes. Other X-linked tumor suppressor gene alterations in high-grade meningiomas include *KDM6A*, *DDX3X*, and *STAG2*, though these are less frequent [49].
- *Loss of H3K27me3:* The nuclear absence of trimethylation of lysine 27 (K27) of histone H3, as identified through immunohistochemistry in meningioma cells, denotes a subset of WHO grade 1 and grade 2 meningiomas with heightened recurrence risk [51–53]. This trend is observed across all meningioma grades, yet the incidence of H3K27me3 loss rises in tandem with tumor grade. Intriguingly, tumors

showing a complete absence of H3K27me3 staining were found to have a higher prevalence of *NF2* and *SUFU* mutations [51]. Whether the identification of H3K27me3 loss, warrants a grade 2 of grade 3 risk prediction is currently unknown and therefore this marker has not yet been included in meningioma WHO grading [53].

3.2 Chromosomal Aberrations

The genomic landscape of high-grade/progressive meningiomas shows Chromosome 22q loss in a staggering 84.1% of the cases [31]. In addition, loss of chromosome 1p is one of the most common copy number alterations in high-grade meningiomas (up to 80%) [31, 54]. Loss of chromosome 1p is identified as a prognostic factor with increased risk for progression in multiple studies [3, 55–57]. Other and 22q and 1p, losses of chromosome 6q, 10, 14q, 18q, and 9p, as well as gains in chromosomes such as 1q, 9q, 12q, 15q, 17q, 19, 20q, and 5, have also been described in multiple studies on meningioma [3, 12]. The independent predictive value of those copy number variations (CNV) remains to be elucidated. Some risk prediction models combine multiple CNVs into one model. One attempt identified the combination of 1p/6q/14q to be most informative, whereas a different model extends these with losses of 3p, 4p/q, 6p, 10p/q, 19p/q, *CDKN2A/B* and the mitotic activity [3, 12, 58].

Some CNVs are associated with specific meningioma subtypes. The deletion of 2p for example, has been reported to be significantly associated with chordoid morphology [59, 60]. Significantly, recurrent tumors from the same patient demonstrated a 75% overlap of arm-level somatic CNAs. In rare cases of fast-growing aggressive meningiomas, complex genomic rearrangements like chromotripsis have been described. Interestingly, high-grade/progressive meningiomas often show sex chromosome losses [49]. Males often present with Y chromosome loss, while some females depict an X chromosome loss [49, 61].

4 Intratumor Heterogeneity

Meningiomas have recently been highlighted for their intratumor heterogeneity. New data points toward distinct regional variations in their genomic, epigenomic, transcriptomic, and histological characteristics [33, 62–64]. This highlights both inter- and intratumor diversity, and underscores the limited understanding of meningioma development, particularly from a monoclonal origin [65, 66]. This lack of depth in research is partly attributed to past analyses often being limited to single tissue samples or recurrent tumors. Single-cell analysis offers invaluable insights into tumor heterogeneity and has emerged as an instrumental tool in understanding meningioma development. Notably, two recent studies have investigated this methodology [[63, 64]. Single-nuclear RNA sequencing was employed on eight tumors, covering all molecular groups and WHO grades. It was found that the majority of the cells were neoplastic, with 14% identified as immune cells, 10% as fibroblasts, and 6% as endothelial cells. Upon further analysis of the neoplastic cells, a dominant cluster and a secondary, less prevalent cluster were revealed, with minimal variability observed in cell copy number profiles. In a similar study, single-cell analysis was conducted on three meningioma samples, revealing heterogeneity in both CNV and gene expression profiles among specific tumor cell populations [63, 64]. Notable variants were identified, including deletions of chromosomes 19q and 22q, and amplifications of chromosomes 7p and 7q.

References

1. Ostrom QT, Price M, Neff C, Cioffi G, Waite KA, Kruchko C, Barnholtz-Sloan JS. CBTRUS statistical report: primary brain and other central nervous system tumors diagnosed in the United States in 2015–2019. Neuro-Oncology. 2022;24:v1–v95.
2. Louis DN, Perry A, Reifenberger G, Deimling A, Figarella-Branger D, Cavenee WK, Ohgaki H, Wiestler OD, Kleihues P, Ellison DW. The 2016 World Health Organization classification of tumors of the central nervous system: a summary. Acta Neuropathol. 2016;131:803–20.

3. Maas SLN, Stichel D, Hielscher T, et al. Integrated molecular-morphologic meningioma classification: a multicenter retrospective analysis, retrospectively and prospectively validated. J Clin Oncol. 2021;39:3839–52.
4. Nassiri F, Liu J, Patil V, et al. A clinically applicable integrative molecular classification of meningiomas. Nature. 2021;597:119–25.
5. Choudhury A, Magill ST, Eaton CD, et al. Meningioma DNA methylation groups identify biological drivers and therapeutic vulnerabilities. Nat Genet. 2022;54:649–59.
6. Wang JZ, Landry AP, Raleigh DR, et al Meningioma: International Consortium on Meningiomas (ICOM) consensus review on scientific advances & treatment paradigms for clinicians, researchers, and patients. Neuro-Oncology (2024) noae082.
7. Louis DN, Perry A, Wesseling P, et al. The 2021 WHO classification of tumors of the central nervous system: a summary. Neuro-Oncology. 2021;23:1231–51.
8. Clark VE, Erson-Omay EZ, Serin A, et al. Genomic analysis of non-NF2 meningiomas reveals mutations in TRAF7, KLF4, AKT1, and SMO. Science. 2013;339:1077–80.
9. Brastianos PK, Horowitz PM, Santagata S, et al. Genomic sequencing of meningiomas identifies oncogenic SMO and AKT1 mutations. Nat Genet. 2013;45:285–9.
10. Berghoff AS, Hielscher T, Ricken G, et al. Prognostic impact of genetic alterations and methylation classes in meningioma. Brain Pathol. 2022;32:e12970.
11. Reuss DE, Piro RM, Jones DTW, et al. Secretory meningiomas are defined by combined KLF4 K409Q and TRAF7 mutations. Acta Neuropathol. 2013;125:351–8.
12. Driver J, Hoffman SE, Tavakol S, et al. A molecularly integrated grade for meningioma. Neuro-Oncology. 2021;24:796–808.
13. Petrilli AM, Fernández-Valle C. Role of Merlin/NF2 inactivation in tumor biology. Oncogene. 2016;35:537–48.
14. Lee Y, Liu J, Patel S, et al. Genomic landscape of meningiomas. Brain Pathol. 2010;20:751–62.
15. Zotti T, Scudiero I, Vito P, Stilo R. The emerging role of TRAF7 in tumor development. J Cell Physiol. 2017;232:1233–8.
16. von Spreckelsen N, Waldt N, Poetschke R, et al. KLF4K409Q–mutated meningiomas show enhanced hypoxia signaling and respond to mTORC1 inhibitor treatment. Acta Neuropathol Commun. 2020;8:1–11.
17. Carpten JD, Faber AL, Horn C, et al. A transforming mutation in the pleckstrin homology domain of AKT1 in cancer. Nature. 2007;448:439–44.
18. Hua L, Alkhatib M, Podlesek D, et al. Two predominant molecular subtypes of spinal meningioma: thoracic NF2-mutant tumors strongly associated with female sex, and cervical AKT1-mutant tumors originating ventral to the spinal cord. Acta Neuropathol. 2022;144:1053–5.
19. Hua L, Alkhatib M, Fujio S, et al. Genetic characterization and mutational profiling of foramen magnum meningiomas: a multi-institutional study. J Neurosurg. 2024;141(1):72–8. PMID: 38277657.
20. Williams SR, Juratli TA, Castro BA, et al. Genomic analysis of posterior fossa meningioma demonstrates frequent AKT1 E17K mutations in foramen magnum meningiomas. J Neurol Surg B Skull Base. 2019;80(6):562–7. PMID: 31750041.
21. Ng JMY, Curran T. The Hedgehog's tale: developing strategies for targeting cancer. Nat Rev Cancer. 2011;11:493–501.
22. Boetto J, Bielle F, Sanson M, Peyre M, Kalamarides M. SMO mutation status defines a distinct and frequent molecular subgroup in olfactory groove meningiomas. Neuro-Oncology. 2017;19:345–51.
23. Strickland MR, Gill CM, Nayyar N, et al. Targeted sequencing of SMO and AKT1 in anterior skull base meningiomas. J Neurosurg. 2017;127(2):438–44. PMID: 27885953
24. Abedalthagafi M, Bi WL, Aizer AA, et al. Oncogenic PI3K mutations are as common as AKT1 and SMO mutations in meningioma. Neuro-Oncology. 2016;18:649–55.
25. Peyre M, Gaillard S, de Marcellus C, et al. Progestin-associated shift of meningioma mutational landscape. Ann Oncol. 2018;29(3):681–6. PMID: 29206892.
26. Clark VE, Harmancı AS, Bai H, et al. Recurrent somatic mutations in POLR2A define a distinct subset of meningiomas. Nat Genet. 2016;48:1253–9.
27. Okano A, Miyawaki S, Teranishi Y, et al. POLR2A mutation is a poor prognostic marker of cerebellopontine angle meningioma. Neurosurgery. 2024;95(2):275–83. PMID: 38380947.
28. Schmitz U, Mueller W, Weber M, Sévenet N, Delattre O, von Deimling A. INI1 mutations in meningiomas at a potential hotspot in exon 9. Br J Cancer. 2001;84:199–201.
29. Christiaans I, Kenter SB, Brink HC, van Os TAM, Baas F, van den Munckhof P, Kidd AMJ, Hulsebos TJM. Germline SMARCB1 mutation and somatic NF2 mutations in familial multiple meningiomas. J Med Genet. 2011;48:93–7.
30. Shay JW. Role of telomeres and telomerase in aging and cancer. Cancer Discov. 2016;6:584–93.
31. Williams EA, Santagata S, Wakimoto H, et al. Distinct genomic subclasses of high-grade/progressive meningiomas: NF2-associated, NF2-exclusive, and NF2-agnostic. Acta Neuropathol Commun. 2020;8:171.
32. Harmancı AS, Youngblood MW, Clark VE, et al. Integrated genomic analyses of de novo pathways underlying atypical meningiomas. Nat Commun. 2017;8:14433.
33. Juratli TA, Thiede C, Koerner MVA, et al. Intratumoral heterogeneity and TERT promoter mutations in progressive/higher-grade meningiomas. Oncotarget. 2017;8:109228–37.

34. Sahm F, Schrimpf D, Olar A, et al. TERT promoter mutations and risk of recurrence in meningioma. JNCIJ. 2016;108:djv377.
35. Mirian C, Duun-Henriksen AK, Juratli T, et al. Poor prognosis associated with TERT gene alterations in meningioma is independent of the WHO classification: an individual patient data meta-analysis. J Neurol Neurosurg Psychiatry. 2020;91:378–87.
36. Mirian C, Grell K, Juratli TA, et al. Implementation of TERT promoter mutations improve prognostication of the WHO classification in meningioma. Neuropathol Appl Neurobio. 2022;48:e12773.
37. Korshunov A, Shishkina L, Golanov A. Immunohistochemical analysis of p16INK4a, p14ARF, p18INK4c, p21CIP1, p27KIP1 and p73 expression in 271 meningiomas correlation with tumor grade and clinical outcome. Int J Cancer. 2003;104:728–34.
38. Sievers P, Hielscher T, Schrimpf D, et al. CDKN2A/B homozygous deletion is associated with early recurrence in meningiomas. Acta Neuropathol. 2020;140:409–13.
39. Shankar GM, Abedalthagafi M, Vaubel RA, et al. Germline and somatic BAP1 mutations in high-grade rhabdoid meningiomas. Neuro-Oncology. 2016;19:535–45.
40. Smith MJ, O'Sullivan J, Bhaskar SS, et al. Loss-of-function mutations in SMARCE1 cause an inherited disorder of multiple spinal meningiomas. Nat Genet. 2013;45:295–8.
41. Smith MJ, Wallace AJ, Bennett C, et al. Germline SMARCE1 mutations predispose to both spinal and cranial clear cell meningiomas. J Pathol. 2014;234:436–40.
42. Sievers P, Sill M, Blume C, et al. Clear cell meningiomas are defined by a highly distinct DNA methylation profile and mutations in SMARCE1. Acta Neuropathol. 2020;141:281–90.
43. Bui C-B, Le HK, Vu DM, Truong K-DD, Nguyen NM, Ho MAN, Truong DQ. ARID1A-SIN3A drives retinoic acid-induced neuroblastoma differentiation by transcriptional repression of TERT. Mol Carcinog. 2019;58:1998–2007.
44. Rahmanto YS, Jung J-G, Wu R-C, Kobayashi Y, Heaphy CM, Meeker AK, Wang T-L, Shih I-M. Inactivating ARID1A tumor suppressor enhances TERT transcription and maintains telomere length in cancer cells*. J Biol Chem. 2016;291:9690–9.
45. Chaluts D, Dullea JT, Ali M, et al. ARID1A mutation associated with recurrence and shorter progression-free survival in atypical meningiomas. J Cancer Res Clin Oncol. 2023;149:5165–72.
46. Williams EA, Wakimoto H, Shankar GM, et al. Frequent inactivating mutations of the PBAF complex gene PBRM1 in meningioma with papillary features. Acta Neuropathol. 2020;140:89–93.
47. Varela I, Tarpey P, Raine K, et al. Exome sequencing identifies frequent mutation of the SWI/SNF complex gene PBRM1 in renal carcinoma. Nature. 2011;469:539–42.
48. Biegel JA, Busse TM, Weissman BE. SWI/SNF chromatin remodeling complexes and cancer. Am J Med Genet C: Semin Med Genet. 2014;166:350–66.
49. Juratli TA, McCabe D, Nayyar N, et al. DMD genomic deletions characterize a subset of progressive/higher-grade meningiomas with poor outcome. Acta Neuropathol. 2018;136:779–92.
50. Paramasivam N, Hübschmann D, Toprak UH, et al. Mutational patterns and regulatory networks in epigenetic subgroups of meningioma. Acta Neuropathol. 2019;138:295–308.
51. Katz LM, Hielscher T, Liechty B, et al. Loss of histone H3K27me3 identifies a subset of meningiomas with increased risk of recurrence. Acta Neuropathol. 2018;135:955–63.
52. Nassiri F, Wang JZ, Singh O, et al. Loss of H3K27me3 in meningiomas. Neuro-Oncology. 2021;23:1282–91.
53. Behling F, Paßlack P, Fodi C-K, Hielscher T, Schittenhelm J, Nassiri F, Wang JZ, Zadeh G, Tabatabai G, Sahm F. Loss of H3K27me3 in meningiomas: an independent marker for CNS WHO grade 2? Neuro-Oncol Adv. 2023;5:vdad112.
54. Lekanne Deprez RH, Riegman PH, van Drunen E, et al. Cytogenetic, molecular genetic and pathological analyses in 126 meningiomas. J Neuropathol Exp Neurol. 1995;54:224–35.
55. Maas SLN, Sievers P, Weber DC, et al. Independent prognostic impact of DNA methylation class and chromosome 1p loss in WHO grade 2 and 3 meningioma undergoing adjuvant high-dose radiotherapy: comprehensive molecular analysis of EORTC 22042–26042. Acta Neuropathol. 2023;146:837–40.
56. Ishino S, Hashimoto N, Fushiki S, Date K, Mori T, Fujimoto M, Nakagawa Y, Ueda S, Abe T, Inazawa J. Loss of material from chromosome arm 1p during malignant progression of meningioma revealed by fluorescent in situ hybridization. Cancer. 1998;83:360–6.
57. Bello MJ, Pestaña A, Rey JA, De Campos JM, Kusak ME, Vaquero J, Sarasay JL. Allelic loss at 1 p is associated with tumor progression of meningiomas. Genes Chromosom Cancer. 1994;9:296–8.
58. Hielscher T, Sill M, Sievers P, Stichel D, Brandner S, Jones DTW, von Deimling A, Sahm F, Maas SLN. Clinical implementation of integrated molecular-morphologic risk prediction for meningioma. Brain Pathol. 2022;33:e13132. https://doi.org/10.1111/bpa.13132.
59. Daoud EV, Zhu K, Mickey B, et al. Epigenetic and genomic profiling of chordoid meningioma: implications for clinical management. Acta Neuropathol Commun. 2022;10:56.
60. Sievers P, Stichel D, Hielscher T, et al. Chordoid meningiomas can be sub-stratified into prognostically distinct DNA methylation classes and are enriched for heterozygous deletions of chromosomal arm 2p. Acta Neuropathol. 2018;136:975–8.
61. Berghaus N, Hielscher T, Savran D, et al. Meningiomas: sex-specific differences and prog-

nostic implications of a chromosome X loss. Neuro Oncol 2024:noae239. PMID: 39535060.
62. Abedalthagafi MS, Bi WL, Merrill PH, et al. ARID1A and TERT promoter mutations in dedifferentiated meningioma. Cancer Genet. 2015;208:345–50.
63. Lucas C-HG, Mirchia K, Seo K, et al. Spatial genomic, biochemical and cellular mechanisms underlying meningioma heterogeneity and evolution. Nat Genet. 2024;56:1121–33.
64. Vasudevan HN, Choudhury A, Hilz S, et al. Intratumor and informatic heterogeneity influence meningioma molecular classification. Acta Neuropathol. 2022;144:579–83.
65. Juratli TA, Prilop I, Saalfeld FC, et al. Sporadic multiple meningiomas harbor distinct driver mutations. Acta Neuropathol Commun. 2021;9:1–4.
66. Islim AI, Lee JX, Mustafa MA, et al. Sporadic multiple intracranial meningioma does not infer worse patient outcomes: results from a case control study. J Neuro-Oncol. 2023;161:287–95.

Location of Meningiomas: Demographic—Pathological and Prognostic Correlations

Francesco Maiuri, Sergio Corvino, Giuseppe Mariniello, and Marialaura Del Basso De Caro

1 Introduction

The location of meningiomas is a well-known factor correlated to clinical presentation and symptoms, extent of surgical resection, and recurrence [1–4]. Several studies [5–16] have also correlated the meningioma location with the tumor pathology and molecular biology.

This chapter discusses the impact of the meningioma location on the tumor pathology and molecular biology, extent of resection, recurrence, and outcome.

2 Classification of Meningioma Locations

Most studies including meningiomas of all locations [5–10, 12–14, 17, 18] classify them in skull base and non-skull base. However, the group of skull base meningiomas includes locations with different potential of growth and recurrence risk. Thus, it is more correct to consider separately medial skull base and lateral skull base tumors. This is supported by the different embryological origin of the meninges [19], depending on the anatomical site. Studies on the meningeal development [20–25] indicate three sources: paraxial mesoderm, dorsal mesoderm, and neural crest. The central skull base develops from the paraxial mesoderm and the posterior skull base from the dorsal mesoderm; the lateral region, including the sphenoid wing, originates from the neural crest. The anterior and middle meningeal convexities, falx cerebri and cerebellar tentorium originate from the neural crest, and the posterior meningeal convexity from the paraxial mesoderm. This suggests classifying intracranial meningiomas in medial skull base, lateral skull base, and non-skull base. Meningiomas of sites of different embryological origin often show different pathological and biomolecular features.

Yasargil in his book [26] classifies intracranial meningiomas in six main groups: basal, fissural, dorsal, intraventricular, orbital, and calvarial. Medial skull base meningiomas are subdivided in midline and paramedian, although of the same embryological origin. Similarly, non-skull base meningiomas are subdivided in dorsal (parasagittal, paramedian, and cerebellar convexities) and fissural (falcine, tentorial, and sylvian). These subdivisions reflect the clinical behavior, patterns of spread, and the involvement of adjacent structures.

F. Maiuri (✉) · S. Corvino · G. Mariniello
Department of Neurosciences, Reproductive and Odontostomatological Sciences, University of Naples "Federico II", Naples, Italy
e-mail: frmaiuri@unina.it

M. Del Basso De Caro
Department of Advanced Biomedical Sciences, Section of Pathology, University of Naples "Federico II", Naples, Italy

F. Maiuri, M. Del Basso De Caro (eds.), *Meningiomas*,
https://doi.org/10.1007/978-3-031-76680-0_8

Table 1 Classification of the meningioma locations

1. Medial Skull Base
Olfactory groove, planum ethmoidalis-sphenoidale
Tuberculum sellae
Parasellar (anterior clinoid, optic canal)
Clival-petroclival
Foramen magnum
2. Lateral Skull Base
Middle and lateral sphenoid wing
Spheno-orbital
Petrous bone
Temporal and occipital fossa
3. Non-Skull Base
Cerebral convexity
Parasagittal
Falx
Tentorium
Cerebellar convexity
4. Intraventricular
5. Spinal

In this chapter, we will use the meningiomas classification into medial skull base, lateral skull base, non-skull base, intraventricular, and spinal (Table 1).

3 Incidence of Meningioma Locations

The rate of meningioma locations has variably been reported. In our previous study [15], meningiomas were in the medial skull base in 24% and in the lateral skull base in 13%; 54% were non-skull base and 9% were in the spinal canal. In the study of Sum et al. [27], non-skull base meningiomas were also prevalent (59%) over the medial (29%) and lateral skull base (19%). On the other hand, in the series of Magill et al. [28] the skull base meningiomas were more numerous.

4 Location and Population-Based Aspects

4.1 Age

The location of intracranial meningiomas in the elderly (≥80 years) is more frequently non-skull base than skull base, as compared to all age groups [29, 30]. The lower surgical risk of non-skull base meningiomas likely favors the indication to surgery of these locations in elderly patients. Spinal meningiomas are also more frequent in the older age.

Meningiomas in children are mainly supratentorial (frontal and parietal convexity), although the posterior fossa locations are more frequent (19%) than all age groups (10%). These tumors frequently occur in unusual locations, such as ventricle and sylvian fissure [31, 32]. Intraventricular meningiomas account for 9–15% versus 5% of adults. Meningiomas in infants aged less than 1 year mainly occur within the brain parenchyma, in intra-axial locations (about 1/3 of the cases).

Spinal meningiomas are very rare in children (2–8% of pediatric meningiomas) and are extradural in up to 15%.

4.2 Sex and Related Conditions

Meningiomas are more frequent in women, with an incidence of about 3:1 [33]. A higher incidence of the female sex is evidenced for spinal (92%) and lateral skull base (88%) locations [15]. No significant difference of the tumor location is observed between premenopausal and postmenopausal women [34].

Meningiomas occurring in pregnancy are mainly supratentorial (95%); medial skull base locations (olfactory groove, tuberculum sellae, anterior clinoid) are significantly more frequent (68%) than non-skull base and infratentorial ones [34–37].

The sex hormone therapies are at risk of occurrence and growth of meningiomas, which are often multiple. Those occurring during therapies with progesterone agonists are more frequently located at the anterior and medial skull base [34, 38–41].

4.3 Familial Genetic Syndromes

Several familial genetic syndromes are associated with the occurrence of meningiomas. These include neurofibromatosis (NF) type 2, basal cell nevus syndrome, Cowed syndrome, Werner syndrome, SMARCE 1 and SMARCB 1 syndromes [42]. Meningiomas occur in 25–40% of NF2 patients [31, 32]; they are mainly located in the brain convexity and also in the lateral and posterior skull base; multiple meningiomas are more frequent than in the general population.

4.4 Radiation Therapy

Radiation therapy for neoplastic or non-tumoral head and neck pathologies may induce the development of intracranial meningiomas; they occur in the field of irradiation after a long latent period (at least 5 years after radiation treatment) [43, 44].

More than 80% of the radiation-induced meningiomas are located over the brain convexity, because of the large doses of radiation received in this region. Irradiation of nasopharyngeal and neck tumors may induce skull base meningiomas. Whole brain radiation therapy for malignant intracranial tumors, mainly medulloblastomas, may cause the development of meningiomas of all locations, mainly at the skull base. Radiation-induced meningiomas often show bone and dural invasion and multiple locations.

The incidence of multiple lesions is high in patients with both high dose and low-dose radiation therapy (4.6–18.7%) [43], and significantly higher than that of sporadic meningiomas.

5 Location and Pathology

5.1 Who Grade and Proliferation Index

The WHO grade and proliferation index are different according to the location of meningiomas [15, 45]. Several studies of the last years have reported significantly higher rates of WHO II and III tumors [6, 8, 10–12, 28, 46] and higher values of Ki67-MIB1 [5–7, 9, 10, 12] in non-skull base versus skull base meningiomas. However, the group of skull base meningiomas includes locations with different rates of atypical forms and different values of Ki67-LI. In our study [15] lateral skull base meningiomas have shown higher rates of WHO II and III tumors and higher values of Ki67-LI than medial ones, and values more similar to the non-skull base ones. This may reflect the above cited different embryological origin of the meninges.

Spinal meningiomas of WHO grade I show lower mean values of Ki67-LI as compared to the intracranial ones [47–52]; their rates of atypical forms and Ki67-LI values are rather similar to those of medial skull base meningiomas and significantly lower than non-skull base ones [6, 15].

5.2 Histological Subtypes of WHO Grade I Meningiomas

The distribution of the histological subtypes of WHO grade I meningiomas in the different locations has been investigated in several studies [15, 53, 54]. Lee et al. [53] found that the meningothelial subtype was significantly more frequent at the medial skull base and spinal canal than lateral skull base and non-skull base locations. In the study of Maiuri et al. [15] the transitional subtype was the most common in the skull base, followed by the meningothelial one; non-skull base meningiomas were mainly of fibrous and transitional types, whereas the meningothelial subtype was rare; besides, all microcystic meningiomas were found at non-skull base locations. Spinal meningiomas were more frequently psammomatous and transitional. Bhat et al. [54] reported no significant differences in the various locations. Thus, it may be suggested that meningiomas of the dural regions originating from the neural crest and dorsal mesoderm are more frequently fibrous; on the contrary, those in the dural locations originating from the paraxial mesoderm are mainly meningothelial [19].

5.3 Progesterone-Receptor Expression

Several studies [38, 55–58] did not find significant correlation between PR status and tumor location, as confirmed in a recent review [59]. However, they consider the overall locations in a unique group. Kuroi et al. [13] reported significantly higher rate of positive PR expression in skull base meningiomas as compared to the non-skull base ones. Another study [59] showed in medial skull base meningiomas significantly higher rate of cases with high PR expression than lateral skull base and non-skull base ones. The inverse correlation between PR expression and Ki 67-LI, reported in several studies of intracranial meningiomas [57, 58, 60–63], is confirmed in groups of different locations, mainly for medial skull base ones.

Spinal meningiomas more often show high PR positivity in high ratings of cases and no significant correlation with the Ki67-LI [15, 48, 49].

6 Location and Biomolecular Profile

Although many genomic studies on meningiomas have been published in the last 25 years, the genomic findings up to 2010 were lacking unanimous consensus and generalizable data for all meningiomas [64].

However, more recent genomic analyses have better clarified the correlation of biomolecular data with the tumor type, location, and behavior [19, 65–72]. NF2 mutations and/or loss of chromosome 22 are predominant in non-skull base meningiomas (cerebral convexities, parasagittal, falcine, cerebellar dura) [65, 66, 68, 69]. Among the different skull base locations, meningiomas of the lateral and posterior skull base harbor more frequently NF2 mutations, loss of chromosome 22 and KLF4, whereas most meningiomas of the medial skull base and anterior fossa are non-NF2 and express AKT1 and SMO genes [66, 67, 72]. The molecular biology and genetics of meningiomas are discussed in chapters "Molecular Biology and Genetics of Meningiomas: Classification and Diagnostic Methods" and "Comprehensive Overview of Genetic Alterations in Meningioma" of this book.

7 Location, Resection, and Recurrence

7.1 Location and Risk of Recurrence

Meningiomas of different locations may show variable and often significant differences of recurrence rates. Involved factors include tumor size, extent of surgical resection, WHO grade, Ki67 MIB1, and biomolecular profile of the meningiomas [4, 73, 74]. However, some of them are differently defined in recent studies [75–78]. Although the Simpson grading scale is still widely used for determining the extent of resection [79], some define "gross-total resection" the macroscopic complete removal of the tumor including Simpson grades I, II, and III [76, 78, 79]. However, grade III resections show different prognostic value and outcome as compared to grades I and II, mainly for atypical meningiomas with significant invasion of the venous sinuses and bone. Besides, defining both WHO grades II and III as "high grade meningiomas" [75] is also questionable; anaplastic WHO III tumors show significantly worse prognosis and shorter survival than atypical WHO grade II ones.

7.2 Medial Skull Base Meningiomas

The group of medial skull base meningiomas includes locations (such as olfactory groove, tuberculum sellae, and anterior clinoid) with often more slow course and low recurrence rates after complete resection, ranging from 0% to 15% [78, 80–92].

In spite of their usually large size, *olfactory groove meningiomas* show low rate of basal bone and paranasal sinus invasion and rare arterial encasement; this allows to obtain Simpson grades I and II resection a very high rate of cases. Several surgical series including more than 60 patients treated by transcranial approaches and with a median follow-up >5 years [82, 83, 92–94] report recurrence rates ≤5%. The recurrences mainly occur at the ethmoid bone and paranasal cavities (Fig. 1). A series of 50 patients treated by endoscopic endonasal approach [85] also reports a 4% recurrence rate.

Tuberculum sellae meningiomas show variable rates of complete resection and recurrence according to their size and relationship with the surrounding structures [88]. In tumors with simple anatomy (diameter ≤17 mm, no or ≤3 mm optic canal invasion, no or <180° growth around the arteries), the rate of complete resection is >90%. In meningiomas with complex anatomy (diameter ≥17 mm, optic canal invasion >3 mm, arterial encasement),

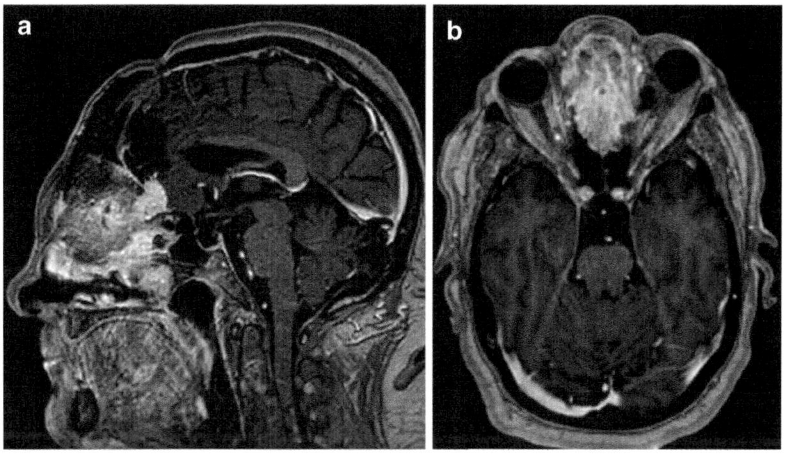

Fig. 1 Post-contrast cranial MRI of a 70-years-old female operated on for large olfactory groove WHO grade I meningioma with complete resection 9 years before: extensive basal recurrence with invasion of the ethmoidal bone and nasal cavity

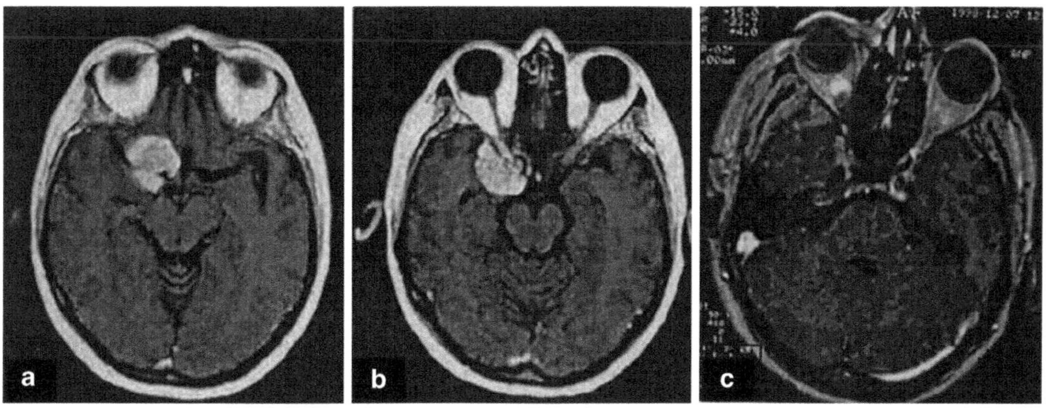

Fig. 2 (**a**–**b**) Post-contrast skull MRI: small right anterior clinoid meningioma on the right side. (**c**) Postoperative MRI: complete tumor resection

the rate of complete resection ranges from 50% to 65%. The recurrence rates range from 10% to 13% [86–88], with no difference between transcranial and transsphenoidal approaches.

Anterior clinoidal meningiomas may be treated by gross-total resection in variable rates according to their size and relationship with the optic nerve and internal carotid artery (Figs. 2 and 3). The total resection rate of Al Mefty [95] type I tumors is low (11–15%) due to the encasement of the internal carotid artery with no arachnoid plane; on the contrary, it is significantly higher for type II (>90%) and type III (80–85%), where a clear arachnoid layer between the tumor and the blood vessels is present [89–91, 96]. The extradural anterior clinoidectomy coupled by unroofing the optic canal improves the surgical exposure and the rate of complete resection [97, 98]. The recurrence–regrowth rates are low (2–11%) after gross-total resection and significantly higher (38–46%) after subtotal resection [89, 90, 96, 99].

In conclusion, medial skull base meningiomas show lower rates of recurrence and longer recurrence free survival than other intracranial locations after complete resection, also because of the lower rates of atypical WHO II forms.

7.3 Lateral Skull Base Meningiomas

The group of the lateral skull base meningiomas includes locations, such as *sphenoid wing* and *spheno-orbital*, with variable and often

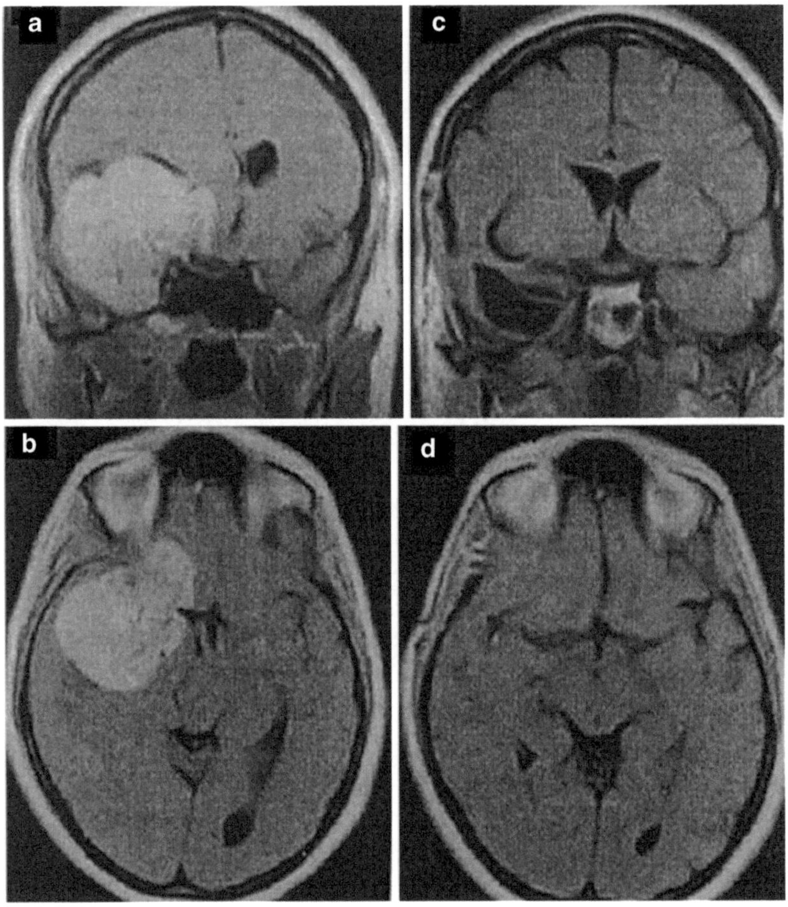

Fig. 3 (**a**–**b**) Post-contrast cranial MRI of a 70-years-old man presenting with right visual loss: large anterior clinoid meningioma on the right side. (**c**–**d**) Postoperative MRI: complete tumor resection

extensive dural, bone, and orbital invasion. In a recent review of 23 surgical series of spheno-orbital meningiomas [100], the recurrence rate was very variable, from 0% to 56.2%; series with complete resection in ≥70% of the cases showed a median recurrence rate of 6.5% versus 29% in studies reporting complete resection in <70% of the patients. Diffuse dural and bone infiltration, involvement of the orbital apex and superior orbital fissure are at higher risk of recurrence [100–103]. Only in the study by Terrier et al. [104], the tumor location and extension and the extent of resection did not predict recurrence.

Thus, lateral skull base meningiomas, as compared to the medial skull base ones, show lower rates of complete resection and higher recurrence rate, also because of the higher amount of WHO grade II forms.

7.4 Non-skull Base Meningiomas

This group includes convexity, parasagittal, falx, and tentorial meningiomas, arising from dural regions derived from the neural crest.

Convexity meningiomas are more favorable from the surgical view point, with rates of Simpson grade I resections of more than 90%; removal of the surrounding dura for more than 2 cm (grade 0 resection) [105] further improves the long-term outcome. Although older studies [106, 107] report high recurrence rates (11–13%), others more recent found recurrence in about 4–6% of the patients [108, 109]. Factors affecting the recurrence rate include large tumor size, loss of arachnoid interface with extrapial and subpial cleavage plane [109], entity of near dural resection [105], location at level of critical areas (motor, speech) [110], and obviously the histological WHO grade.

Parasagittal meningiomas have variable but high recurrence rate ranging from 13 to 40% [111]. Surgical findings that hinder the gross-total resection include the entity of sinus invasion [112], the pial invasion and involvement of the rolandic vein at the motor area [110]. Benign WHO grade I lesions, attached to or invading the ipsilateral sinus wall or lateral recess, and those totally resected have a lower risk of recurrence than atypical tumors invading or occluding the sinus lumen and incompletely resected, as confirmed in a recent meta-analysis [112]. Thus, the WHO grading (grades II–III versus I) is the strongest predictive factor of recurrence, followed by extent of sinus invasion (Sindou [112] types IV–V versus I–III) and extent of resection (partial versus gross total).

Falcine meningiomas (approximately 5% of all intracranial meningiomas [106, 113]) may undergo gross-total resection in about 90% of the cases. Factors that may hinder the complete resection include the proximity with the superior sagittal sinus and cortical veins and the encasement of the pericallosal arteries. The recurrence rates for patients who underwent Simpson grade I resection range from 7.3% [114] to 18.1% [106] after follow-up of more than 5 years. Mooney et al. [115] report a median progression-free survival of 84 months after gross-total resection of "de novo" falcine meningiomas versus 30 months after subtotal resection. Thus, the extent of resection and the WHO grade are the main factors at risk of recurrence, also because of the high rate of atypical WHO grade II tumors in this location [115, 116].

The falx anatomically originates by the reflection of the meningeal layers of the neural crest on the midline; thus, it is formed by two leaflets of dura. It is possible that tumor nodules may diffusely spread along the falx between the two layers; this may explain late and distant recurrences after Simpson grade I resection. In this view, additional 2 cm (grade 0) resection anterior and posterior to the initial attachment may reduce the recurrence rate [115] (Fig. 4).

Tentorial meningiomas (3–5% of all intracranial meningiomas) may arise at different tentorial areas and may be classified according to their direction of growth (supratentorial, infratentorial or both), relationship with the transverse sinuses and extension in the tentorial notch. The rates of gross-total resection are different according to the involved tentorial area [117–120]; the most common reasons for subtotal resection include involvement of neurovascular structures at the tentorial incisure, invasion of torcula and nearby venous sinus (mainly transverse) and very large size with multicompartmental growth, sometimes requiring separate approaches, The reported recurrence rates of tentorial meningiomas are very variable, from 8.6% to 31.7% [120]; this reflects the different tentorial locations with different rates of gross-total resection. Thus, the tumor location along the tentorium and the relationship with the neurovascular structures are the main predictors of recurrence.

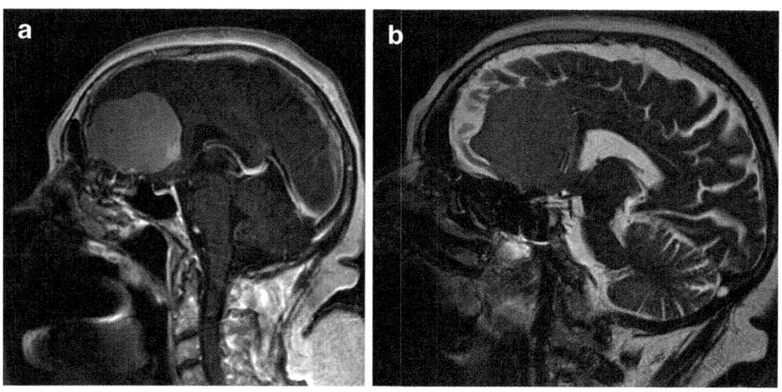

Fig. 4 Falcine Meningioma. Preoperative sagittal MRI (**a**–**b**) showing large meningioma of the anterior third of the falx cerebri

7.5 Intraventricular Meningiomas

Although intraventricular meningiomas are included in most studies in the non-skull base group, they must be treated as a separate entity because of their embryological origin different from other locations. These meningiomas (0.5–3%) mainly occur in the lateral ventricles (80%), more rarely in the third (15%) and fourth (4%) ventricles [121]. They arise from arachnoid cap cells present in the choroid plexus and in the tela coroidea of the vellum interpositum. A recent review of 98 studies from the literature [121] reports gross-total resection in 94.7% of the patients. Intraventricular meningiomas are rather indolent tumors with a low rate of atypical WHO grade I histology. Recurrences occur in about 5% of the cases, almost exclusively after subtotal resection [121, 122]. From the biomolecular view point, these meningiomas frequently harbor NF2 mutations but lack other common genetic alterations [123, 124].

7.6 Spinal Meningiomas

Spinal meningiomas are known to have favorable prognosis. The complete surgical resection is almost always possible; it results in gratifying improvement of the neurological function and low recurrence rates. In a review of 19 series of spinal meningiomas including different grades of surgical resection and different histology [52] the recurrence rate was ≤5% in 9 and between 5% and 18% in 10; the role of the different length of the follow-up was less significant and correlated to the rate of complete resection [47, 49–51, 125]. The WHO grade (II, III) and the incomplete resection are the main factors at risk of recurrence. In benign WHO grade I meningiomas, the invasion of the pia-arachnoid and the Ki67-MIB1 are significant. On the other hand, the tumor size, its location along the spinal axis, the distribution along the circumference of the spinal canal, the resection of the infiltrated dura (Simpson grade I versus II) are not correlated with the recurrence [52].

8 Conclusion

The meningioma location is relevant not only from the surgical but also from the pathological and prognostic viewpoint, according to the different embryological origin of the meninges. Medial skull base and spinal meningiomas show significantly lower incidence of atypical WHO II forms, lower rates of cases with Ki 67 Li values >4%, and lower recurrence rate than lateral skull base and non-skull base meningiomas. NF2 mutations and loss of chromosome 22 are the most frequent biomolecular findings in non-skull base and lateral skull base tumors. Although complete resection is the most relevant factor to prevent recurrences, the evaluation of the long-term outcome and mainly the risk of late recurrences must also consider the different pathological and biomolecular aspects according to the different locations.

References

1. Mirimanoff RO, Dosoretz DE, Linggood RM, Ojemann RG, Martuza RL. Meningioma: analysis of recurrence and progression following neurosurgical resection. J Neurosurg. 1985;62(1):18–24. https://doi.org/10.3171/jns.1985.62.1.0018.
2. Rohringer M, Sutherland GR, Louw DF, Sima AA. Incidence and clinicopathological features of meningioma. J Neurosurg. 1989;71(5 Pt 1):665–72. https://doi.org/10.3171/jns.1989.71.5.0665.
3. Nakasu S, Nakasu Y, Nakajima M, Matsuda M, Handa J. Preoperative identification of meningiomas that are highly likely to recur. J Neurosurg. 1999;90(3):455–62. https://doi.org/10.3171/jns.1999.90.3.0455.
4. Maiuri F, Mariniello G, Peca C, Guadagno E, Corvino S, d'Avanzo S, et al. Multicentric and diffuse recurrences of meningiomas. Br J Neurosurg. 2020;34:1–8. https://doi.org/10.1080/02688697.2020.1754335.
5. Kasuya H, Kubo O, Tanaka M, Amano K, Kato K, Hori T. Clinical and radiological features related to the growth potential of meningioma. Neurosurg Rev. 2006;29(4):293–6.; discussion 6–7. https://doi.org/10.1007/s10143-006-0039-3.
6. Sade B, Chahlavi A, Krishnaney A, Nagel S, Choi E, Lee JH. World Health Organization Grades II and III meningiomas are rare in the cranial base and spine. Neurosurgery. 2007;61(6):1194–8. ; discussion 8. https://doi.org/10.1227/01.neu.0000306097.38141.65.

7. McGovern SL, Aldape KD, Munsell MF, Mahajan A, DeMonte F, Woo SY. A comparison of World Health Organization tumor grades at recurrence in patients with non-skull base and skull base meningiomas. J Neurosurg. 2010;112(5):925–33. https://doi.org/10.3171/2009.9.JNS09617.
8. Kane AJ, Sughrue ME, Rutkowski MJ, Shangari G, Fang S, McDermott MW, et al. Anatomic location is a risk factor for atypical and malignant meningiomas. Cancer. 2011;117(6):1272–8. https://doi.org/10.1002/cncr.25591.
9. Hashimoto N, Rabo CS, Okita Y, Kinoshita M, Kagawa N, Fujimoto Y, et al. Slower growth of skull base meningiomas compared with non-skull base meningiomas based on volumetric and biological studies. J Neurosurg. 2012;116(3):574–80. https://doi.org/10.3171/2011.11.JNS11999.
10. Cornelius JF, Slotty PJ, Steiger HJ, Hänggi D, Polivka M, George B. Malignant potential of skull base versus non-skull base meningiomas: clinical series of 1,663 cases. Acta Neurochir. 2013;155(3):407–13. https://doi.org/10.1007/s00701-012-1611-y.
11. Zhou P, Ma W, Yin S, Li Y, Jiang S. Three risk factors for WHO grade II and III meningiomas: a study of 1737 cases from a single center. Neurol India. 2013;61(1):40–4. https://doi.org/10.4103/0028-3886.107928.
12. Mansouri A, Klironomos G, Taslimi S, Kilian A, Gentili F, Khan OH, et al. Surgically resected skull base meningiomas demonstrate a divergent postoperative recurrence pattern compared with non-skull base meningiomas. J Neurosurg. 2016;125(2):431–40. https://doi.org/10.3171/2015.7.JNS15546.
13. Kuroi Y, Matsumoto K, Shibuya M, Kasuya H. Progesterone receptor is responsible for benign biology of skull base meningioma. World Neurosurg. 2018;118:e918–e24. https://doi.org/10.1016/j.wneu.2018.07.100.
14. Savardekar AR, Patra DP, Bir S, Thakur JD, Mohammed N, Bollam P, et al. Differential tumor progression patterns in skull base versus non-skull base meningiomas: a critical analysis from a long-term follow-up study and review of literature. World Neurosurg. 2018;112:e74–83. https://doi.org/10.1016/j.wneu.2017.12.035.
15. Maiuri F, Mariniello G, Guadagno E, Barbato M, Corvino S, Del Basso De Caro M. WHO grade, proliferation index, and progesterone receptor expression are different according to the location of meningioma. Acta Neurochir. 2019;161(12):2553–61. https://doi.org/10.1007/s00701-019-04084-z.
16. Maiuri F, Mariniello G, Barbato M, Corvino S, Guadagno E, Chiariotti L, et al. Malignant intraventricular meningioma: literature review and case report. Neurosurg Rev. 2022;45(1):151–66. https://doi.org/10.1007/s10143-021-01585-3.
17. Magill ST, Dalle Ore CL, Diaz MA, Jalili DD, Raleigh DR, Aghi MK, et al. Surgical outcomes after reoperation for recurrent non-skull base meningiomas. J Neurosurg. 2018;131:1–9. https://doi.org/10.3171/2018.6.JNS18118.
18. Magill ST, Lee DS, Yen AJ, Lucas CG, Raleigh DR, Aghi MK, et al. Surgical outcomes after reoperation for recurrent skull base meningiomas. J Neurosurg. 2018;130(3):876–83. https://doi.org/10.3171/2017.11.JNS172278.
19. Okano A, Miyawaki S, Hongo H, Dofuku S, Teranishi Y, Mitsui J, et al. Associations of pathological diagnosis and genetic abnormalities in meningiomas with the embryological origins of the meninges. Sci Rep. 2021;11(1):6987. https://doi.org/10.1038/s41598-021-86298-9.
20. O'Rahilly R, Müller F. The meninges in human development. J Neuropathol Exp Neurol. 1986;45(5):588–608.
21. Jiang X, Iseki S, Maxson RE, Sucov HM, Morriss-Kay GM. Tissue origins and interactions in the mammalian skull vault. Dev Biol. 2002;241(1):106–16. https://doi.org/10.1006/dbio.2001.0487.
22. Noden DM, Trainor PA. Relations and interactions between cranial mesoderm and neural crest populations. J Anat. 2005;207(5):575–601. https://doi.org/10.1111/j.1469-7580.2005.00473.x.
23. McBratney-Owen B, Iseki S, Bamforth SD, Olsen BR, Morriss-Kay GM. Development and tissue origins of the mammalian cranial base. Dev Biol. 2008;322(1):121–32. https://doi.org/10.1016/j.ydbio.2008.07.016.
24. Wu T, Chen G, Tian F, Liu HX. Contribution of cranial neural crest cells to mouse skull development. Int J Dev Biol. 2017;61(8–9):495–503. https://doi.org/10.1387/ijdb.170051gc.
25. Dasgupta K, Jeong J. Developmental biology of the meninges. Genesis. 2019;57(5):e23288. https://doi.org/10.1002/dvg.23288.
26. Yasargil MG. Chapter 9A. In: Microneurosurgery, vol. IV B. Stuttgart: Georg Thieme Verlag; 1996. p. 134–65.
27. Sun C, Dou Z, Wu J, Jiang B, Iranmanesh Y, Yu X, et al. The preferred locations of meningioma according to different biological characteristics based on voxel-wise analysis. Front Oncol. 2020;10:1412. https://doi.org/10.3389/fonc.2020.01412.
28. Magill ST, Young JS, Chae R, Aghi MK, Theodosopoulos PV, McDermott MW. Relationship between tumor location, size, and WHO grade in meningioma. Neurosurg Focus. 2018;44(4):E4. https://doi.org/10.3171/2018.1.FOCUS17752.
29. Joubert C, Sellier A, Sahuc P, Beucler N, Desse N, Bernard C, et al. Neurosurgery for intracranial meningioma in patients aged more than 80 years: benefits and rationale. Br J Neurosurg. 2021;35(4):470–5. https://doi.org/10.1080/02688697.2021.1875397.
30. Maiuri F, Corvino S, Lorenzetti M, Franca RA, Esposito F, Del Basso De Caro M. Intracranial meningiomas in patients aged ≥80 years: pathological features and surgical problems. World Neurosurg.

2023;173:e498–508. https://doi.org/10.1016/j.wneu.2023.02.086.
31. Arivazhagan A, Devi BI, Kolluri SV, Abraham RG, Sampath S, Chandramouli BA. Pediatric intracranial meningiomas—do they differ from their counterparts in adults? Pediatr Neurosurg. 2008;44(1):43–8. https://doi.org/10.1159/000110661.
32. Phillips D, Auguste KI, Gupta N. Meningiomas in children. Handb Clin Neurol. 2020;169:253–9. https://doi.org/10.1016/B978-0-12-804280-9.00016-0.
33. Wiemels J, Wrensch M, Claus EB. Epidemiology and etiology of meningioma. J Neuro-Oncol. 2010;99(3):307–14. https://doi.org/10.1007/s11060-010-0386-3.
34. Maiuri F, Mariniello G, Somma T, Guadagno E, Corvino S, Pagano S, et al. Meningiomas in premenopausal women: role of the hormone related conditions. Front Oncol. 2020;10:556701. https://doi.org/10.3389/fonc.2020.556701.
35. Lusis EA, Scheithauer BW, Yachnis AT, Fischer BR, Chicoine MR, Paulus W, et al. Meningiomas in pregnancy: a clinicopathologic study of 17 cases. Neurosurgery. 2012;71(5):951–61. https://doi.org/10.1227/NEU.0b013e31826adf65.
36. Laviv Y, Bayoumi A, Mahadevan A, Young B, Boone M, Kasper EM. Meningiomas in pregnancy: timing of surgery and clinical outcomes as observed in 104 cases and establishment of a best management strategy. Acta Neurochir. 2018;160(8):1521–9. https://doi.org/10.1007/s00701-017-3146-8.
37. Chakravarthy V, Kaplan B, Gospodarev V, Myers H, De Los RK, Achiriloaie A. Houdini tumor: case report and literature review of pregnancy-associated meningioma. World Neurosurg. 2018;114:e1261–e5. https://doi.org/10.1016/j.wneu.2018.03.187.
38. Portet S, Banor T, Bousquet J, Simonneau A, Flores M, Ingrand P, et al. New insights into expression of hormonal receptors by meningiomas. World Neurosurg. 2020;140:e87–96. https://doi.org/10.1016/j.wneu.2020.04.168.
39. Champeaux-Depond C, Weller J, Froelich S, Sartor A. Cyproterone acetate and meningioma: a nationwide-wide population based study. J Neuro-Oncol. 2021;151(2):331–8. https://doi.org/10.1007/s11060-020-03672-9.
40. Samarut E, Lugat A, Amelot A, Scharbarg E, Hadjadj S, Primot C, et al. Meningiomas and cyproterone acetate: a retrospective, monocentric cohort of 388 patients treated by surgery or radiotherapy for intracranial meningioma. J Neuro-Oncol. 2021;152(1):115–23. https://doi.org/10.1007/s11060-020-03683-6.
41. Lee KS, Zhang JJY, Kirollos R, Santarius T, Nga VDW, Yeo TT. A systematic review and meta-analysis of the association between cyproterone acetate and intracranial meningiomas. Sci Rep. 2022;12(1):1942. https://doi.org/10.1038/s41598-022-05773-z.
42. Kerr K, Qualmann K, Esquenazi Y, Hagan J, Kim DH. Familial syndromes involving meningiomas provide mechanistic insight into sporadic disease. Neurosurgery. 2018;83(6):1107–18. https://doi.org/10.1093/neuros/nyy121.
43. Umansky F, Shoshan Y, Rosenthal G, Fraifeld S, Spektor S. Radiation-induced meningioma. Neurosurg Focus. 2008;24(5):E7. https://doi.org/10.3171/FOC/2008/24/5/E7.
44. Yamanaka R, Hayano A, Kanayama T. Radiation-induced meningiomas: an exhaustive review of the literature. World Neurosurg. 2017;97:635–44.e8. https://doi.org/10.1016/j.wneu.2016.09.094.
45. Franca RA, Della Monica R, Corvino S, Chiariotti L, Del Basso De Caro M. WHO grade and pathological markers of meningiomas: clinical and prognostic role. Pathol Res Pract. 2023;243:154340. https://doi.org/10.1016/j.prp.2023.154340.
46. Meling TR, Da Broi M, Scheie D, Helseth E. Meningiomas: skull base versus non-skull base. Neurosurg Rev. 2019;42(1):163–73. https://doi.org/10.1007/s10143-018-0976-7.
47. Klekamp J, Samii M. Surgical results for spinal meningiomas. Surg Neurol. 1999;52(6):552–62. https://doi.org/10.1016/s0090-3019(99)00153-6.
48. Roser F, Nakamura M, Bellinzona M, Ritz R, Ostertag H, Tatagiba MS. Proliferation potential of spinal meningiomas. Eur Spine J. 2006;15(2):211–5. https://doi.org/10.1007/s00586-005-0937-z.
49. Barresi V, Alafaci C, Caffo M, Barresi G, Tuccari G. Clinicopathological characteristics, hormone receptor status and matrix metallo-proteinase-9 (MMP-9) immunohistochemical expression in spinal meningiomas. Pathol Res Pract. 2012;208(6):350–5. https://doi.org/10.1016/j.prp.2012.02.013.
50. Nakamura M, Tsuji O, Fujiyoshi K, Hosogane N, Watanabe K, Tsuji T, et al. Long-term surgical outcomes of spinal meningiomas. Spine (Phila Pa 1976). 2012;37(10):E617–23. https://doi.org/10.1097/BRS.0b013e31824167f1.
51. Maiti TK, Bir SC, Patra DP, Kalakoti P, Guthikonda B, Nanda A. Spinal meningiomas: clinicoradiological factors predicting recurrence and functional outcome. Neurosurg Focus. 2016;41(2):E6. https://doi.org/10.3171/2016.5.FOCUS16163.
52. Maiuri F, Del Basso De Caro M, de Divitiis O, Guadagno E, Mariniello G. Recurrence of spinal meningiomas: analysis of the risk factors. Br J Neurosurg. 2019;34:1–6. https://doi.org/10.1080/02688697.2019.1638886.
53. Lee JH, Sade B, Choi E, Golubic M, Prayson R. Meningothelioma as the predominant histological subtype of midline skull base and spinal meningioma. J Neurosurg. 2006;105(1):60–4. https://doi.org/10.3171/jns.2006.105.1.60.
54. Bhat AR, Wani MA, Kirmani AR, Ramzan AU. Histological-subtypes and anatomical location correlated in meningeal brain tumors (meningiomas). J Neurosci Rural Pract. 2014;5(3):244–9. https://doi.org/10.4103/0976-3147.133568.

55. Gursan N, Gundogdu C, Albayrak A, Kabalar ME. Immunohistochemical detection of progesterone receptors and the correlation with Ki-67 labeling indices in paraffin-embedded sections of meningiomas. Int J Neurosci. 2002;112(4):463–70. https://doi.org/10.1080/00207450290025581.
56. Konstantinidou AE, Korkolopoulou P, Mahera H, Kotsiakis X, Hranioti S, Eftychiadis C, et al. Hormone receptors in non-malignant meningiomas correlate with apoptosis, cell proliferation and recurrence-free survival. Histopathology. 2003;43(3):280–90. https://doi.org/10.1046/j.1365-2559.2003.01712.x.
57. Roser F, Nakamura M, Bellinzona M, Rosahl SK, Ostertag H, Samii M. The prognostic value of progesterone receptor status in meningiomas. J Clin Pathol. 2004;57(10):1033–7. https://doi.org/10.1136/jcp.2004.018333.
58. Wolfsberger S, Doostkam S, Boecher-Schwarz HG, Roessler K, van Trotsenburg M, Hainfellner JA, et al. Progesterone-receptor index in meningiomas: correlation with clinico-pathological parameters and review of the literature. Neurosurg Rev. 2004;27(4):238–45. https://doi.org/10.1007/s10143-004-0340-y.
59. Maiuri F, Mariniello G, de Divitiis O, Esposito F, Guadagno E, Teodonno G, et al. Progesterone receptor expression in meningiomas: pathological and prognostic implications. Front Oncol. 2021;11:611218. https://doi.org/10.3389/fonc.2021.611218.
60. Roser F, Nakamura M, Ritz R, Bellinzona M, Dietz K, Samii M, et al. Proliferation and progesterone receptor status in benign meningiomas are not age dependent. Cancer. 2005;104(3):598–601. https://doi.org/10.1002/cncr.21192.
61. Pravdenkova S, Al-Mefty O, Sawyer J, Husain M. Progesterone and estrogen receptors: opposing prognostic indicators in meningiomas. J Neurosurg. 2006;105(2):163–73. https://doi.org/10.3171/jns.2006.105.2.163.
62. Maiuri F, De Caro MB, Esposito F, Cappabianca P, Strazzullo V, Pettinato G, et al. Recurrences of meningiomas: predictive value of pathological features and hormonal and growth factors. J Neuro-Oncol. 2007;82(1):63–8. https://doi.org/10.1007/s11060-005-9078-9.
63. Metellus P, Nanni I, Dussert C, Trinkhaus M, Fuentes S, Chinot O, et al. [Prognostic implications of biologic markers in intracranial meningiomas: 120 cases]. Neurochirurgie. 2008;54(6):750–6. https://doi.org/10.1016/j.neuchi.2008.07.005.
64. Aarhus M, Lund-Johansen M, Knappskog PM. Gene expression profiling of meningiomas: current status after a decade of microarray-based transcriptomic studies. Acta Neurochir. 2011;153(3):447–56. https://doi.org/10.1007/s00701-010-0906-0.
65. Brastianos PK, Horowitz PM, Santagata S, Jones RT, McKenna A, Getz G, et al. Genomic sequencing of meningiomas identifies oncogenic SMO and AKT1 mutations. Nat Genet. 2013;45(3):285–9. https://doi.org/10.1038/ng.2526.
66. Clark VE, Erson-Omay EZ, Serin A, Yin J, Cotney J, Ozduman K, et al. Genomic analysis of non-NF2 meningiomas reveals mutations in TRAF7, KLF4, AKT1, and SMO. Science. 2013;339(6123):1077–80. https://doi.org/10.1126/science.1233009.
67. Reuss DE, Piro RM, Jones DT, Simon M, Ketter R, Kool M, et al. Secretory meningiomas are defined by combined KLF4 K409Q and TRAF7 mutations. Acta Neuropathol. 2013;125(3):351–8. https://doi.org/10.1007/s00401-013-1093-x.
68. Bi WL, Abedalthagafi M, Horowitz P, Agarwalla PK, Mei Y, Aizer AA, et al. Genomic landscape of intracranial meningiomas. J Neurosurg. 2016;125(3):525–35. https://doi.org/10.3171/2015.6.JNS15591.
69. Clark VE, Harmancı AS, Bai H, Youngblood MW, Lee TI, Baranoski JF, et al. Recurrent somatic mutations in POLR2A define a distinct subset of meningiomas. Nat Genet. 2016;48(10):1253–9. https://doi.org/10.1038/ng.3651.
70. Sahm F, Schrimpf D, Stichel D, Jones DTW, Hielscher T, Schefzyk S, et al. DNA methylation-based classification and grading system for meningioma: a multicentre, retrospective analysis. Lancet Oncol. 2017;18(5):682–94. https://doi.org/10.1016/S1470-2045(17)30155-9.
71. Karsy M, Azab MA, Abou-Al-Shaar H, Guan J, Eli I, Jensen RL, et al. Clinical potential of meningioma genomic insights: a practical review for neurosurgeons. Neurosurg Focus. 2018;44(6):E10. https://doi.org/10.3171/2018.2.FOCUS1849.
72. Youngblood MW, Duran D, Montejo JD, Li C, Omay SB, Özduman K, et al. Correlations between genomic subgroup and clinical features in a cohort of more than 3000 meningiomas. J Neurosurg. 2019;133(5):1345–54. https://doi.org/10.3171/2019.8.JNS191266.
73. Maiuri F, Corvino S, Corazzelli G, Berardinelli J, Di Crescenzo RM, Del Basso De Caro M. Time to recurrence of intracranial meningiomas from a monoinstitutional surgical series. World Neurosurg. 2024;185:e612. https://doi.org/10.1016/j.wneu.2024.02.087.
74. Maiuri F, Corvino S, Corazzelli G, Del Basso De Caro M. Single versus multiple reoperations for recurrent intracranial meningiomas. J Neuro-Oncol. 2024;168:527.
75. Goldbrunner R, Minniti G, Preusser M, Jenkinson MD, Sallabanda K, Houdart E, et al. EANO guidelines for the diagnosis and treatment of meningiomas. Lancet Oncol. 2016;17(9):e383–91. https://doi.org/10.1016/S1470-2045(16)30321-7.
76. Otero-Rodriguez A, Tabernero MD, Munoz-Martin MC, Sousa P, Orfao A, Pascual-Argente D, et al. Re-evaluating Simpson Grade I, II, and III resections in neurosurgical treatment of World Health Organization Grade I meningio-

mas. World Neurosurg. 2016;96:483–8. https://doi.org/10.1016/j.wneu.2016.09.007.
77. Schwartz TH, McDermott MW. The Simpson grade: abandon the scale but preserve the message. J Neurosurg. 2020;135:1–8. https://doi.org/10.3171/2020.6.JNS201904.
78. Traylor JI, Plitt AR, Hicks WH, Mian TM, Mickey BE, Barnett SL. Evaluating risk of recurrence in patients with meningioma. J Neurosurg. 2023;138(3):621–8. https://doi.org/10.3171/2022.6.JNS221162.
79. Przybylowski CJ, Hendricks BK, Frisoli FA, Zhao X, Cavallo C, Borba Moreira L, et al. Prognostic value of the Simpson grading scale in modern meningioma surgery: Barrow Neurological Institute experience. J Neurosurg. 2020;135:1–9. https://doi.org/10.3171/2020.6.JNS20374.
80. Nanda A, Vannemreddy P. Recurrence and outcome in skull base meningiomas: do they differ from other intracranial meningiomas? Skull Base. 2008;18(4):243–52. https://doi.org/10.1055/s-2007-1016956.
81. Lam Shin Cheung V, Kim A, Sahgal A, Das S. Meningioma recurrence rates following treatment: a systematic analysis. J Neuro-Oncol. 2018;136(2):351–61. https://doi.org/10.1007/s11060-017-2659-6.
82. Spektor S, Valarezo J, Fliss DM, Gil Z, Cohen J, Goldman J, et al. Olfactory groove meningiomas from neurosurgical and ear, nose, and throat perspectives: approaches, techniques, and outcomes. Neurosurgery. 2005;57(4 Suppl):268–80.; discussion 80. https://doi.org/10.1227/01.neu.0000176409.70668.eb.
83. Nakamura M, Struck M, Roser F, Vorkapic P, Samii M. Olfactory groove meningiomas: clinical outcome and recurrence rates after tumor removal through the frontolateral and bifrontal approach. Neurosurgery. 2007;60(5):844–52.; discussion 52. https://doi.org/10.1227/01.NEU.0000255453.20602.80.
84. Pallini R, Fernandez E, Lauretti L, Doglietto F, D'Alessandris QG, Montano N, et al. Olfactory groove meningioma: report of 99 cases surgically treated at the Catholic University School of Medicine, Rome. World Neurosurg. 2015;83(2):219–31.e1–3. https://doi.org/10.1016/j.wneu.2014.11.001.
85. Koutourousiou M, Fernandez-Miranda JC, Wang EW, Snyderman CH, Gardner PA. Endoscopic endonasal surgery for olfactory groove meningiomas: outcomes and limitations in 50 patients. Neurosurg Focus. 2014;37(4):E8. https://doi.org/10.3171/2014.7.focus14330.
86. Makarenko S, Carreras EM, Akagami R. Craniotomy for perisellar meningiomas: comparison of simple (appropriate for endoscopic approach) versus complex anatomy and surgical outcomes. J Neurosurg. 2017;126(4):1191–200. https://doi.org/10.3171/2016.3.JNS152307.
87. Bander ED, Singh H, Ogilvie CB, Cusic RC, Pisapia DJ, Tsiouris AJ, et al. Endoscopic endonasal versus transcranial approach to tuberculum sellae and planum sphenoidale meningiomas in a similar cohort of patients. J Neurosurg. 2018;128(1):40–8. https://doi.org/10.3171/2016.9.JNS16823.
88. Magill ST, Morshed RA, Lucas CG, Aghi MK, Theodosopoulos PV, Berger MS, et al. Tuberculum sellae meningiomas: grading scale to assess surgical outcomes using the transcranial versus transsphenoidal approach. Neurosurg Focus. 2018;44(4):E9. https://doi.org/10.3171/2018.1.FOCUS17753.
89. Bassiouni H, Asgari S, Sandalcioglu IE, Seifert V, Stolke D, Marquardt G. Anterior clinoidal meningiomas: functional outcome after microsurgical resection in a consecutive series of 106 patients. Clinical article. J Neurosurg. 2009;111(5):1078–90. https://doi.org/10.3171/2009.3.17685.
90. Xu T, Yan Y, Evins AI, Gong Z, Jiang L, Sun H, et al. Anterior clinoidal meningiomas: meningeal anatomical considerations and surgical implications. Front Oncol. 2020;10:634. https://doi.org/10.3389/fonc.2020.00634.
91. Chen LH, Xia Y, Wei F, Sun K, Huang HZ, Xu RX. The factors influencing postoperative efficacy of anterior clinoidal meningioma treatment and an analysis of best-suited surgical strategies. Front Neurol. 2023;14:1097686. https://doi.org/10.3389/fneur.2023.1097686.
92. Solero CL, Giombini S, Morello G. Suprasellar and olfactory meningiomas. Report on a series of 153 personal cases. Acta Neurochir. 1983;67(3–4):181–94. https://doi.org/10.1007/BF01401420.
93. Bitter AD, Stavrinou LC, Ntoulias G, Petridis AK, Dukagjin M, Scholz M, et al. The role of the pterional approach in the surgical treatment of olfactory groove meningiomas: a 20-year experience. J Neurol Surg B Skull Base. 2013;74(2):97–102. https://doi.org/10.1055/s-0033-1333618.
94. Guduk M, Yener U, Sun HI, Hacihanefioglu M, Ozduman K, Pamir MN. Pterional and unifrontal approaches for the microsurgical resection of olfactory groove meningiomas: experience with 61 consecutive patients. Turk Neurosurg. 2017;27(5):707–15. https://doi.org/10.5137/1019-5149.JTN.17154-16.1.
95. Al-Mefty O. Clinoidal meningiomas. J Neurosurg. 1990;73(6):840–9. https://doi.org/10.3171/jns.1990.73.6.0840.
96. Giammattei L, Starnoni D, Levivier M, Messerer M, Daniel RT. Surgery for clinoidal meningiomas: case series and meta-analysis of outcomes and complications. World Neurosurg. 2019;129:e700–e17. https://doi.org/10.1016/j.wneu.2019.05.253.
97. Mariniello G, de Divitiis O, Seneca V, Maiuri F. Classical pterional compared to the extended skull base approach for the removal of clinoidal meningiomas. J Clin Neurosci. 2012;19(12):1646–50. https://doi.org/10.1016/j.jocn.2011.12.033.
98. Mariniello G, de Divitiis O, Bonavolontà G, Maiuri F. Surgical unroofing of the optic canal and visual outcome in basal meningiomas. Acta Neurochir.

2013;155(1):77–84. https://doi.org/10.1007/s00701-012-1485-z.
99. Kim JH, Jang WY, Jung TY, Kim IY, Lee KH, Kang WD, et al. Predictive factors for surgical outcome in anterior clinoidal meningiomas: analysis of 59 consecutive surgically treated cases. Medicine (Baltimore). 2017;96(15):e6594. https://doi.org/10.1097/MD.0000000000006594.
100. Mariniello G, de Divitiis O, Corvino S, Strianese D, Iuliano A, Bonavolontà G, et al. Recurrences of spheno-orbital meningiomas: risk factors and management. World Neurosurg. 2022;161:e514. https://doi.org/10.1016/j.wneu.2022.02.048.
101. Shrivastava RK, Sen C, Costantino PD, Della RR. Sphenoorbital meningiomas: surgical limitations and lessons learned in their long-term management. J Neurosurg. 2005;103(3):491–7. https://doi.org/10.3171/jns.2005.103.3.0491.
102. Bloss HG, Proescholdt MA, Mayer C, Schreyer AG, Brawanski A. Growth pattern analysis of sphenoid wing meningiomas. Acta Neurochir. 2010;152(1):99–103; discussion. https://doi.org/10.1007/s00701-009-0556-2.
103. Leroy HA, Leroy-Ciocanea CI, Baroncini M, Bourgeois P, Pellerin P, Labreuche J, et al. Internal and external spheno-orbital meningioma varieties: different outcomes and prognoses. Acta Neurochir. 2016;158(8):1587–96. https://doi.org/10.1007/s00701-016-2850-0.
104. Terrier LM, Bernard F, Fournier HD, Morandi X, Velut S, Hénaux PL, et al. Spheno-orbital meningiomas surgery: multicenter management study for complex extensive tumors. World Neurosurg. 2018;112:e145–e56. https://doi.org/10.1016/j.wneu.2017.12.182.
105. Kinjo T, al-Mefty O, Kanaan I. Grade zero removal of supratentorial convexity meningiomas. Neurosurgery. 1993;33(3):394–9.; discussion 9. https://doi.org/10.1227/00006123-199309000-00007.
106. Giombini S, Solero CL, Morello G. Late outcome of operations for supratentorial convexity meningiomas. Report on 207 cases. Surg Neurol. 1984;22(6):588–94. https://doi.org/10.1016/0090-3019(84)90436-1.
107. Borovich B, Doron Y. Recurrence of intracranial meningiomas: the role played by regional multicentricity. J Neurosurg. 1986;64(1):58–63. https://doi.org/10.3171/jns.1986.64.1.0058.
108. Morokoff AP, Zauberman J, Black PM. Surgery for convexity meningiomas. Neurosurgery. 2008;63(3):427–33.; discussion 33–4. https://doi.org/10.1227/01.NEU.0000310692.80289.28.
109. Alvernia JE, Dang ND, Sindou MP. Convexity meningiomas: study of recurrence factors with special emphasis on the cleavage plane in a series of 100 consecutive patients. J Neurosurg. 2011;115(3):491–8. https://doi.org/10.3171/2011.4.JNS101922.
110. Maiuri F, Corvino S. Meningiomas of the rolandic region: risk factors for motor deficit and role of intra-operative monitoring. Acta Neurochir. 2023;165(7):1707–16. https://doi.org/10.1007/s00701-023-05630-6.
111. Balik V, Kourilova P, Sulla I, Vrbkova J, Srovnal J, Hajduch M, et al. Recurrence of surgically treated parasagittal meningiomas: a meta-analysis of risk factors. Acta Neurochir. 2020;162(9):2165–76. https://doi.org/10.1007/s00701-020-04336-3.
112. Sindou MP, Alvernia JE. Results of attempted radical tumor removal and venous repair in 100 consecutive meningiomas involving the major dural sinuses. J Neurosurg. 2006;105(4):514–25. https://doi.org/10.3171/jns.2006.105.4.514.
113. Chung SB, Kim CY, Park CK, Kim DG, Jung HW. Falx meningiomas: surgical results and lessons learned from 68 cases. J Korean Neurosurg Soc. 2007;42(4):276–80. https://doi.org/10.3340/jkns.2007.42.4.276.
114. Murrone D, De Paulis D, di Norcia V, Di Vitantonio H, Galzio RJ. Surgical management of falcine meningiomas: experience of 95 patients. J Clin Neurosci. 2017;37:25–30. https://doi.org/10.1016/j.jocn.2016.11.002.
115. Mooney MA, Abolfotoh M, Bi WL, Tavanaiepour D, Almefty RO, Bassiouni H, et al. Is Falcine meningioma a diffuse disease of the Falx? Case series and analysis of a "Grade Zero" resection. Neurosurgery. 2020;87(5):900–9. https://doi.org/10.1093/neuros/nyaa038.
116. Sughrue ME, Rutkowski MJ, Shangari G, Parsa AT, Berger MS, McDermott MW. Results with judicious modern neurosurgical management of parasagittal and falcine meningiomas. Clinical article. J Neurosurg. 2011;114(3):731–7. https://doi.org/10.3171/2010.9.JNS10646.
117. Bassiouni H, Hunold A, Asgari S, Stolke D. Tentorial meningiomas: clinical results in 81 patients treated microsurgically. Neurosurgery. 2004;55(1):108–16. ; discussion 16–8. https://doi.org/10.1227/01.neu.0000126886.48372.49.
118. Shukla D, Behari S, Jaiswal AK, Banerji D, Tyagi I, Jain VK. Tentorial meningiomas: operative nuances and perioperative management dilemmas. Acta Neurochir. 2009;151(9):1037–51. https://doi.org/10.1007/s00701-009-0421-3.
119. Hashemi M, Schick U, Hassler W, Hefti M. Tentorial meningiomas with special aspect to the tentorial fold: management, surgical technique, and outcome. Acta Neurochir. 2010;152(5):827–34. https://doi.org/10.1007/s00701-009-0591-z.
120. Nanda A, Patra DP, Savardekar A, Maiti TK, Konar SK, Notarianni C, et al. Tentorial meningiomas: reappraisal of surgical approaches and their outcomes. World Neurosurg. 2018;110:e177–e96. https://doi.org/10.1016/j.wneu.2017.10.115.
121. Pereira BJA, de Almeida AN, Paiva WS, de Aguiar PHP, Teixeira MJ, Marie SKN. Natural history of intraventricular meningiomas: systematic review. Neurosurg Rev. 2020;43(2):513–23. https://doi.org/10.1007/s10143-018-1019-0.

122. Grujicic D, Cavallo LM, Somma T, Illic R, Milicevic M, Raicevic S, et al. Intraventricular meningiomas: a series of 42 patients at a single institution and literature review. World Neurosurg. 2017;97:178–88. https://doi.org/10.1016/j.wneu.2016.09.068.
123. Jungwirth G, Warta R, Beynon C, Sahm F, von Deimling A, Unterberg A, et al. Intraventricular meningiomas frequently harbor NF2 mutations but lack common genetic alterations in TRAF7, AKT1, SMO, KLF4, PIK3CA, and TERT. Acta Neuropathol Commun. 2019;7(1):140. https://doi.org/10.1186/s40478-019-0793-4.
124. Ammendola S, Simbolo M, Ciaparrone C, Rizzo PC, Caffo M, Pinna G, et al. Intraventricular meningiomas: clinical-pathological and genetic features of a monocentric series. Curr Oncol. 2022;29(1):178–85. https://doi.org/10.3390/curroncol29010017.
125. Sandalcioglu IE, Hunold A, Müller O, Bassiouni H, Stolke D, Asgari S. Spinal meningiomas: critical review of 131 surgically treated patients. Eur Spine J. 2008;17(8):1035–41. https://doi.org/10.1007/s00586-008-0685-y.

Brain–Tumor Interface, Brain Invasion, and Peritumoral Edema in Meningiomas: Pathology—Clinical Implications

Felix Behling

1 The Brain–Meningioma Interface

What makes meningioma such a fascinating tumor entity is its clinical and radiographic wide range of presentations as well as its neuropathological variety. While most tumors are considered benign, some have a higher tendency to recur and require a multimodal treatment. Incidentally, meningiomas can take a malignant course and in very rare cases even develop extraneural metastases [1–4]. Meningiomas also display some hallmarks of cancer [5], and invasive growth has especially gained more attention since its introduction in the WHO classification of central nervous system tumors in 2016 [1, 6].

Based on autopsy reports of 27 non-operated meningiomas the brain–meningioma interface was first classified in 1989 by Nakasu and colleagues. Four subtypes were described based on the contact of the meningioma with the brain and its subarachnoid vessels: smooth, lobular, and finger-like expansion as well as invasive growth [7]. Later the same study group reported on the different intraoperative characteristics at the brain–tumor interface of 50 meningiomas together with immunohistochemical assessments. Collagen type 4 was identified in the basal membrane that demarcated benign meningiomas at the brain–tumor interface but was missing in atypical and anaplastic meningiomas. Partial or complete disruption of the arachnoid membrane was observed in the majority of cases. True brain invasion revealed immunopositivity for the glial fibrillary acidic protein (GFAP) within the tumor and axonal degenerative changes such as ballooning or varicose swelling in the invaded brain parenchyma [8].

The mechanisms underlying these differing behaviors at the brain–meningioma contact zone remain poorly understood. However, some potentially driving factors have been identified. Based on the knowledge on other cancer types cytoskeletal alterations as well as interactions with other cells and the extracellular matrix are important prerequisites for invasive growth [5]. In order to invade adjacent healthy tissue, tumor cells need to adopt different capabilities that enable disengagement from cell–cell and cell–extra cellular matrix integration [9]. Merlin as the product of the *NF2* gene is a cytoskeletal protein and tumor suppressor [10], which is inactivated in the majority of meningiomas [11]. It is widely

F. Behling (✉)
Department of Neurosurgery and Neurotechnology, University Hospital Tübingen, Eberhard Karls-University Tübingen, Tübingen, Germany

Hertie Institute for Clinical Brain Research, Tübingen, Germany

Center for CNS Tumors, Comprehensive Cancer Center Tübingen-Stuttgart, University Hospital Tübingen, Eberhard-Karls-University, Tübingen, Germany
e-mail: felix.behling@med.uni-tuebingen.de

F. Maiuri, M. Del Basso De Caro (eds.), *Meningiomas*,
https://doi.org/10.1007/978-3-031-76680-0_9

accepted that the loss of its tumor suppressor function is decisive for meningioma development [12], but furthermore, it has been shown in schwannomas that this seems to be independent from its cytoskeletal role [13]. Many factors that potentially influence the integrity of the brain–meningioma interface have been suggested [9]. Especially matrix metalloproteinases (MMPs) have been in the focus of cancer research for several years [14], including meningiomas. Their protease function degrades extracellular matrix architecture and enables invasive growth [15]. High expression of MMP-9 was identified as an independent prognostic factor resulting in a higher risk of meningioma recurrence [16, 17]. Factors involved in the regulation of MMP expression have been described in meningiomas as well and suggested as an indicator for the risk of recurrence [18]. However, the true significance of brain invasion in meningiomas is still unclear. If we learn to better understand how the integrity of the brain–meningioma interface is maintained and disrupted, new possibilities for diagnostic risk-stratification or even therapeutic targeting may be revealed.

2 Brain Invasion

2.1 The Evolution of the Role of Brain Invasion in Meningioma

With the introduction of the updated WHO classification of central nervous system tumors in 2016 the histopathological evidence of brain invasion became a stand-alone criterion for atypia in meningiomas [1, 6]. This change is largely based on the works of Perry and colleagues, who described brain invasion as a potential marker for higher grading in meningiomas. It is worth mentioning, that they had to enlarge their prior cohort by 116 malignant meningioma cases in order to obtain statistically significant results [19, 20]. However, several retrospective cohort studies have since questioned the prognostic relevance of brain invasion and its current integration in the histopathological diagnostic work up. In 2016, Baumgarten et al. published their results of a selective cohort of 229 meningiomas from two neurosurgical centers. They suggested to exclude brain invasion from the grading of meningiomas based on their assessment of 61 meningiomas that were otherwise histopathologically benign but showed signs of brain invasion. Compared to atypical meningiomas, due to established histopathological signs of atypia like increased mitotic count, the authors were able to show a favorable progression-free survival of brain invasive but otherwise benign meningiomas [21]. In the same year, Spille et al. reviewed the prognostic impact of brain invasion in 467 meningiomas. While 32 cases showed signs of brain invasion, an independent prognostic impact was not evident [22]. Another two-institutional retrospective analysis confirmed the lack of prognostic impact of brain invasion in 875 meningiomas of which only 28 were invasive [23]. Overall, the growing evidence that especially meningiomas with brain invasion, but otherwise benign features (BIOB) may have a more favorable clinical course similar to grade 1 tumors, led to an ongoing discussion about the WHO classification of meningiomas. In 2021, a meta-analysis included studies from 2000 until 2020 and concluded that there was currently not enough evidence to support the application of brain invasion as a standalone criterion especially due to the lack of significance in BIOBs [24]. It must be mentioned at this point that the incidence of BIOBs seems to have increased unproportionally with the introduction of the 2016 classification, presumably because of most pathologists' inclination not to miss a potentially decisive prognostic factor [25]. While the topic remains controversial, several studies have followed, with some in support of the prognostic role of brain invasion [26, 27]. A mentionable study assessed 1006 meningiomas and found 282 cases to be brain invasive, which seems an exceptionally high rate. They proposed that otherwise benign meningiomas with signs of brain invasion may have an intermediate prognosis compared to atypical and benign meningiomas [28]. Recently, a higher rate of proliferation expressed by the immunohistochemical marker MIB1 was described in meningiomas with brain invasion in a cohort of 1718 tumors [29]. Overall, the debate about the prognostic impact of brain invasion is still not resolved. There are several problems that may underlie the conflicting results of the mentioned retrospective studies such as the application of non-uniform histopathological criteria for brain invasion and non-standardized meningioma tissue sampling.

2.2 The Histopathological Assessment of Brain Invasion

Brain invasion has been addressed in the WHO classification of central nervous system tumors dating back to the second edition in 1993. Back then, it was mentioned as a criterion for malignancy for anaplastic meningiomas [30]. Over the following years, the focus was shifted to otherwise benign meningiomas resulting in the first evidence of a potential prognostic impact [19, 20]. Since 2007, the WHO classification recommended to consider brain invasive meningiomas prognostically as grade II [31] but it took until the 4th edition in 2016 to finally incorporate this characteristic as a stand-alone criterion for atypia [6]. Since then, "irregular tongue-like protrusions of tumor cells infiltrating underlying parenchyma, without an intervening layer of leptomeninges" define brain invasion. Typically, GFAP positive islands can be appreciated within the periphery of the tumor. It is important to note that extension into Virchow-Robin spaces does not count as brain invasion because there is no breach of the pia mater. Additionally, perivascular spread and hyalinaztion are also not considered as brain invasion but rather mimic meningoangiomatosis in children [6].

By gradually lifting the importance of brain invasion for the histopathological assessment of meningiomas, the topic was handed differently among neuropathological departments in their routine work up. The definite change made in the WHO classification in 2016 resulted in the identification of exceptionally high rates of brain invasion in otherwise benign meningiomas (BIOBs) in some pathology departments as recently reported by Perry. It is assumed that this phenomenon may be caused by the inclination of pathologists to be on the safe site and grade borderline cases as invasive in order to not miss a potentially decisive prognostic factor [25]. In the latest WHO classification of 2021, the criteria have not been changed but a clear statement about the controversial role in BIOB meningiomas has been included [1, 25]. In a personal statement by Perry, recommendations for BIOB meningioma cases were given. Obtaining further immunohistochemical stainings, especially GFAP, and ordering molecular studies to identify prognostically relevant biomarkers such as the homozygous deletion of CDKN2A or a TERT promoter mutation [32, 33] were suggested in assisting diagnostic certainty in borderline cases. Additionally, a clear statement should be given in the pathology report if a case was graded as borderline invasive, resulting in closer follow-up [25]. Taken together these diagnostic uncertainties, there is a need and potential for a reliable diagnostic biomarker that may assist in assuring brain invasion detection in borderline cases.

2.3 The Sampling Dilemma

It is obviously not advisable to sample the adjacent central nervous system tissue of each meningioma to have a complete picture of the brain–meningioma interface. As described above, GFAP positive cells can usually be appreciated within the meningioma border in cases of true invasion [6], making sampling of the parenchyma unnecessary. However, there are some aspects of meningioma tissue sampling that were discussed during the last years, which have resulted in a clearer picture of potential sampling problems and possibilites of optimization. In 2014 Pizem et al. demonstrated that extensive sampling of meningiomas can lead to a higher detection rate of brain invasion [34]. Assessing meningioma samples from eight different neurosurgical centers, Timme et al. found marked differences in the characteristics of provided samples during resection, resulting in differences in the detection of brain invasion, ranging from 5% to 28% [35]. Both studies provide important evidence for the influence of non-standardized sampling during meningioma surgery on the detection rate of brain invasion.

A recent survey evaluated how the methodology of intraoperative meningioma sampling currently varied among European neurosurgical centers and if the new role of brain invasion since the 2016 classification had any influence on routine sampling practice. Only 47.2% of respondents stated that they applied a standardized sampling methodology during meningioma resection and only 54.9% obtain a complete sampling of the meningioma surface in contact with the brain. 77.5% of respondents did not change their sampling method after the WHO classification of 2016 included brain invasion as a standalone criterion for atypia [36].

A differentiated view was directed at brain invasion and its prognostic value in a comparative retrospective analysis of intraoperative and histopathological features of invasive growth in 1517 meningiomas in a single neurosurgical institution. Intraoperative signs of possible brain invasion were seen in 23.7% but only 4.8% of cases were histopathologically graded as CNS-invasive. Furthermore, it was shown that the intraoperative detection of infiltrative growth of meningiomas into central nervous system tissue can complement the histopathological assessment of CNS invasion. The combined assessment of brain invasion was an independent factor for tumor recurrence. This study showed that incorporating the intraoperative impression regarding invasive meningioma growth can influence the prognostic impact of brain invasion, suggesting that sampling may be incomplete in some cases [37]. However, the intraoperative impression of the surgeon regarding brain invasion is a non-uniform criterion and there are no established intraoperative signs that are reported in a structured way. There is also a high suspicion that the gross impression intraoperatively may not represent true invasion but more frequently simple adhesion of the meningioma to the pial surface, resulting in overreporting of potential brain invasion when reviewing operative reports [25]. An example of intraoperative impressions that may be incorrectly interpreted as potential invasive growth is illustrated in Fig. 1.

Regarding the marked differences of brain invasion detection rates that have been reported, this is most likely due to non-standardized sampling. It is obvious that a complete sampling of the meningioma-tumor contact area is necessary to give the neuropathologist the chance to confidently rule out or detect brain invasion, as recently discussed in a review on the prognostic potential of brain invasion. There, the authors furthermore critically discuss the difficulty of achieving a complete sampling in certain tumor locations, depending also on eloquence and especially the technical approach by the surgeon, which can vary based on experience and training. Figure 2 illustrates different meningiomas and highlights the area of interest regarding the detection of CNS invasion, differentiated into eloquent and non-eloquent contact areas [38].

The insights that have thus far been obtained on the topic of brain invasion call for a standardized

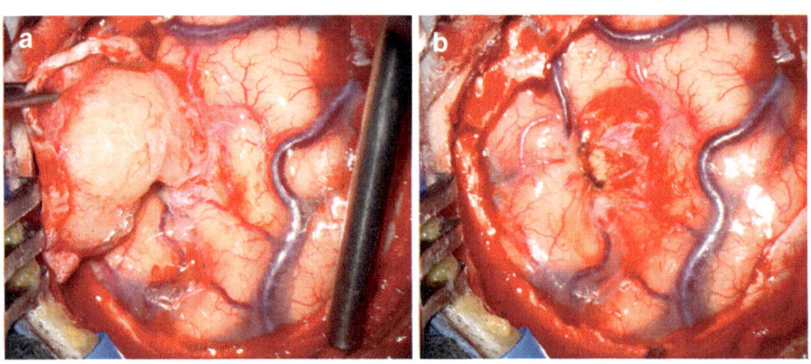

Fig. 1 Intraoperative impressions of a convexity meningioma. After resecting the dura together with the meningioma, marked adhesion to the pial surface was seen (**a**). After further removal pial surface breach was evident and thus suggestive of invasive growth (**b**). The complete contact area of the meningioma was sent for histopathological evaluation with a separate focused sample of the potentially invasive area of the pial breach. However, a chordoid meningioma was diagnosed without histopathological evidence of brain invasion

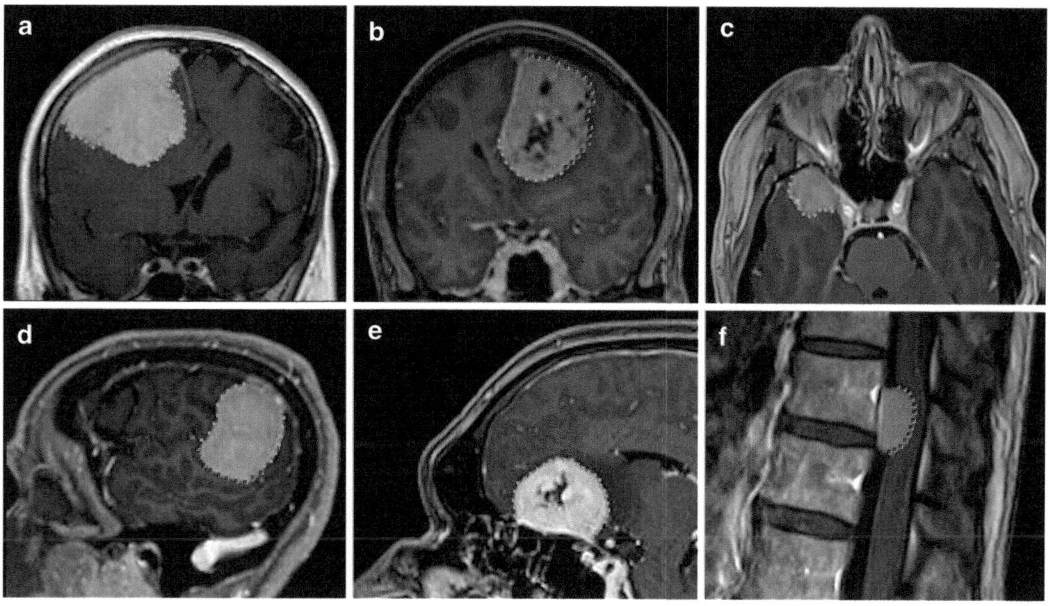

Fig. 2 Illustration of the area of interest for meningioma sampling regarding the detection of CNS invasion. Dotted lines represent contact of the meningioma surface with non-eloquent (green) and eloquent (red) parenchyma ((**a**): right frontal convexity meningioma; (**b**): left frontoparietal falcine meningioma; (**c**): right-sided sphenoid wing meningioma; (**d**): left temporo-parietal convexity meningioma; (**e**): olfactory groove meningioma; (**f**): ventral thoracic meningioma). (Unchanged from Behling et al. 2021 [38], https://creativecommons.org/licenses/by/4.0/deed.en)

sampling recommendation. This is an important step to get a better understanding of the real incidence of brain invasion, especially since varying detection rates have been reported in retrospective studies, ranging from 4.8% to 28.0% [21, 22, 26, 28, 37]. Therefore, it is crucial to learn more about how the neurosurgical sampling influences the diagnostic yield and what factors may lead to incomplete sampling. Recently a multicenter study that evaluates the intraoperative sampling and assessment as well as the histopathological evaluation has been launched and will hopefully formulate a basis for discussing a standardized sampling recommendation (Tübingen Invader Study).

3 Brain Invasion and Imaging

3.1 Radiographic Features of Invasive Growth

It seems obvious that disruption of the pial surface and invasion into the brain parenchyma should be somehow visible on preoperative imaging of meningiomas. There have been several reports about multiple preoperative imaging characteristics that supposedly correlate with invasive growth into brain parenchyma. In 1988 Go et al. described that the size and disruption of the cortical layer were associated with edema in preoperative computer tomographies of 38 meningiomas. Additionally, edema was reported to be more common in meningothelial and transitional meningioma [39]. Ide et al. experienced meningothelial meningiomas to be more commonly associated with preoperative edema in a series of 39 tumors but observed no impact of tumor size [40]. Later Mantle et al. published a series of 135 patients. Assessing the edema in preoperative computer tomographies revealed that the chance of brain invasion increased by 20% for each centimeter of edema [41].

With the introduction of the MRI, the correlation between brain invasion and edema was confirmed and further linked to preoperative seizures in a retrospective cohort of 176 meningiomas [42]. Of note, other clinical manifestation such as the risk of postoperative hemorrhage has also been shown to be related to invasive growth of meningiomas [43], underlining that the disruption of the pial surface and intrusion into the parenchyma has marked clinical implications. Adeli et al. reviewed preoperative MR images of 617 meningiomas of which 24 were brain invasive. They found a significant correlation of irregular tumor shape, heterogeneous contrast enhancement and peritumoral edema with brain invasion, while only increasing edema volume remained significant in the multivariate analysis [44]. Although the cohort was underpowered with only 24 brain invasive cases, the results demonstrated the great potential in further exploring how invasive growth may be displayed in preoperative imaging (Fig. 3).

3.2 The Potential of Radiomics

Radiomics offer an expansive view into the raw imaging data that provide more information than the human eye can perceive in a common routine MR or CT imaging. In oncology research radiomics have shown to provide new prognostic criteria in different cancer types [45]. The evolution of this new technology has proven to be of great potential when combined with clinical data. This has also been explored in meningiomas with a surge of publications that claim to predict different characteristics like WHO grade [46], risk of recurrence [47], response to radiosurgery [48], or the expression of different markers such as MIB1, P53, and progesterone receptor or even the presence of a *TERT* promoter mutation [49–51]. While it is overwhelming to learn about the possibilities of radiomics and the great asset it potentially provides for patient management when combined with clinical data, caution is recommended when taking a closer look at the construction of some models, especially regarding what clinical data is included and what in some cases may have been overlooked. Nonetheless, the potential seems to be infinite and in the future physicians will have a great asset in radiomics.

Regarding brain invasion, there have also been some noteworthy radiomics studies. Kademirli et al. compared filtration-based first-order histogram derived texture parameters from postcontrast T1 MR-sequences of 108 meningiomas of which 56 were brain invasive. They described good accuracy of the applied model and underlined the potential of radiomics in the prediction of brain invasion [52]. A more detailed approach was presented by Joo et al. in 2021. They presented an imaging-based model that focused on radiomics of the meningioma-brain interface in 454 patients of which 88 cases had evidence of brain invasion. The volume of parenchymal edema was revealed as an independent predictor for brain invasion [53], which is interestingly in line with the main result of imaging studies on brain invasion before the advent of radiomics.

Overall, the strength of radiomics undoubtedly depends on a firm basis of high-quality clinical and pathological data. Considering that the reported incidence of brain invasion varies throughout different study cohorts, it can be assumed that this will also impact future

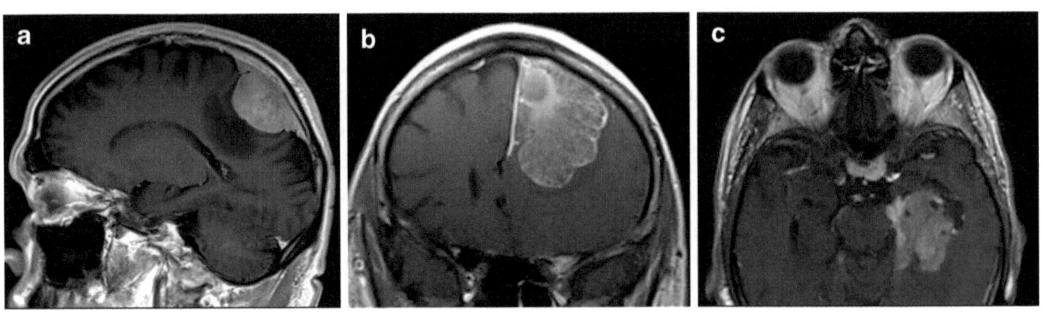

Fig. 3 Three meningiomas of the Tübingen Meningioma Cohort are shown as representative examples of MR-Imaging characteristics that have been associated with invasive growth (peritumoral edema (**a**); irregular contrast enhancement (**b**); irregular tumor shape (**c**)). In all three cases, histopathological evidence of invasive growth into CNS parenchyma was detected. (Unchanged from Behling et al. 2021 [38], https://creativecommons.org/licenses/by/4.0/deed.en)

radiomics studies that incorporate invasive growth. However, with high-quality and reliable data, radiomics has the chance to unfold its high potential, which especially holds true for brain invasion in meningiomas.

4 Future Directions

Taken together the insights that have thus far been obtained on the topic of brain invasion in meningioma, it is understandable that some may remain skeptical of tis integration into the WHO classification. However, the numerous retrospective analyses that argue for and against the prognostic value of brain invasion have some variations and weaknesses. It is therefore necessary to design structured multicenter trials to further evaluate brain invasion in a large cohort of meningiomas. For this purpose, comprehensive high-quality clinical data are important to explore the full potential of brain invasion.

Possibly, identifying a marker that can reliably detect invasive growth would be of great help, regarding the histopathological uncertainty in borderline and BIOB cases and potentially solve the sampling dilemma. Until then, it is recommended to pursuit a complete intraoperative sampling of the brain-meningioma contact area if technically feasible.

Furthermore, gaining more insights into the molecular mechanism of brain invasion may reveal potential therapeutical targets or assisting methodologies for intraoperative visualization of invasive tumor residues that allow for more certainty of complete resection.

References

1. Louis DN, Perry A, Wesseling P, Brat DJ, Cree IA, Figarella-Branger D, et al. The 2021 WHO classification of tumors of the central nervous system: a summary. Neuro-Oncology. 2021; https://doi.org/10.1093/neuonc/noab106.
2. Ostrom QT, Price M, Neff C, Cioffi G, Waite KA, Kruchko C, et al. CBTRUS statistical report: primary brain and other central nervous system tumors diagnosed in the United States in 2016–2020. Neuro-Oncology. 2023;25(12 Suppl 2):iv1–iv99. https://doi.org/10.1093/neuonc/noad149.
3. Goldbrunner R, Minniti G, Preusser M, Jenkinson MD, Sallabanda K, Houdart E, et al. EANO guidelines for the diagnosis and treatment of meningiomas. Lancet Oncol. 2016;17(9):e383–91. https://doi.org/10.1016/S1470-2045(16)30321-7.
4. Watanabe G, Young K, Rauber E, Khan MF, Suzuki R, Riestenberg R, et al. A systematic review of extraneural meningioma metastasis: timing, evolution and outlook. J Neuro-Oncol. 2024;168:187. https://doi.org/10.1007/s11060-024-04659-6.
5. Hanahan D, Weinberg RA. Hallmarks of cancer: the next generation. Cell. 2011;144(5):646–74. https://doi.org/10.1016/j.cell.2011.02.013.
6. Louis DN, Perry A, Reifenberger G, von Deimling A, Figarella-Branger D, Cavenee WK, et al. The 2016 World Health Organization classification of tumors of the central nervous system: a summary. Acta Neuropathol. 2016;131(6):803–20. https://doi.org/10.1007/s00401-016-1545-1.
7. Nakasu S, Hirano A, Llena JF, Shimura T, Handa J. Interface between the meningioma and the brain. Surg Neurol. 1989;32(3):206–12. https://doi.org/10.1016/0090-3019(89)90180-8.
8. Nakasu S, Fukami T, Jito J, Matsuda M. Microscopic anatomy of the brain-meningioma interface. Brain Tumor Pathol. 2005;22(2):53–7. https://doi.org/10.1007/s10014-005-0187-0.
9. von Spreckelsen N, Kesseler C, Brokinkel B, Goldbrunner R, Perry A, Mawrin C. Molecular neuropathology of brain-invasive meningiomas. Brain Pathol. 2022;32(2):e13048. https://doi.org/10.1111/bpa.13048.
10. Hovens CM, Kaye AH. The tumour suppressor protein NF2/merlin: the puzzle continues. J Clin Neurosci. 2001;8(1):4–7. https://doi.org/10.1054/jocn.2000.0784.
11. Ruttledge MH, Sarrazin J, Rangaratnam S, Phelan CM, Twist E, Merel P, et al. Evidence for the complete inactivation of the NF2 gene in the majority of sporadic meningiomas. Nat Genet. 1994;6(2):180–4. https://doi.org/10.1038/ng0294-180.
12. Lee S, Karas PJ, Hadley CC, Bayley VJ, Khan AB, Jalali A, et al. The role of Merlin/NF2 loss in meningioma biology. Cancers (Basel). 2019;11(11) https://doi.org/10.3390/cancers11111633.
13. Lallemand D, Saint-Amaux AL, Giovannini M. Tumor-suppression functions of merlin are independent of its role as an organizer of the actin cytoskeleton in Schwann cells. J Cell Sci. 2009;122(Pt 22):4141–9. https://doi.org/10.1242/jcs.045914.
14. Pickup MW, Mouw JK, Weaver VM. The extracellular matrix modulates the hallmarks of cancer. EMBO Rep. 2014;15(12):1243–53. https://doi.org/10.15252/embr.201439246.
15. Stamenkovic I. Extracellular matrix remodelling: the role of matrix metalloproteinases. J Pathol. 2003;200(4):448–64. https://doi.org/10.1002/path.1400.

16. Barresi V, Vitarelli E, Tuccari G, Barresi G. MMP-9 expression in meningiomas: a prognostic marker for recurrence risk? J Neuro-Oncol. 2011;102(2):189–96. https://doi.org/10.1007/s11060-010-0312-8.
17. Nordqvist AC, Smurawa H, Mathiesen T. Expression of matrix metalloproteinases 2 and 9 in meningiomas associated with different degrees of brain invasiveness and edema. J Neurosurg. 2001;95(5):839–44. https://doi.org/10.3171/jns.2001.95.5.0839.
18. Okuducu AF, Zils U, Michaelis SA, Mawrin C, von Deimling A. Increased expression of avian erythroblastosis virus E26 oncogene homolog 1 in World Health Organization grade 1 meningiomas is associated with an elevated risk of recurrence and is correlated with the expression of its target genes matrix metalloproteinase-2 and MMP-9. Cancer. 2006;107(6):1365–72. https://doi.org/10.1002/cncr.22130.
19. Perry A, Scheithauer BW, Stafford SL, Lohse CM, Wollan PC. "Malignancy" in meningiomas: a clinicopathologic study of 116 patients, with grading implications. Cancer. 1999;85(9):2046–56. https://doi.org/10.1002/(sici)1097-0142(19990501)85:9<2046::aid-cncr23>3.0.co;2-m.
20. Perry A, Stafford SL, Scheithauer BW, Suman VJ, Lohse CM. Meningioma grading: an analysis of histologic parameters. Am J Surg Pathol. 1997;21(12):1455–65. https://doi.org/10.1097/00000478-199712000-00008.
21. Baumgarten P, Gessler F, Schittenhelm J, Skardelly M, Tews DS, Senft C, et al. Brain invasion in otherwise benign meningiomas does not predict tumor recurrence. Acta Neuropathol. 2016;132(3):479–81. https://doi.org/10.1007/s00401-016-1598-1.
22. Spille DC, Hess K, Sauerland C, Sanai N, Stummer W, Paulus W, et al. Brain invasion in meningiomas: incidence and correlations with clinical variables and prognosis. World Neurosurg. 2016;93:346–54. https://doi.org/10.1016/j.wneu.2016.06.055.
23. Biczok A, Jungk C, Egensperger R, von Deimling A, Suchorska B, Tonn JC, et al. Microscopic brain invasion in meningiomas previously classified as WHO grade I is not associated with patient outcome. J Neuro-Oncol. 2019;145(3):469–77. https://doi.org/10.1007/s11060-019-03312-x.
24. Nakasu S, Nakasu Y. Prognostic significance of brain invasion in meningiomas: systematic review and meta-analysis. Brain Tumor Pathol. 2021;38(2):81–95. https://doi.org/10.1007/s10014-020-00390-y.
25. Perry A. The definition and role of brain invasion in meningioma grading: still controversial after all these years. Free Neuropathol. 2021:2. https://doi.org/10.17879/freeneuropathology-2021-3276.
26. Banan R, Abbetmeier-Basse M, Hong B, Dumitru CA, Sahm F, Nakamura M, et al. The prognostic significance of clinicopathological features in meningiomas: microscopic brain invasion can predict patient outcome in otherwise benign meningiomas. Neuropathol Appl Neurobiol. 2021;47:724. https://doi.org/10.1111/nan.12700.
27. Picart T, Dumot C, Guyotat J, Pavlov V, Streichenberger N, Vasiljevic A, et al. Clinical and pathological impact of an optimal assessment of brain invasion for grade 2 meningioma diagnosis: lessons from a series of 291 cases. Neurosurg Rev. 2022;45(4):2797–809. https://doi.org/10.1007/s10143-022-01792-6.
28. Li HY, Ying YZ, Zheng D, Dong GH, Zhang GB, Liu XM, et al. Is brain invasion sufficient as a stand-alone criterion for grading atypical meningioma? J Neurosurg. 2023;139(4):953–64. https://doi.org/10.3171/2023.2.JNS222751.
29. Behling F, Fodi C, Wang S, Hempel JM, Hoffmann E, Tabatabai G, et al. Increased proliferation is associated with CNS invasion in meningiomas. J Neuro-Oncol. 2021;155(3):247–54. https://doi.org/10.1007/s11060-021-03892-7.
30. Kleihues P, Burger PC, Scheithauer BW. The new WHO classification of brain tumours. Brain Pathol. 1993;3(3):255–68. https://doi.org/10.1111/j.1750-3639.1993.tb00752.x.
31. Louis DN, Ohgaki H, Wiestler OD, Cavenee WK, Burger PC, Jouvet A, et al. The 2007 WHO classification of tumours of the central nervous system. Acta Neuropathol. 2007;114(2):97–109. https://doi.org/10.1007/s00401-007-0243-4.
32. Sahm F, Schrimpf D, Olar A, Koelsche C, Reuss D, Bissel J, et al. TERT promoter mutations and risk of recurrence in meningioma. J Natl Cancer Inst. 2016;108(5):djv377. https://doi.org/10.1093/jnci/djv377.
33. Sievers P, Hielscher T, Schrimpf D, Stichel D, Reuss DE, Berghoff AS, et al. CDKN2A/B homozygous deletion is associated with early recurrence in meningiomas. Acta Neuropathol. 2020;140(3):409–13. https://doi.org/10.1007/s00401-020-02188-w.
34. Pizem J, Velnar T, Prestor B, Mlakar J, Popovic M. Brain invasion assessability in meningiomas is related to meningioma size and grade, and can be improved by extensive sampling of the surgically removed meningioma specimen. Clin Neuropathol. 2014;33(5):354–63. https://doi.org/10.5414/NP300750.
35. Timme M, Thomas C, Spille DC, Stummer W, Ebel H, Ewelt C, et al. Brain invasion in meningiomas: does surgical sampling impact specimen characteristics and histology? Neurosurg Rev. 2020;43(2):793–800. https://doi.org/10.1007/s10143-019-01125-0.
36. Behling F, Bruneau M, Honegger J, Berhouma M, Jouanneau E, Cavallo L, et al. Differences in intraoperative sampling during meningioma surgery regarding CNS invasion—results of a survey on behalf of the EANS skull base section. Brain Spine. 2023;3:101740. https://doi.org/10.1016/j.bas.2023.101740.
37. Behling F, Fodi C, Gepfner-Tuma I, Machetanz K, Renovanz M, Skardelly M, et al. CNS invasion in meningioma-how the intraoperative assessment can improve the prognostic evaluation of tumor recurrence. Cancers (Basel). 2020;12(12) https://doi.org/10.3390/cancers12123620.

38. Behling F, Hempel JM, Schittenhelm J. Brain invasion in meningioma-A prognostic potential worth exploring. Cancers (Basel). 2021;13(13) https://doi.org/10.3390/cancers13133259.
39. Go KG, Wilmink JT, Molenaar WM. Peritumoral brain edema associated with meningiomas. Neurosurgery. 1988;23(2):175–9. https://doi.org/10.1227/00006123-198808000-00008.
40. Ide M, Jimbo M, Kubo O, Yamamoto M, Imanaga H. Peritumoral brain edema associated with meningioma—histological study of the tumor margin and surrounding brain. Neurol Med Chir (Tokyo). 1992;32(2):65–71. https://doi.org/10.2176/nmc.32.65.
41. Mantle RE, Lach B, Delgado MR, Baeesa S, Belanger G. Predicting the probability of meningioma recurrence based on the quantity of peritumoral brain edema on computerized tomography scanning. J Neurosurg. 1999;91(3):375–83. https://doi.org/10.3171/jns.1999.91.3.0375.
42. Hess K, Spille DC, Adeli A, Sporns PB, Brokinkel C, Grauer O, et al. Brain invasion and the risk of seizures in patients with meningioma. J Neurosurg. 2018;130(3):789–96. https://doi.org/10.3171/2017.11.JNS172265.
43. Brokinkel B, Sicking J, Spille DC, Hess K, Paulus W, Stummer W. Letter to the editor. Brain invasion and the risk for postoperative hemorrhage and neurological deterioration after meningioma surgery. J Neurosurg. 2018;129(3):849–51. https://doi.org/10.3171/2018.5.JNS181287.
44. Adeli A, Hess K, Mawrin C, Streckert EMS, Stummer W, Paulus W, et al. Prediction of brain invasion in patients with meningiomas using preoperative magnetic resonance imaging. Oncotarget. 2018;9(89):35974–82. https://doi.org/10.18632/oncotarget.26313.
45. Aerts HJ, Velazquez ER, Leijenaar RT, Parmar C, Grossmann P, Carvalho S, et al. Decoding tumour phenotype by noninvasive imaging using a quantitative radiomics approach. Nat Commun. 2014;5:4006. https://doi.org/10.1038/ncomms5006.
46. Zhao Z, Nie C, Zhao L, Xiao D, Zheng J, Zhang H, et al. Multi-parametric MRI-based machine learning model for prediction of WHO grading in patients with meningiomas. Eur Radiol. 2024;34(4):2468–79. https://doi.org/10.1007/s00330-023-10252-8.
47. Ren L, Chen J, Deng J, Qing X, Cheng H, Wang D, et al. The development of a combined clinico-radiomics model for predicting post-operative recurrence in atypical meningiomas: a multicenter study. J Neuro-Oncol. 2024;166(1):59–71. https://doi.org/10.1007/s11060-023-04511-3.
48. Speckter H, Radulovic M, Trivodaliev K, Vranes V, Joaquin J, Hernandez W, et al. MRI radiomics in the prediction of the volumetric response in meningiomas after gamma knife radiosurgery. J Neuro-Oncol. 2022;159(2):281–91. https://doi.org/10.1007/s11060-022-04063-y.
49. Akkurt BH, Spille DC, Peetz-Dienhart S, Kiolbassa NM, Mawrin C, Musigmann M, et al. Radiomics-based prediction of TERT promotor mutations in intracranial high-grade meningiomas. Cancers (Basel). 2023;15(17) https://doi.org/10.3390/cancers15174415.
50. Duan C, Li N, Li Y, Cui J, Xu W, Liu X. Prediction of progesterone receptor expression in high-grade meningioma by using radiomics based on enhanced T1WI. Clin Radiol. 2023;78(10):e752–e7. https://doi.org/10.1016/j.crad.2023.06.006.
51. Moon CM, Lee YY, Kim DY, Yoon W, Baek BH, Park JH, et al. Preoperative prediction of Ki-67 and p53 status in meningioma using a multiparametric MRI-based clinical-radiomic model. Front Oncol. 2023;13:1138069. https://doi.org/10.3389/fonc.2023.1138069.
52. Kandemirli SG, Chopra S, Priya S, Ward C, Locke T, Soni N, et al. Presurgical detection of brain invasion status in meningiomas based on first-order histogram based texture analysis of contrast enhanced imaging. Clin Neurol Neurosurg. 2020;198:106205. https://doi.org/10.1016/j.clineuro.2020.106205.
53. Joo L, Park JE, Park SY, Nam SJ, Kim YH, Kim JH, et al. Extensive peritumoral edema and brain-to-tumor interface MRI features enable prediction of brain invasion in meningioma: development and validation. Neuro-Oncology. 2021;23(2):324–33. https://doi.org/10.1093/neuonc/noaa190.

Malignant Meningiomas

Lorenzo Chiariotti, Rosa Della Monica, Rosa Maria Di Crescenzo, Sergio Corvino, Francesco Maiuri, and Marialaura Del Basso De Caro

1 Definition and Origin

Malignant meningiomas correspond to the Grade III of the WHO Classification [1]. The term malignant meningioma is often used as synonymous of anaplastic meningioma; however, anaplastic meningiomas account for about 95% of all WHO grade III tumors, the other being represented by rhabdoid and papillary types [2].

About half of WHO grade III meningiomas are diagnosed de novo, while the others occur from the anaplastic transformation of lower grade tumors. However, only 2% of benign WHO grade I meningiomas [3] and 15% of atypical grade II ones [4] progress to malignant forms. This mainly occurs at further and often late recurrences, whereas it is rarer at the first reoperation.

L. Chiariotti · R. D. Monica
Department of Molecular Medicine and Medical Biotechnology, University of Naples "Federico II", Naples, Italy

R. M. Di Crescenzo · M. Del Basso De Caro (✉)
Department of Advanced Biomedical Sciences, Section of Pathology, University of Naples "Federico II", Naples, Italy
e-mail: delbasso@unina.it

S. Corvino · F. Maiuri
Department of Neurosciences, Reproductive and Odontostomatological Sciences, University of Naples "Federico II", Naples, Italy

2 Incidence

Malignant meningiomas account for 1–3% of all meningiomas [2, 5–9]. They are slightly more common in men than in women [6, 10]. According to the age, the peak of incidence is between 75 and 84 years; however, it is different according to the sex: the incidence is slightly higher in young females (aged 35–64 years) and in males aged >75 years [10]. On the other hand, Cao et al. [11] have reported decreased incidence of malignant meningiomas in middle-aged females. Radiation-induced meningiomas show significantly higher rates of malignant WHO grade III forms than the sporadic ones [12].

3 Pathology

The last WHO Classification [1] defines the anaplastic (malignant) meningioma as a high-grade meningioma with clear malignant cytomorphology (anaplasia) that can 1. resemble carcinoma, high-grade sarcoma or melanoma; 2. display markedly elevated mitotic index (≥20 mitoses/10HPF) [13]; 3. harbor TERT promoter mutation, and/or 4 homozygous CDKN2A/B deletion. Most of these tumors show extensive necrosis and brain invasion. However, in borderline cases determining the cut-off between atypical and anaplastic meningioma based on anaplasia alone may be difficult [14].

F. Maiuri, M. Del Basso De Caro (eds.), *Meningiomas*,
https://doi.org/10.1007/978-3-031-76680-0_10

The papillary and rhabdoid meningiomas are two rare subtypes associated to aggressiveness and worse prognosis. Focal papillary and/or rhabdoid architecture is often present in atypical meningiomas in the absence of other features of high grades; this is not sufficient for designating the tumor as malignant [15, 16]. On the other hand, the finding of areas with papillary or rhabdoid architectures is sometimes found in tumors which otherwise fulfill the criteria to be classified as WHO grade anaplastic malignant meningiomas [17, 18].

The papillary meningioma is defined by the presence of a predominant perivascular pseudopapillary pattern; the tumor cells surround thin-walled blood vessels in perivascular, pseudorosette-like pattern. The rhabdoid subtype is formed by rhabdoid plump cells with eccentric nuclei and eosinophilic nuclei. Both papillary and rhabdoid areas are sometimes present in the same tumor specimen [15, 16, 19, 20]. Besides, both rhabdoid and papillary subtypes, although found at initial diagnosis, are more often evident at the recurrences.

4 Genetic and Epigenetic Characteristics

Molecular analysis is now considered an important step toward definition of main types and subtypes of meningiomas and in some cases of great help to design a correct therapeutic plan. Molecular lesions may be in common among different grades. However, some specific features may define malignant meningiomas [21].

In particular, genetic alterations involving TERT1 and CDKN2A/B exclusive for grade III (WHO 2021). *TERT promoter mutations* are associated to rapid tumor progression and biological aggressiveness. *CDKN2A/B deletion* are frequently present in recurrent and progressive meningiomas [1].

Based on WHO 2021 brain tumors classification the diagnosis of anaplastic meningioma (grade III) can be performed in the presence of one of the above genetic lesions or histological features such as frank anaplasia or high mitotic index.

However, many other genetic alterations may contribute to malignant phenotype. These include:

Histone 3 mutation at lysine 27 (K27), associated with loss of K27 tri-methylation and with chromatin rearrangement. This alteration is present in about 20% of malignant meningioma and it correlates with poor prognosis.

BAP1 inactivation is common in meningiomas with rhabdoid features and its inactivation is associated to rapid recurrence [19].

TRAF7, AKT1, KLF4, SMARCE1 mutations are associated to specific histological subtypes but the histological grade is not predictable by the presence of mutations in these genes [22].

Copy Number variations (CNVs): 22q; 1p, 6p, 9q, 10q, 14q, and 18q chromosome losses are often present but are not unique of malignant meningioma.

Loss of function mutations of NF2 are associated with inherited meningioma while somatic alteration of this gene is often represented by loss of chromosome 22q material where NF2 is located.

Epigenetic profiling of brain tumors is essential to define diagnosis of different tumor especially tumor of glial origin [23]. Several studies have been conducted on meningiomas, in particular, Felix Sahm et al. analyzed 497 meningiomas divided into three groups based on histopathological grading [22]. Epigenetic profiling was able to distinguish five different methylation groups. Among these groups, malignant meningiomas include meningiomas with different genetic alterations (such as 22q; 1p, 10 chromosome deletions, TERT mutations, and CDKN2A/B deletions). However, it is possible to diagnose a malignant meningioma based on specific genome-wide arrangement of methylated CpG (methylome profiles) independently from genetic and histological features doing on the power of epigenetic analysis that can help to solve challenging cases.

5 Diagnostic Imaging

Magnetic Resonance Imaging (MRI) evidence several features suggesting the malignant nature of meningioma: on T1W the tumor margins are indistinct, with tumors interdigitations extending in the brain. Although marked peritumoral edema is often evident on FLAIR sequences, this finding is also found in the more frequent WHO grade I and II meningiomas [24]. Diffusion Weighted Images show marked hyperintensity and restricted diffusion. T1-contrast-enhanced images show inhomogeneous enhancement of the tumor due to necrotic areas; plaque-like tumor digitation may involve the dura and the brain. Skull and scalp invasion is often present. MR spectroscopy evidences elevated levels of alanine. Variable and often marked increase of the blood perfusion is found on perfusion MRI.

The nuclear medicine imagines techniques, such as the positron emission tomography (PET) and the single-photon emission computed tomography (SPECT), although seldom used in the current clinical practice for meningioma diagnosis, may provide useful information on the histologic grade, outcome after surgical resection and differentiation between fibrous tissue and residual-recurrent tumor [25, 26]. In PET studies with 18F-FDG, the uptake of FDG is significantly higher in high-grade meningiomas. Also, in SPECT studies, the intensity of the 99mTc-TF uptake is significantly higher in anaplastic than in typical meningiomas.

6 Natural History

Malignant meningiomas are aggressive tumors with rapid growth. Rapid increase of the tumor size and frequent brain invasion result in early occurrence of neurological symptoms. Spread along the meningeal surface through the cerebrospinal fluid may result in even distant intracranial tumor nodules [2]. Distant metastases are reported in 6–43% of the cases [27, 28]; they mainly occur in the lungs, but also in the liver, bone, and skin. Malignant meningiomas associated to distant metastases are mainly "de novo" WHO grade III and located in the convexity and parasagittal region [27].

Malignant meningiomas show higher rates of recurrence and death. The median survival is 2–3 years [6].

7 Surgery

Maximal resection and conformed radiotherapy are considered the best management of malignant meningiomas. A recent review [29] including 14 studies [3, 6, 27–40] has examined the correlation between rates of gross-total resection and survival. The reported rates of gross-total resection are very variable, ranging from 15% to 100%, but the 5-years survival is less than 50%. Some studies suggest that greater extent of resection significantly prolongs overall survival [5, 18, 27, 30, 31, 34, 36, 40]. However, single-center retrospective studies are less reliable, because of the different tumor extension and aggressiveness. Two large population-based studies from the National Cancer Database [37, 39] found that gross-total resection in combination with radiotherapy, provides a survival benefit, but the result was not statistically significant.

In other studies, neither progression-free survival nor overall survival was affected by the extent of resection [32, 41–44]. Peyre et al. [38] have shown statistically significative difference between GTR and STR only in primary anaplastic meningiomas.

However, the extent of resection must be balanced with the risk of neurological complications. Sun et al. [45] recommend in a review the maximal safe resection of malignant meningiomas. Sughrue et al. [6] found in their series a neurosurgical complications rate of 21% due to the surgical trauma related to tumor resection with pial transgression. They concluded that attempting to resect a small tumor remnant (<5% of the men) in areas with significant morbidity may be deleterious.

The patient age and preoperative health conditions are important for surgical indication. Younger patents without significant comorbidities show improved overall survival after surgery compared to older ones [32, 33]. According to the location, one study [30] found better survival in anaplastic meningiomas of the brain convexity.

8 Adjuvant Treatments

The use of adjuvant radiotherapy after the surgical resection of malignant meningiomas is always necessary, irrespective of the extent of resection.

The association of the external radiotherapy, as compared to surgery alone, results in improvement of the overall survival [9, 32], progression-free survival [9, 31], and local tumor control [4, 5, 18, 37, 38, 45].

The mostly used radiation dose is 60 Gy. Several studies [46, 47] have reported better progression-fee survival in patients treated with doses >60 Gy; however, they were published before the introduction of the CT-based planning [2].

There is no consensus about the security of the margins (4 cm in the study of Dziuk [31]). It varies up to 1 cm among the studies according to the tumor extension and postoperative changes in the meninges and brain tissue on MRI.

The benefit of the radiation treatment is most pronounced when administered immediately after the resection; on the other hand, savage radiation therapy at progression results in little or no improvement [41].

Stereotactic radiosurgery has sometimes been used for anaplastic meningiomas; on the other hand, the optimal timing is still not defined [48, 49]. It should be considered in patients who cannot afford additional radiosurgery or when external radiotherapy has failed to obtain tumor control. However, it does not result in a better outcome [6], also because of the difficulty in defining the target volume.

The role of radiotherapy at initial diagnosis is widely discussed in chapter "Meningioma Pathology and Radiotherapy at Initial Diagnosis".

The medical therapy of malignant meningiomas is still controversial and provides scarce results. Despite the numerous genes alterations, target therapies are not yet applicable.

Currently, the most widely used pharmacological approaches are based on anti-VEGF drugs and mTOR inhibitors (REF) VEGF expression is frequently associated with the grade of meningiomas. Phase II clinical trial in 2016 suggests that patients with refractory meningioma could benefit from a combined therapy with bevacizumab (anti-VEGF) and everolimus [50–52].

Another target drug used in clinical trials is Sunitinib, a small molecule able to inhibit different targets, such as VEGFR, PDGFR, KIT. In 2015, a prospective and multicentric phase II trial for recurrent and progressive atypical/anaplastic meningiomas was conducted. It is been demonstrated that Sunitinib is active in recurrent atypical/malignant meningioma patients. In particular, an improvement in PFS of 42% at 6 months was observed compared to PFS of 5–30% at 6 months reported in the natural history [53].

BAP1 alteration is frequent in malignant meningioma, Tazemetostat is a molecule which interferes with the molecular pathway activated by BAP1. Rhabdoid grade III meningioma presents often BAP1 genetic alteration. Tazemetostat targets increased level of PCR2 complex protein, EZH2, due to loss of BAP1. Some mutations may drive epigenetic changes, leading to altered methylation profiles which might lead to disease formation or progression of disease [54].

9 Recurrent Malignant Meningiomas

Although malignant meningiomas are rare, they recur in almost all cases in a short period. Surgery is the first management option of recurrent malignant meningiomas. Thus, reoperation and re-resection are a relatively common problem. However, obtaining a gross-total resection is often very difficult. Dural adhesions to the cortex can lead to disruption of the pial surface and injury to the cortical vein. The tumor–arachnoid interface is difficult to identify in presence of scar tissue and radiation-induced changes. This results in high complication rate [29, 55], mainly for meningiomas of the middle third of the sagittal plane, where injury of the superior sagittal sinus, rolandic vein and motor cortex may occur. However, even when complete resection is not possible, maximal cytoreduction at resurgery and even serial re-resections in appropriate surgical candidates, while preserving the functional structures, may have significant impact on the clinical outcome [29].

Recurrent malignant meningiomas have mostly been irradiated at the initial diagnosis or at malignant progression from a low-grade tumor. Thus, external re-irradiation is not advisable. In those cases, the brachytherapy has been used [56]. Although the complication rate, including radiation necrosis, wound breakdown and infection, is high (34%), the median overall survival (2.3 years) is better than after resurgery alone [56].

A high rate (up to ¾) of patients with recurrent malignant meningiomas show intracranial dissemination with even distant tumor nodules; leptomeningeal seeding is observed in about 10% of the cases [9]. This significantly limits the surgical resection at the disease progression.

Approximately, 50% of malignant meningiomas recur between 2 and 3 years after the initial surgery. Champeaux et al. [57] report a median time to recurrence of 1.3 years between the first and the second surgery. Although an average patient with malignant meningiomas undergo three surgeries [6, 27], the survival benefit of more than two operations is not defined.

10 Outcome and Predictors of Survival

Malignant meningiomas are tumors with bad prognosis, with an overall survival of 3 years after surgical resection [6]. In USA, a population-based study showed a 5-years survival of 41.4%. Gross-total resection seems to be associated to a better prognosis. However, the adjuvant radiotherapy administered soon after the resection is the most important prognostic factor affecting the progression-free survival and overall survival.

Several predicting factors of survival have been suggested. Those include patient sex and age, primary versus secondary tumor origin, location, postoperative KPS, proliferation and biomolecular markers.

Data on patient sex and age are controversial. Several studies report higher rates of malignant meningiomas and worse outcome in men [27, 58, 59] or in women [28]; however, in most studies [7, 32, 41, 60, 61], the difference between genders could not be confirmed as an independent predictive and prognostic factor. According to the age, patients older than 60 years were reported to present worse outcome [32, 43, 62]; on the other hand, in other studies [7, 60] age was not a prognostic factor for a better outcome.

Secondary malignant meningiomas (recurrent tumors that progressed from a lower grade) were found to be associated with a worse prognosis than "de novo" anaplastic meningiomas in recent studies [27, 28, 36, 38]. Variance of histological and biomolecular features during the anaplastic transformation, various patterns of relapse and previous radiotherapy and chemotherapy treatments in secondary anaplastic tumors may account for the different outcome [29].

The tumor location was not significantly correlated to the outcome [6, 7, 63, 64], although others found a non-skull base location to be associated to anaplastic histology [58] or better survival [27].

The postoperative KPS does not significantly impact on the survival [6, 7, 41].

Although proliferation markers are suggested an additional support to predict survival of meningiomas, data on the malignant WHO grade III tumor are scarce and not conclusive.

As previously discussed, the TERT promoter mutation and methylation are associated to worse progression-free survival of a secondary anaplastic meningiomas; their absence is also associated with better overall survival [38].

11 Conclusion

Malignant meningiomas are a major problem for neurosurgeons and for neuro-oncologist. The maximal safe resection with preservation of the neurological functions is the first management; adjuvant external conformal radiotherapy soon after the surgical resection is essential to improve the survival. However, the prognosis is poor, with a median survival of about 3 years, also because of the intracranial dissemination and recurrence in most patients. Future direction of the studies includes the identification of genetic and epigenetic profiles and appropriate medical target therapies.

References

1. Cancer WCoTEBWHOCoTotCNSteLIAfRo, 2021.
2. Fountain DM, Young AMH, Santarius T. Malignant meningiomas. Handb Clin Neurol. 2020;170:245–50. https://doi.org/10.1016/B978-0-12-822198-3.00044-6.

3. Yang SY, Park CK, Park SH, Kim DG, Chung YS, Jung HW. Atypical and anaplastic meningiomas: prognostic implications of clinicopathological features. J Neurol Neurosurg Psychiatry. 2008;79(5):574–80. https://doi.org/10.1136/jnnp.2007.121582.
4. Ding D, Starke RM, Hantzmon J, Yen CP, Williams BJ, Sheehan JP. The role of radiosurgery in the management of WHO Grade II and III intracranial meningiomas. Neurosurg Focus. 2013;35(6):E16. https://doi.org/10.3171/2013.9.FOCUS13364.
5. Modha A, Gutin PH. Diagnosis and treatment of atypical and anaplastic meningiomas: a review. Neurosurgery. 2005;57(3):538–50; discussion 50. https://doi.org/10.1227/01.neu.0000170980.47582.a5.
6. Sughrue ME, Sanai N, Shangari G, Parsa AT, Berger MS, McDermott MW. Outcome and survival following primary and repeat surgery for World Health Organization Grade III meningiomas. J Neurosurg. 2010;113(2):202–9. https://doi.org/10.3171/2010.1.JNS091114.
7. Masalha W, Heiland DH, Delev D, Fennell JT, Franco P, Scheiwe C, et al. Survival and prognostic predictors of anaplastic meningiomas. World Neurosurg. 2019;131:e321–e8. https://doi.org/10.1016/j.wneu.2019.07.148.
8. Ogasawara C, Philbrick BD, Adamson DC. Meningioma: a review of epidemiology, pathology, diagnosis, treatment, and future directions. Biomedicines. 2021;9(3) https://doi.org/10.3390/biomedicines9030319.
9. Seo DO, Song SW, Kim YH, Hong CK, Kim JH. Anaplastic meningioma: clinical characteristics, prognostic factors and survival outcome. Brain Tumor Res Treat. 2022;10(4):244–54. https://doi.org/10.14791/btrt.2022.0030.
10. Kshettry VR, Ostrom QT, Kruchko C, Al-Mefty O, Barnett GH, Barnholtz-Sloan JS. Descriptive epidemiology of World Health Organization grades II and III intracranial meningiomas in the United States. Neuro-Oncology. 2015;17(8):1166–73. https://doi.org/10.1093/neuonc/nov069.
11. Cao J, Yan W, Hong X, Yan H. Epidemiology and survival of non-malignant and malignant meningiomas in middle-aged females, 2004–2018. Front Oncol. 2023;13:1157182. https://doi.org/10.3389/fonc.2023.1157182.
12. Umansky F, Shoshan Y, Rosenthal G, Fraifeld S, Spektor S. Radiation-induced meningioma. Neurosurg Focus. 2008;24(5):E7. https://doi.org/10.3171/FOC/2008/24/5/E7.
13. Perry A, Scheithauer BW, Stafford SL, Lohse CM, Wollan PC. "Malignancy" in meningiomas: a clinicopathologic study of 116 patients, with grading implications. Cancer. 1999;85(9):2046–56. https://doi.org/10.1002/(sici)1097-0142(19990501)85:9<2046::aid-cncr23>3.0.co;2-m.
14. Gauchotte G, Peyre M, Pouget C, Cazals-Hatem D, Polivka M, Rech F, et al. Prognostic value of histopathological features and loss of H3K27me3 immunolabeling in anaplastic meningioma: a multicenter retrospective study. J Neuropathol Exp Neurol. 2020;79(7):754–62. https://doi.org/10.1093/jnen/nlaa038.
15. Vaubel RA, Chen SG, Raleigh DR, Link MJ, Chicoine MR, Barani I, et al. Meningiomas with rhabdoid features lacking other histologic features of malignancy: a study of 44 cases and review of the literature. J Neuropathol Exp Neurol. 2016;75(1):44–52. https://doi.org/10.1093/jnen/nlv006.
16. Wu YT, Ho JT, Lin YJ, Lin JW. Rhabdoid papillary meningioma: a clinicopathologic case series study. Neuropathology. 2011;31(6):599–605. https://doi.org/10.1111/j.1440-1789.2011.01201.x.
17. Perry A, Stafford SL, Scheithauer BW, Suman VJ, Lohse CM. Meningioma grading: an analysis of histologic parameters. Am J Surg Pathol. 1997;21(12):1455–65.
18. Zhang GJ, Zhang YS, Zhang GB, Li D, Zhang LW, Wu Z, et al. Prognostic factors and the management of anaplastic meningioma. Clin Neurol Neurosurg. 2018;170:13–9. https://doi.org/10.1016/j.clineuro.2018.03.028.
19. Shankar GM, Santagata S. BAP1 mutations in high-grade meningioma: implications for patient care. Neuro-Oncology. 2017;19(11):1447–56. https://doi.org/10.1093/neuonc/nox094.
20. Williams EA, Santagata S, Wakimoto H, Shankar GM, Barker FG, Sharaf R, et al. Distinct genomic subclasses of high-grade/progressive meningiomas: NF2-associated, NF2-exclusive, and NF2-agnostic. Acta Neuropathol Commun. 2020;8(1):171. https://doi.org/10.1186/s40478-020-01040-2.
21. Salles D, Santino SF, Malinverni ACM, Stávale JN. Meningiomas: a review of general, histopathological, clinical and molecular characteristics. Pathol Res Pract. 2021;223:153476. https://doi.org/10.1016/j.prp.2021.153476.
22. Sahm F, Schrimpf D, Stichel D, Jones DTW, Hielscher T, Schefzyk S, et al. DNA methylation-based classification and grading system for meningioma: a multicentre, retrospective analysis. Lancet Oncol. 2017;18(5):682–94. https://doi.org/10.1016/S1470-2045(17)30155-9.
23. Capper D, Jones DTW, Sill M, Hovestadt V, Schrimpf D, Sturm D, et al. DNA methylation-based classification of central nervous system tumours. Nature. 2018;555(7697):469–74. https://doi.org/10.1038/nature26000.
24. Kunimatsu A, Kunimatsu N, Kamiya K, Katsura M, Mori H, Ohtomo K. Variants of meningiomas: a review of imaging findings and clinical features. Jpn J Radiol. 2016;34(7):459–69. https://doi.org/10.1007/s11604-016-0550-6.
25. Valotassiou V, Leondi A, Angelidis G, Psimadas D, Georgoulias P. SPECT and PET imaging of meningiomas. ScientificWorldJournal. 2012;2012:412580. https://doi.org/10.1100/2012/412580.
26. Filis P, Alexiou GA, Zigouris A, Sioka C, Filis N, Voulgaris S. Meningioma grading based on posi-

tron emission tomography: a systematic review and meta-analysis. World Neurosurg X. 2023;18:100167. https://doi.org/10.1016/j.wnsx.2023.100167.
27. Moliterno J, Cope WP, Vartanian ED, Reiner AS, Kellen R, Ogilvie SQ, et al. Survival in patients treated for anaplastic meningioma. J Neurosurg. 2015;123(1):23–30. https://doi.org/10.3171/2014.10.JNS14502.
28. Zhao P, Hu M, Zhao M, Ren X, Jiang Z. Prognostic factors for patients with atypical or malignant meningiomas treated at a single center. Neurosurg Rev. 2015;38(1):101–7; discussion 7. https://doi.org/10.1007/s10143-014-0558-2.
29. Pacult MA, Przybylowski CJ, Raza SM, DeMonte F. Surgical management of high-grade meningiomas. Cancers (Basel). 2024;16(11) https://doi.org/10.3390/cancers16111978.
30. Palma L, Celli P, Franco C, Cervoni L, Cantore G. Long-term prognosis for atypical and malignant meningiomas: a study of 71 surgical cases. J Neurosurg. 1997;86(5):793–800. https://doi.org/10.3171/jns.1997.86.5.0793.
31. Dziuk TW, Woo S, Butler EB, Thornby J, Grossman R, Dennis WS, et al. Malignant meningioma: an indication for initial aggressive surgery and adjuvant radiotherapy. J Neuro-Oncol. 1998;37(2):177–88. https://doi.org/10.1023/a:1005853720926.
32. Durand A, Labrousse F, Jouvet A, Bauchet L, Kalamaridès M, Menei P, et al. WHO grade II and III meningiomas: a study of prognostic factors. J Neuro-Oncol. 2009;95(3):367–75. https://doi.org/10.1007/s11060-009-9934-0.
33. Pasquier D, Bijmolt S, Veninga T, Rezvoy N, Villa S, Krengli M, et al. Atypical and malignant meningioma: outcome and prognostic factors in 119 irradiated patients. A multicenter, retrospective study of the Rare Cancer Network. Int J Radiat Oncol Biol Phys. 2008;71(5):1388–93. https://doi.org/10.1016/j.ijrobp.2007.12.020.
34. Aizer AA, Bi WL, Kandola MS, Lee EQ, Nayak L, Rinne ML, et al. Extent of resection and overall survival for patients with atypical and malignant meningioma. Cancer. 2015;121(24):4376–81. https://doi.org/10.1002/cncr.29639.
35. Kumar N, Kumar R, Khosla D, Salunke PS, Gupta SK, Radotra BD. Survival and failure patterns in atypical and anaplastic meningiomas: a single-center experience of surgery and postoperative radiotherapy. J Cancer Res Ther. 2015;11(4):735–9. https://doi.org/10.4103/0973-1482.151426.
36. Champeaux C, Dunn L. World Health Organization Grade II meningioma: a 10-year retrospective study for recurrence and prognostic factor assessment. World Neurosurg. 2016;89:180–6. https://doi.org/10.1016/j.wneu.2016.01.055.
37. Orton A, Frandsen J, Jensen R, Shrieve DC, Suneja G. Anaplastic meningioma: an analysis of the National Cancer Database from 2004 to 2012. J Neurosurg. 2018;128(6):1684–9. https://doi.org/10.3171/2017.2.JNS162282.
38. Peyre M, Gauchotte G, Giry M, Froehlich S, Pallud J, Graillon T, et al. De novo and secondary anaplastic meningiomas: a study of clinical and histomolecular prognostic factors. Neuro-Oncology. 2018;20(8):1113–21. https://doi.org/10.1093/neuonc/nox231.
39. Rydzewski NR, Lesniak MS, Chandler JP, Kalapurakal JA, Pollom E, Tate MC, et al. Gross total resection and adjuvant radiotherapy most significant predictors of improved survival in patients with atypical meningioma. Cancer. 2018;124(4):734–42. https://doi.org/10.1002/cncr.31088.
40. Champeaux C, Jecko V, Houston D, Thorne L, Dunn L, Fersht N, et al. Malignant meningioma: an international multicentre retrospective study. Neurosurgery. 2019;85(3):E461–E9. https://doi.org/10.1093/neuros/nyy610.
41. Rosenberg LA, Prayson RA, Lee J, Reddy C, Chao ST, Barnett GH, et al. Long-term experience with World Health Organization grade III (malignant) meningiomas at a single institution. Int J Radiat Oncol Biol Phys. 2009;74(2):427–32. https://doi.org/10.1016/j.ijrobp.2008.08.018.
42. Adeberg S, Hartmann C, Welzel T, Rieken S, Habermehl D, von Deimling A, et al. Long-term outcome after radiotherapy in patients with atypical and malignant meningiomas—clinical results in 85 patients treated in a single institution leading to optimized guidelines for early radiation therapy. Int J Radiat Oncol Biol Phys. 2012;83(3):859–64. https://doi.org/10.1016/j.ijrobp.2011.08.010.
43. Detti B, Scoccianti S, Di Cataldo V, Monteleone E, Cipressi S, Bordi L, et al. Atypical and malignant meningioma: outcome and prognostic factors in 68 irradiated patients. J Neuro-Oncol. 2013;115(3):421–7. https://doi.org/10.1007/s11060-013-1239-7.
44. Balasubramanian SK, Sharma M, Silva D, Karivedu V, Schmitt P, Stevens GH, et al. Longitudinal experience with WHO Grade III (anaplastic) meningiomas at a single institution. J Neuro-Oncol. 2017;131(3):555–63. https://doi.org/10.1007/s11060-016-2321-8.
45. Sun SQ, Hawasli AH, Huang J, Chicoine MR, Kim AH. An evidence-based treatment algorithm for the management of WHO Grade II and III meningiomas. Neurosurg Focus. 2015;38(3):E3. https://doi.org/10.3171/2015.1.FOCUS14757.
46. Goldsmith BJ, Wara WM, Wilson CB, Larson DA. Postoperative irradiation for subtotally resected meningiomas. A retrospective analysis of 140 patients treated from 1967 to 1990. J Neurosurg. 1994;80(2):195–201. https://doi.org/10.3171/jns.1994.80.2.0195.
47. Hug EB, Devries A, Thornton AF, Munzenride JE, Pardo FS, Hedley-Whyte ET, et al. Management of atypical and malignant meningiomas: role of high-dose, 3D-conformal radiation therapy. J Neuro-

Oncol. 2000;48(2):151–60. https://doi.org/10.1023/a:1006434124794.
48. Helis CA, Hughes RT, Cramer CK, Tatter SB, Laxton AW, Bourland JD, et al. Stereotactic radiosurgery for atypical and anaplastic meningiomas. World Neurosurg. 2020;144:e53–61. https://doi.org/10.1016/j.wneu.2020.07.211.
49. Shepard MJ, Xu Z, Kearns K, Li C, Chatrath A, Sheehan K, et al. Stereotactic radiosurgery for atypical (World Health Organization II) and anaplastic (World Health Organization III) meningiomas: results from a multicenter, international cohort study. Neurosurgery. 2021;88(5):980–8. https://doi.org/10.1093/neuros/nyaa553.
50. Caruso G, Ferrarotto R, Curcio A, Metro L, Pasqualetti F, Gaviani P, et al. Novel advances in treatment of meningiomas: prognostic and therapeutic implications. Cancers (Basel). 2023;15(18) https://doi.org/10.3390/cancers15184521.
51. Hou J, Kshettry VR, Selman WR, Bambakidis NC. Peritumoral brain edema in intracranial meningiomas: the emergence of vascular endothelial growth factor-directed therapy. Neurosurg Focus. 2013;35(6):E2. https://doi.org/10.3171/2013.8.FOCUS13301.
52. Shih KC, Chowdhary S, Rosenblatt P, Weir AB, Shepard GC, Williams JT, et al. A phase II trial of bevacizumab and everolimus as treatment for patients with refractory, progressive intracranial meningioma. J Neuro-Oncol. 2016;129(2):281–8. https://doi.org/10.1007/s11060-016-2172-3.
53. Kaley TJ, Wen P, Schiff D, Ligon K, Haidar S, Karimi S, et al. Phase II trial of sunitinib for recurrent and progressive atypical and anaplastic meningioma. Neuro-Oncology. 2015;17(1):116–21. https://doi.org/10.1093/neuonc/nou148.
54. Kim L. A narrative review of targeted therapies in meningioma. Chin Clin Oncol. 2020;9(6):76. https://doi.org/10.21037/cco-2020-mbt-01.
55. Magill ST, Dalle Ore CL, Diaz MA, Jalili DD, Raleigh DR, Aghi MK, et al. Surgical outcomes after reoperation for recurrent non-skull base meningiomas. J Neurosurg. 2018;131:1–9. https://doi.org/10.3171/2018.6.JNS18118.
56. Magill ST, Lau D, Raleigh DR, Sneed PK, Fogh SE, McDermott MW. Surgical resection and interstitial Iodine-125 brachytherapy for high-grade meningiomas: a 25-year series. Neurosurgery. 2017;80(3):409–16. https://doi.org/10.1227/NEU.0000000000001262.
57. Champeaux C, Dunn L. World Health Organization grade II meningiomas. Acta Neurochir. 2016;158(5):921–9; discussion 9. https://doi.org/10.1007/s00701-016-2771-y.
58. Kane AJ, Sughrue ME, Rutkowski MJ, Shangari G, Fang S, McDermott MW, et al. Anatomic location is a risk factor for atypical and malignant meningiomas. Cancer. 2011;117(6):1272–8. https://doi.org/10.1002/cncr.25591.
59. Cain SA, Smoll NR, Van Heerden J, Tsui A, Drummond KJ. Atypical and malignant meningiomas: considerations for treatment and efficacy of radiotherapy. J Clin Neurosci. 2015;22(11):1742–8. https://doi.org/10.1016/j.jocn.2015.03.054.
60. Cao X, Hao S, Wu Z, Wang L, Jia G, Zhang L, et al. Survival rates, prognostic factors and treatment of anaplastic meningiomas. J Clin Neurosci. 2015;22(5):828–33. https://doi.org/10.1016/j.jocn.2014.11.022.
61. Pisćević I, Villa A, Milićević M, Ilić R, Nikitović M, Cavallo LM, et al. The influence of adjuvant radiotherapy in atypical and anaplastic meningiomas: a series of 88 patients in a single institution. World Neurosurg. 2015;83(6):987–95. https://doi.org/10.1016/j.wneu.2015.02.021.
62. Colli BO, Carlotti Junior CG, Assirati Junior JA, VeP CJ, Neder L. Atypical and anaplastic meningiomas in a public hospital in São Paulo State, Brazil. Arq Neuropsiquiatr. 2015;73(9):770–8. https://doi.org/10.1590/0004-282X20150106.
63. Maiuri F, Mariniello G, Barbato M, Corvino S, Guadagno E, Chiariotti L, et al. Malignant intraventricular meningioma: literature review and case report. Neurosurg Rev. 2022;45(1):151–66. https://doi.org/10.1007/s10143-021-01585-3.
64. Maiuri F, Mariniello G, Guadagno E, Barbato M, Corvino S, Del Basso De Caro M. WHO grade, proliferation index, and progesterone receptor expression are different according to the location of meningioma. Acta Neurochir. 2019;161(12):2553–61. https://doi.org/10.1007/s00701-019-04084-z.

Part III

Meningioma Pathology and Imaging

Pathology: Imaging Correlations of Meningiomas—Imaging Mimics

Sirio Cocozza, Giuseppe Pontillo, Alessandra Scaravilli, Alessandra Alfieri, Donatella Franco, Renata Conforti, Ferdinando Caranci, and Andrea Elefante

1 Introduction

Meningiomas are the most common primary intracranial tumors; although typically benign and slow-growing tumors, these lesions are characterized by an extensive histological diversity with numerous microscopic subtypes described. This histologic variability often leads to a diagnostic challenge with other tumor types and with other different conditions sharing imaging characteristics mimicking meningiomas, with the most common represented by metastasis, lymphomas, solitary fibrous tumors (SFT), and schwannomas [1–3].

Accurate identification of meningioma mimics is therefore critical for diagnosis and appropriate management and it is based on a comprehensive diagnostic approach that integrates clinical, neuroradiological, and histopathological findings. In this setting, advanced imaging techniques, including diffusion-weighted imaging (DWI), perfusion-weighted imaging (PWI), and magnetic resonance spectroscopy (MRS), play an adjunctive role in helping to unravel these diagnostic dilemmas, providing insights into tissue microstructure, vascularity, and metabolism [4].

In this chapter on meningioma mimics, we examined the diverse spectrum of the most common entities that can enter the differential diagnosis with meningiomas at imaging, exploring their computed tomography (CT) and magnetic resonance imaging (MRI) features. Through a systematic analysis of their typical and atypical imaging characteristics, we aim to provide physicians and radiologists with the knowledge and the tools needed to confidently address these diagnostic challenges [5].

2 Typical and Atypical Imaging Features

2.1 Typical Imaging Features of Meningioma

Understanding the typical and atypical imaging features of meningiomas is essential for achieving an accurate diagnosis and excluding imaging mimics. MRI stands out as the optimal imaging modality in the characterization of meningiomas, allowing exploration of different findings [3].

As previously described, meningiomas usually present as rounded well-defined extra-axial masses

S. Cocozza · G. Pontillo · A. Scaravilli · A. Elefante
Department of Advanced Biomedical Sciences, University of Naples "Federico II", Naples, Italy

A. Alfieri
Department of Neurosurgery, "S. Anna & S. Sebastiano" Hospital, Caserta, Italy

D. Franco · R. Conforti · F. Caranci (✉)
Department of Medicine of Precision, University "Luigi Vanvitelli", Naples, Italy
e-mail: ferdinando.caranci@unicampania.it

F. Maiuri, M. Del Basso De Caro (eds.), *Meningiomas*,
https://doi.org/10.1007/978-3-031-76680-0_11

hyperdense on CT and hypo-isointense to the cortex on MRI, with homogeneous contrast enhancement. They are often associated with a dural tail, which is a dural thickening at the tumor margins thought to be caused by the presence of vascularized connective tissue near the meningioma and not by a proper dural invasion. Although highly suggestive, the dural tail sign is not a pathognomonic feature of meningioma and can occur in many mimickers. Furthermore, an "en plaque" pattern with diffuse sheet-like dural thickening can be also observed. Despite their generally benign nature, meningiomas can present different imaging characteristics depending on factors such as histological subtype, location, and size of the tumor [3]. For instance, atypical meningiomas present a more heterogeneous contrast enhancement due to the presence of areas of necrosis. Calcifications, usually diffuse and punctate, best seen on CT or susceptibility-weighted imaging (SWI), are associated with lower grade and slow growth as well as adjacent bony hyperostosis. Calvarial hyperostosis, better visualized on CT, is a pathognomonic feature of meningiomas and is more common in skull base locations, as the sphenoid wing, and in "en plaque" meningiomas. An additional typical imaging feature, the identification of a cerebrospinal fluid (CSF) cleft between the tumor and the brain parenchyma, often containing displaced vessels, is also highly suggestive of a benign nature of the meningioma, excluding brain invasion. On the other hand, loss of the CSF cleft with invasion of surrounding structures is a typical feature of malignant meningiomas displaying aggressive features. The presence of perilesional vasogenic edema is more commonly associated with certain subtypes (e.g., angiomatous, secretory, lymphoplasmacyte-rich, microcystic, and high-grade meningiomas). The occurrence of this phenomenon can result from compression of the parenchyma and cerebral vessels, sometimes resulting in local ischemia and recruitment of pial vascular supply, or from a direct invasion of brain tissue.

Meningiomas can grow encasing vessels, around cranial nerves and through foramina, causing arterial insufficiency and cranial palsy. However, invasion of the dural sinuses is much more common. Vascular supply is usually provided by dural branches originating from the external carotid artery (ECA), with a lesser contribution from the internal carotid artery (ICA) and the posterior circulation. Meningiomas are highly vascularized lesions, with a complete lack of a blood-brain barrier (BBB), and therefore they exhibit avid contrast enhancement on T1-weighted images and a hyperperfusion on PWI, with relative cerebral blood volume (rCBV) values reported between 6 and 9 compared to the normal brain tissue. The highly vascular nature of meningioma is also reflected in its angiographic vascular pattern, showing a hypointense linear signal that radiates from the center to the periphery in a sunburst pattern and in persistent blush.

MRS typically shows a typical alanine peak at 1.48 ppm and a glutamine/glutamate peak, with a high alanine/creatine ratio that can be useful in distinguishing them from other intracranial tumors. Higher-grade meningiomas can present high lipid and lactate peaks and are usually associated with lower apparent diffusion coefficient (ADC) values on DWI.

2.2 Imaging Red Flags for Mimics

Hyperintensity on T2-weighted images is an uncommon finding in meningiomas, indicating high water content. A high T2-weighted signal may suggest various features, such as a cartilaginous matrix in chondrosarcoma, the presence of fluid-rich material as seen in chordoma, cystic degeneration typical of schwannoma, or necrotic areas in rapidly growing tumors. Solitary fibrous tumors (SFT) may exhibit high T2 signal due to their distinctive vascular histopathology. It is also important to differentiate this high signal from intratumoral cysts, a characteristic also observed in meningiomas.

Similarly, a T2-weighted hypointensity is also atypical in meningiomas, possibly indicating increased cellularity (characteristic of lymphoma), the presence of fibrous tissue (as seen in SFT), or extensive calcification (observed in chondrosarcoma). Additionally, some of these mimics may present with both hyperintense and hypointense T2 components, as in the case of chondrosarcoma or SFT.

Leptomeningeal or subarachnoid extension is generally absent in meningiomas and may raise suspicion for potential mimics. Lesions originating in the brain parenchyma and subsequently extending into the leptomeningeal space and the dura can resemble dural-based masses, as in metastasis, lymphoma, schwannoma, and hemangioblastoma. Similarly, hemangiopericytoma can invade the dura and extend into the leptomeningeal space or the brain parenchyma.

Bone destruction or erosion is an aggressive feature, uncommon in typical low-grade meningioma, while it can be observed in a large number of conditions including but not limited to chondrosarcoma, chordoma, aggressive SFT, lymphomas, or metastases.

Finally, it is noteworthy to mention that, although meningioma typically exhibits a high rCBV compared to most of its mimics, hypervascular metastases (i.e., from renal carcinoma or melanoma) could be indistinguishable from meningiomas on PWI.

3 Imaging Mimics

Meningioma mimickers can be divided, basing on the etiologic condition, into neoplastic and non-neoplastic causes, that will be here therefore treated separately.

3.1 Neoplastic Conditions Mimicking Meningiomas

The largest group of meningioma mimics is certainly represented by neoplastic conditions. An accurate differential diagnosis between meningioma and other neoplastic lesions is crucial for precise diagnosis and adequate treatment planning, including surgery and radiotherapy. Neoplastic lesions that can mimic a meningioma encompass a wide spectrum of conditions, from intra-axial lesions to extra-axial lesions originating from adjacent structures, such as the skull base or cranial nerves. Achieving an accurate neuroradiological differentiation relies on an in-depth knowledge of the anatomy, as well as their distinct neuroradiological characteristics across imaging modalities.

3.1.1 Metastasis

Metastases are typically located at the interface of gray and white matter, as well as within the transitional areas between arterial territories, associated with varying degrees of vasogenic edema. A characteristic ring enhancement pattern with a central region of necrosis and hemorrhages is often observed. Metastatic lesions involving the dura mater, particularly from breast and lung carcinomas, often mimic meningiomas due to their extra-axial location, dural-based growth, and avid and inhomogeneous contrast enhancement [6, 7]. Dural metastases are more frequently associated with invasion of adjacent skull and soft tissues (Fig. 1).

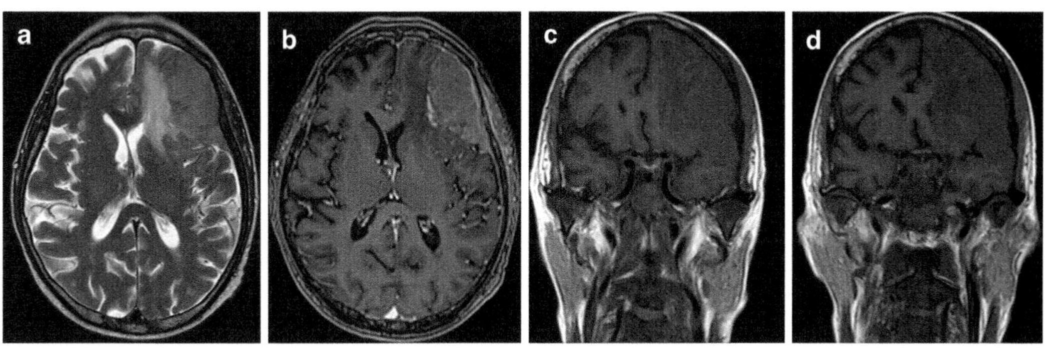

Fig. 1 Prostate carcinoma metastasis simulating meningioma. (**a**) T2-weighted axial section; (**b**) post-contrast T1-weighted axial section; (**c**, **d**) T1-weighted coronal sections. Presence of a left frontal dural-based extra-axial lesion, with irregular margins and peripheral edema; it shows a moderate enhancement after contrast medium injection (**b**), contiguous dural enhancement and significative osseous involvement (**c**, **d**)

Metastases usually exhibit a lower rCBV compared to meningioma on PWI. However, hypervascular metastases, as those found in patients with renal carcinoma and melanoma, represent an exception and therefore a challenging differential diagnosis [7].

Careful evaluation of the clinical history, presence of systemic malignancy, and multiplicity of lesions can aid in differentiation (Fig. 2).

3.1.2 Solitary Fibrous Tumor

Formerly known also as hemangiopericytoma, SFT are rare fibroblastic tumors arising from pericytes surrounding capillaries. They share several imaging features with meningiomas, including dural attachment, avid contrast enhancement and most common locations, as skull base, falcine, and parasagittal, while cerebellopontine angle and sellar regions are less common (Fig. 3). Although no specific radiological features can be used in differential diagnosis with meningiomas, SFT tend to have more heterogeneous enhancement, high or mixed intensity on T2-weighted images, due to hypointense region of collagen, with a propensity for bone destruction [8]. Furthermore, T2-weighted hypointense regions usually show contrast enhancement. Intralesional flow-voids may be present, especially in high-grade SFT, while bone hyperostosis is usually absent [8]. Finally, SFT may have high rCBVvalues similarly to those observed in meningiomas.

3.1.3 Schwannoma

Distinguishing between meningiomas and schwannomas can pose a challenge to neuroradiologists, especially when considering the cerebellopontine angle (CPA), where schwannomas and meningiomas represent the most common tumors affecting this area, the middle cranial fossa and spinal regions. Nonetheless, some features that are commonly found in typical meningiomas and schwannomas can aid to a correct diagnosis (1). These include a high and heterogeneous signal on T2-weighted images in schwannomas, often with areas of cystic degeneration (particularly in larger tumors), less commonly found in meningiomas (2). Similarly, contrast enhancement in meningiomas is typically uniform, whereas it can vary in schwannomas (Fig. 4). Meningiomas of the CPA usually originate from the posterior surface of the petrous bone and may extend secondarily into the internal auditory canal (IAC), while primary IAC meningiomas are very rare. An obstructive vestibular schwannoma often demonstrates a more noticeable decrease in perilymphatic signal on T2-weighted images, aiding in the differentiation from IAC meningiomas. Finally, when evaluating the spine, it is noteworthy to remember that the dural tail sign is commonly seen in meningiomas, while neural foraminal extension with widening of the intervertebral foramen and cystic changes are highly suggestive of schwannomas, given that meningiomas with neural foraminal extension are uncommon.

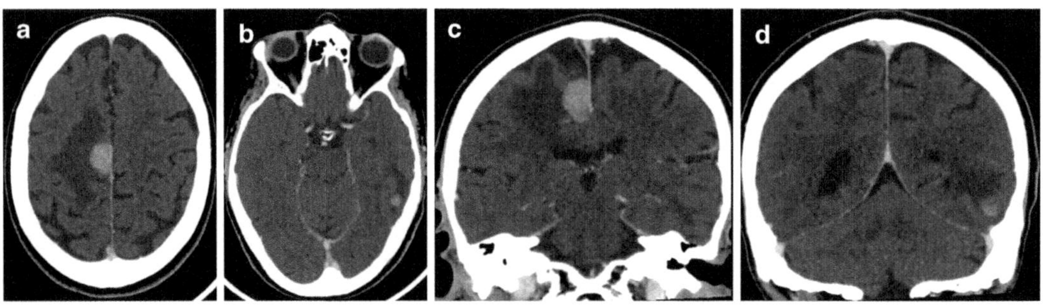

Fig. 2 Brain metastases. (**a**, **b**). Axial post-contrast CT images; (**c**, **d**). Coronal post-contrast CT images. Presence of an avid enhancing mass adherent to the right side of the falx cerebri (**a**, **c**), simulating a meningioma, surrounded by extensive vasogenic edema; the coexistence of a left temporal enhancing nodular lesion, surrounded by edema (**b**, **d**) orient toward a diagnosis of brain metastases

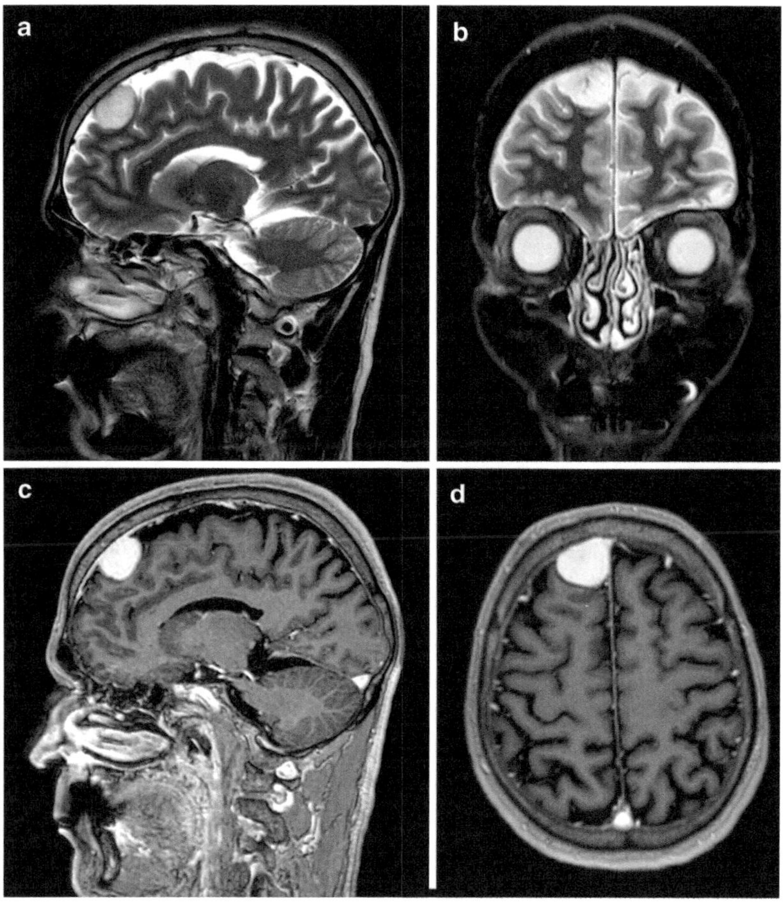

Fig. 3 Solitary fibrous tumor. (**a**) T2-weighted axial section; (**b**) STIR coronal section; (**c**, **d**) post-contrast T1-weighted sagittal and axial sections. Presence of an extra-axial lesion located at the right frontal convexity, showing hyperintense signal with internal flow-voids on T2-weighted images (**a**, **b**), and avid enhancement after contrast medium injection (**c**, **d**)

3.1.4 Glioblastoma/Gliosarcoma

Although significantly different [9, 10], in literature there are cases of documented glioblastomas exhibiting imaging characteristics mimicking meningioma. Glioblastomas are intra-axial masses with irregular margins, showing heterogeneous contrast enhancement patterns with a peripheral enhancement indicative of central necrotic/hemorrhagic regions; unlike meningiomas, bony changes are typically not present. Gliosarcoma (a variant of glioblastoma with IDH-wild type phenotype) may have some radiological characteristics that can help to distinguish it from meningiomas [11]. These encompass clearly defined margins, solid-cystic elements, the "salt and pepper" sign (a crescent-shaped region of enhancement at the interface of solid and cystic components), and an irregular rim or peropheral enhancement.

3.1.5 Pleomorphic Xanthoastrocytoma

Pleomorphic xanthoastrocytoma (PXA) shares some MRI characteristics with meningioma, mostly represented by the presence of a dural tail. The typical presentation of PXA comprises a cystic mass situated superficially in the supratentorial region, with a mural nodule positioned close to the leptomeninges. However, up to half of PXAs may not exhibit cystic changes, and intratumoral calcifications are infrequent.

Nonetheless, PXA is a differential diagnosis to be considered in cases involving young patients [2].

3.1.6 Skull Base Chordomas

Chordomas, arising from remnants of the notochord, can mimic meningiomas on imaging when located at the skull base, due to their dural attachment.

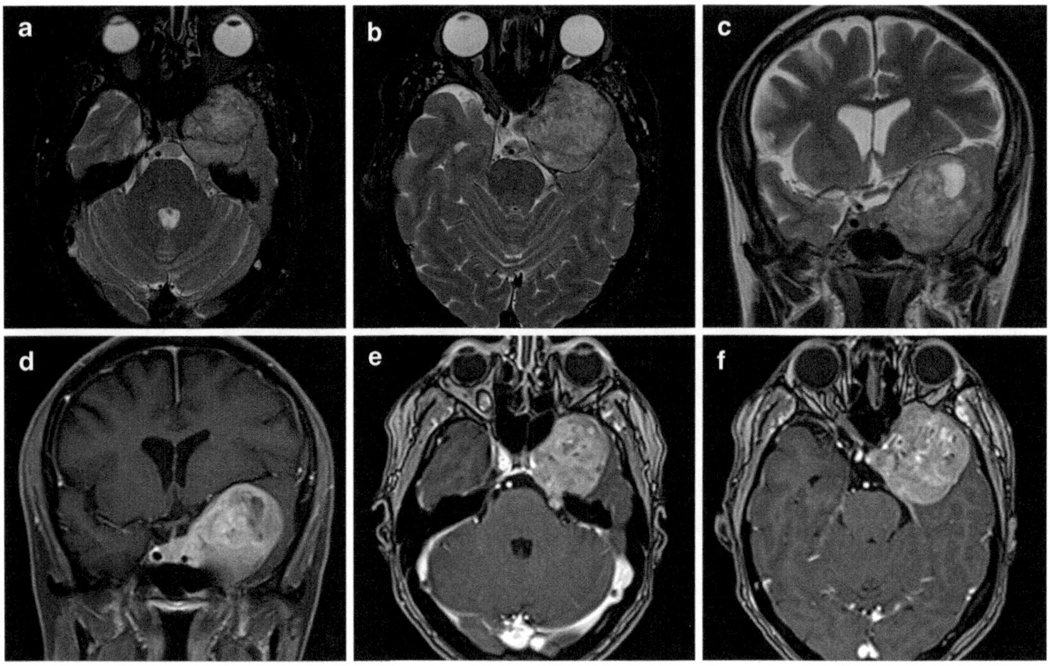

Fig. 4 Trigeminal schwannoma. (**a–c**) T2-weighted axial and coronal sections; **d–f**) post-contrast T1-weighted coronal and axial sections. Large extra-axial tumor in the left middle cranial fossa, extending posteriorly along the cisternal trigeminal nerve and anteriorly toward the orbit, showing heterogeneous signal on T2-weighted images (**a**, **b**), and enhancement after contrast medium injection (**c**, **d**), without dural tail sign (**c**, **d**)

The clival region represents the second most common location, accounting for approximatively 30–35% of cases [12]. Chordomas present as well-circumscribed, midline masses with osseous erosion and heterogeneous enhancement. Furthermore, they typically exhibit very high T2 signal intensity and bone erosion and irregular intratumoral calcifications, thought to represent sequestra of normal bone, and variable intralesional hemorrhage, all characteristics aiding in the differentiation from meningiomas [2].

3.1.7 Lymphoma

Primary central nervous system (CNS) lymphomas can occasionally mimic meningiomas, particularly when involving the dura mater or leptomeninges, presenting as a solitary mass or plaque-like thickening [13, 14]. Lymphomas usually appear as CT hyperdense, T1-and T2-weighted hypointense lobulated mass, with a lower apparent diffusion coefficient (ADC) compared to other CNS neoplasm, with a strong contrast enhancement, because of their high cellularity and high nuclear-to-cytoplasm ratio [14]. Calcification and bone hyperostosis are usually absent.

3.1.8 Erdheim-Chester Disease

Erdheim-Chester disease is a clonal histiocytosis with chronic inflammation and variable fibrosis. It may involve the CNS and meninges with or without systemic lesions. Dura is involved in 15–25% of cases, presenting with dural thickening, a meningioma-like mass, or a combination of both [15, 16]. Lesions are usually hyperdense on CT, with nodular or diffuse appearance, iso-to hypointense on T2-weighted images, and show a strong and homogeneous contrast enhancement. Other involved CNS regions include the brainstem and the cerebellum, spinal cord, and hypothalamus. A diffuse calvarial or paranasal sinus wall osteosclerosis or thickening is almost always present in patients with intracranial disease [15, 16].

3.1.9 Rosai-Dorfman Disease

Rosai-Dorfman disease similarly may involve the CNS or the meninges with or without systemic lesions [17, 18]. Lesions are usually well-circumscribed, CT hyperdense, iso/hypointense on T1-weighted images, hypointense on T2-weighted images, show homogeneous contrast enhancement and are more likely to have erosion of adjacent bone in absence of calcification or hyperostosis [17, 18]. Lesions are usually hypovascular.

3.1.10 Meningeal Sarcomas

Primary meningeal sarcomas are rare malignant tumors arising from mesenchymal tissue within the meninges. They can mimic meningiomas on imaging, particularly when presenting as enhancing extra-axial masses. However, meningeal sarcomas tend to demonstrate infiltrative growth, with ill-defined margins and surrounding edema [2].

3.1.11 Plasmacytoma

Solitary plasmacytomas are rare neoplasms that originate from plasma cells. Intracranial involvement in plasmacytoma is rare and may involve the cranial vault or the brain parenchyma. These tumors may mimic meningiomas since they present as extra-axial masses with dural attachment on imaging studies, hyperdense on CT, with a variable signal intensity on T2-weighted MRI imaging, and strong and homogeneous contrast enhancement. Plasmacytomas can present as highly vascular mass with pathological vessels; a tumor blush may be seen on angiography [2].

3.1.12 Circumscribed Meningeal Melanocytic Neoplasms

Circumscribed meningeal melanocytic neoplasms arise from the leptomeningeal melanocytes, with a predilection for the spinal canal, mostly cervical and thoracic, and posterior fossa. They include well-differentiated (e.g., melanocytoma), and aggressive malignant (e.g., meningeal melanoma) neoplasms. They usually show contrast enhancement and appear T1-weighted hyperintense and T2-weighted hypointense, due to high content of melanin or hemorrhage, and hyperdense on CT [19]. Bone hyperostosis and calcification are usually absent.

3.1.13 Other Neoplastic Mimics

Pituitary adenomas and sometimes craniopharyngiomas can mimic parasellar meningiomas on imaging.

Pituitary macroadenomas should be differentiated from meningiomas in cases of dominant suprasellar extension, homogeneous enhancement, cavernous sinus extension, and, rarely, dural tail sign; the key imaging differentiating features include the sellar epicenter of the tumor with sellar expansion, and nonvisualization of the pituitary gland apart from the mass [20].

Sinonasal tumors could be considered in the differential diagnosis of anterior skull base meningiomas. Esthesioneuroblastomas originate from the olfactory epithelia, and may present as dumbbell-shaped intracranial and extracranial nasal masses through the cribriform plate [2]; they show homogeneous intense enhancement and should be differentiated from olfactory groove or extracranial nasal meningiomas (Fig. 5).

3.2 Non-neoplastic Conditions

3.2.1 Neurosarcoidosis

Neurosarcoidosis, a rare presentation of systemic sarcoidosis impacting CNS, can display imaging features that resemble those of meningiomas [2]. These features may include single or multiple enhancing, tumor-like extra-axial dural-based masses, along with perilesional edema and the dural tail sign. However, dural involvement without coexisting extraneural manifestations is less frequent. Meningeal lesions typically exhibit a very low T2-weighted signal, likely due to the presence of fibrocollagenous tissue. Additional MRI findings associated with neurosarcoidosis may encompass intraparenchymal lesions, leptomeninges, hypothalamus, cranial nerve, and pituitary involvement, as well as vasculitis-like lesions.

3.2.2 IgG4-Related Inflammatory Pseudotumors

Inflammatory pseudotumors are part of the broader spectrum of systemic IgG4-related disease [21, 22]. Initially thought to be a type of pancreatitis, IgG4-related disease has expanded to include disorders affecting various organs, such as the CNS, with cases of hypophysitis, pachymeningitis, and inflammatory pseudotumor

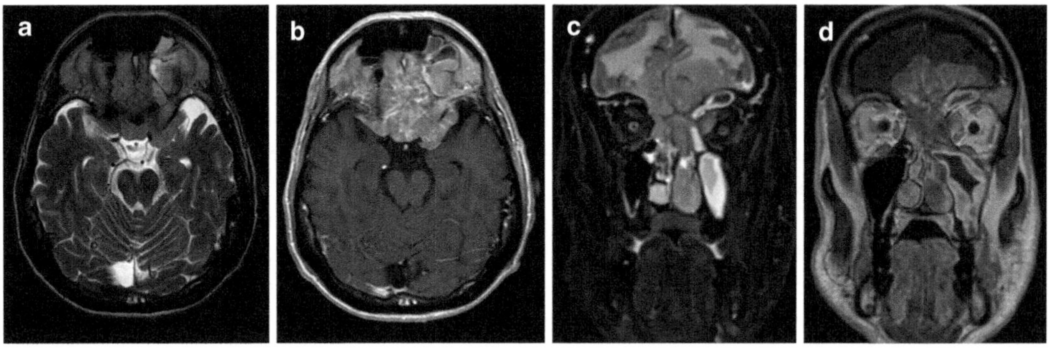

Fig. 5 Esthesioneuroblastoma. (**a**) T2-weighted axial section; (**b**) post-contrast T1-weighted axial sections; (**c**) STIR coronal section; (**d**) post-contrast T1-weighted coronal section. Strongly enhancing extra-axial mass centered on the fovea ethmoidalis, and extending toward the sphenoid wings; coronal sections show the extension of the tumor from the left nasal cavity and the ethmoid bone through the cribriform plate into the anterior cranial fossa, and the invasion of the left orbital cavity

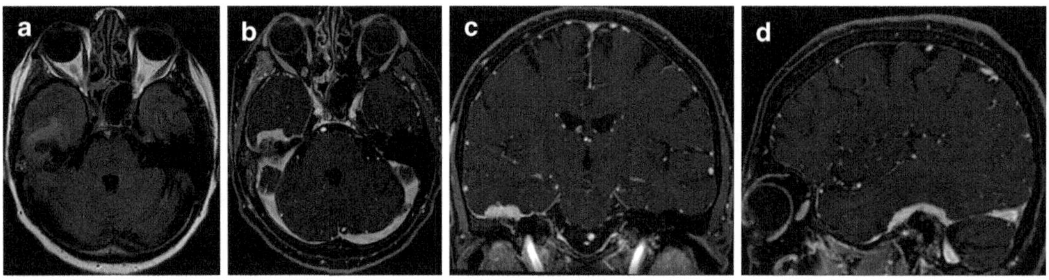

Fig. 6 Inflammatory pseudotumor of the temporal bone. (**a**) Flair axial section; (**b**–**d**) post-contrast T1-weighted axial, coronal, and sagittal sections. Strong and pseudonodular enhancement of dura surrounding the petrous portion of the right temporal bone, associated with vasogenic edema

of the trigeminal nerve documented in the literature [23, 24]. Inflammatory pseudotumors can mimic meningiomas on imaging [21, 22], appearing as extra-axial contrast-enhancing masses (Fig. 6).

3.2.3 Intracranial Foreign Body

Intracranial foreign body granuloma (FBG) is a rare inflammatory response to foreign material, which can manifest either shortly after or months to years following neurosurgical procedures. It appears to primarily stem from synthetic materials intentionally left, including nonabsorbable substances like Teflon, suture threads, or cotton, as well as absorbable hemostatic materials. Similar to other treatment-related effects such as pseudoprogression and treatment-induced necrosis, the clinical and radiographic characteristics associated with intracranial FBG are difficult to distinguish from those of progressive disease [25]. This underscores its relevance as a rare but crucial differential diagnosis in neuro-oncology, with important therapeutic implications. On MRI, cases of intracranial FBG demonstrating a peripherally enhancing lesion with central diffusion restriction have been described.

3.2.4 Tuberculosis and Fungal Diseases

Tuberculosis can manifest in various extrapulmonary sites, including the CNS, where the most common lesions are leptomeningitis and tuberculomas [26]. Tuberculomas, in particular, can resemble meningiomas on imaging; they should be suspected when encountering solitary extra-axial lesions in individuals from regions with high tuberculosis prevalence.

Clinically, they often present with progressive neurological symptoms, sometimes accompanied by signs of meningeal involvement, and are more frequently seen in immunocompromised patients with signs of tuberculosis spread. These lesions tend to localize primarily at the periventricular gray-white matter junction and within the posterior fossa. Tuberculomas exhibit different appearances on CT and MRI scans, influenced by their developmental stage and the presence of caseous necrosis [27–29]. Non-caseating tuberculomas appear hypointense on T1-weighted and hyperintense on T2-weighted images, displaying intense and homogeneous contrast enhancement without restricted diffusion. Caseating tuberculomas, on the other hand, typically show hypo-or iso-signal on both T1-and T2-weighted images and are often associated with perilesional edema; they do not show restricted diffusion and exhibit homogeneous or ring enhancement. Spectroscopy can reveal a decrease in NAA/Cr with elevated lipid-lactate peaks [30]. Caseating granulomas with central liquefaction present as hypointense on T1-weighted images with a hyperintense rim and hyperintense on T2-weighted images with a hypointense rim, surrounded by vasogenic edema and exhibiting ring enhancement with variable diffusion restriction. Finally, calcified tuberculomas are usually hypointense on both T1-and T2-weighted images, without diffusion restriction or contrast enhancement.

Dural-based masses due to fungal infection are rare but can resemble meningiomas; clinical information is the key to diagnosis.

3.2.5 Cavernous Venous Malformations

Cavernous venous malformations are benign vascular lesions that account for 0.4–2% of all intracranial vascular malformations. These lesions appear as well-defined, extra-axial masses with dural attachment, and therefore resemble in appearance meningiomas [31]. Possible localizations are the cavernous sinus (Fig. 7), convexity, tentorium, cerebellopontine angle, Meckel's cave, sella turcica, and IAC, although occasionally these have been described as arising from the dura mater [32].

These lesions typically demonstrate hypo-or isointensity on T1-weighted images and hyperintensity on T2-weighted images, appearing more hyperintense than neningiomas, with some intratumor linear hypointensity that is generally washed out by homogenous enhancement after intravenous contrast medium injection, likely representing thin walled vessels within the vascular malformation [31]. Some studies have observed a hyperintense peripheral enhancement leading to a central hyperintensity pattern with timed contrast administration that is classically seen in hepatic cavernous hemangiomas.

3.2.6 Meningioangiomatosis

Meningioangiomatosis is a rare benign hamartomatous lesion primarily affecting the leptomeninges and superficial cerebral cortex [33]. It typically originates from the cortex and may manifest as a single or multiple lesions. The most common clinical manifestations include headaches and seizures. Radiologically, meningioangiomatosis can resemble meningioma, presenting as a dural-based mass with a dural tail sign and contrast enhancement. Typically, it appears as a gyriform cortical/subcortical hyperdense lesion with punctate calcifications on CT scans, while on MRI it typically shows a gyriform T2 hypointense signal with minimal T2/FLAIR hyperintensity. It displays varying degrees of calcification, cystic degeneration, and enhancement, with a preference for occurring in the frontal and temporal lobes [34].

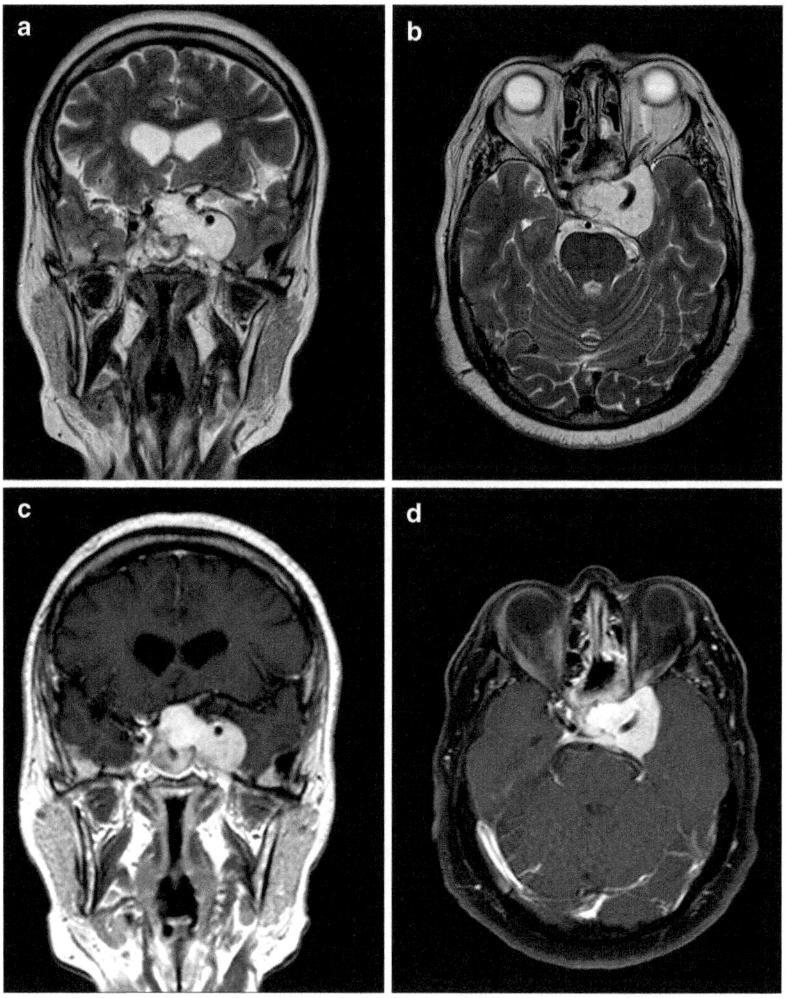

Fig. 7 Cavernous sinus hemangioma. (**a**, **b**) T2-weighted coronal and axial sections; (**c**, **d**) post-contrast T1-weighted coronal and axial sections. Presence of a well-limited lobulated mass centered at the left cavernous sinus, extending to the sellar and suprasellar region, showing a high, homogeneous signal on T2-weighted images (**a**, **b**), and intense and homogeneous contrast medium enhancement

4 Conclusions

While meningiomas are the most common extra-axial tumors encountered in clinical practice, several entities can mimic their imaging appearance, posing diagnostic challenges not only radiologically but also intraoperatively [35]. Careful evaluation of clinical history and imaging characteristics is essential for accurate diagnosis and appropriate management. Collaboration between neuroradiologists, neurosurgeons, and oncologists is crucial in navigating these diagnostic dilemmas and ensuring optimal patient care.

References

1. Lyndon D, Lansley JA, Evanson J, Krishnan AS. Dural masses: meningiomas and their mimics. Insights Imaging. 2016;10(1):11. https://doi.org/10.1186/s13244-019-0697-7. PMID: 30725238; PMCID: PMC6365311.
2. Ahmed M, Lee JH, Masaryk TJ. Meningiomas: Imaging Mimics. In: Lee JH, editor. Meningiomas. London: Springer; 2009. https://doi.org/10.1007/978-1-84628-784-8_8.
3. Starr CJ, Cha S. Meningioma mimics: five key imaging features to differentiate them from meningiomas. Clin Radiol. 2017;72(9):722–8.

4. Di Salle F, Esposito F, Elefante A, Scarabino T, Volpicelli A, Cirillo S, Elefante R, Seifritz E. High field functional MRI. Eur J Radiol. 2003;48(2):138–45. https://doi.org/10.1016/j.ejrad.2003.08.010.
5. Cocozza S, Russo C, Pontillo G, Ugga L, Macera A, Cervo A, De Liso M, Di Paolo N, Ginocchio MI, Giordano F, Leone G, Rusconi G, Stanzione A, Briganti F, Quarantelli M, Caranci F, D'Amico A, Elefante A, Tedeschi E, Brunetti A. Is advanced neuroimaging for neuroradiologists? A systematic review of the scientific literature of the last decade. Neuroradiology. 2016;58(12):1233–9. https://doi.org/10.1007/s00234-016-1761-3. Epub 2016 Nov 8.
6. Nayak L, Abrey LE, Iwamoto FM. Intracranial dural metastases. Cancer. 2009;115(9):1947–53. https://doi.org/10.1016/j.crad.2017.05.002.
7. Da Silva AN, Schiff D. Dural and skull base metastases. Cancer Treat Res. 2007;136:117–41.
8. Weon YC, Kim EY, Kim HJ, Byun HS, Park K, Kim JH. Intracranial solitary fibrous tumors: imaging findings in 6 consecutive patients. AJNR Am J Neuroradiol. 2007;28(8):1466–9.
9. de Robles P, Fiest KM, Frolkis AD, Pringsheim T, Atta C, St Germaine-Smith C, Day L, Lam D, Jette N. The worldwide incidence and prevalence of primary brain tumors: a systematic review and meta-analysis. Neuro Oncol. 2015;17(6):776–83. https://doi.org/10.1093/neuonc/nou283. Epub 2014 Oct.
10. Ostrom QT, Cioffi G, Gittleman H, Patil N, Waite K, Kruchko C, Barnholtz-Sloan JS. CBTRUS statistical report: primary brain and other central nervous system tumors diagnosed in the United States in 2012-2016. Neuro-Oncology. 2019;21(Suppl 5):v1–v100.
11. Zhang BY, Chen H, Geng DY, et al. Computed tomography and magnetic resonance features of gliosarcoma: a study of 54 cases. J Comput Assist Tomogr. 2011;35(6):667–73.
12. Elefante A, Caranci F, Del Basso De Caro ML, Peca C, Guadagno E, Severino R, Mariniello G, Maiuri F. Paravertebral high cervical chordoma. A case report. Neuroradiol J. 2013;26(2):227–32. https://doi.org/10.1177/197140091302600214. Epub 2013 May 10. PMID: 23859247; PMCID: PMC5228733.
13. Jahnke K, Schilling A, Heidenreich J, et al. Radiologic morphology of low-grade primary central nervous system lymphoma in immunocompetent patients. AJNR Am J Neuroradiol. 2005;26(10):2446–54.
14. Iwamoto FM, Abrey LE. Primary dural lymphomas: a review. Neurosurg Focus. 2006;21(5):E5.
15. Lachenal F, Cotton F, Desmurs-Clavel H, et al. Neurological manifestations and neuroradiological presentation of Erdheim-Chester disease: report of 6 cases and systematic review of the literature. J Neurol. 2006;253(10):1267–77.
16. Drier A, Haroche J, Savatovsky J, et al. Cerebral, facial, and orbital involvement in Erdheim-Chester disease: CT and MR imaging findings. Radiology. 2010;255(2):586–94.
17. Huang BY, Zong M, Zong WJ, Sun YH, Zhang H, Zhang HB. Intracranial Rosai-Dorfman disease. J Clin Neurosci. 2016;32:133–6.
18. Raslan OA, Schellingerhout D, Fuller GN, Ketonen LM. Rosai-Dorfman disease in neuroradiology: imaging findings in a series of 10 patients. AJR Am J Roentgenol. 2011;196(2):W187–93.
19. Chen CJ, Hsu YI, Ho YS, Hsu YH, Wang LJ, Wong YC. Intracranial meningeal melanocytoma: CT and MRI. Neuroradiology. 1997;39(11):811–4.
20. Cappabianca P, Cirillo S, Alfieri A, D'Amico A, Maiuri F, Mariniello G, Caranci F, de Divitiis E. Pituitary macroadenoma and diaphragma sellae meningioma: differential diagnosis on MRI. Neuroradiology. 1999;41:22–6. https://doi.org/10.1007/s002340050698.
21. Thompson A, Whyte A. Imaging of IgG4-related disease of the head and neck. Clin Radiol. 2018;73(1):106–20.
22. Toyoda K, Oba H, Kutomi K, et al. MR imaging of IgG4-related disease in the head and neck and brain. AJNR Am J Neuroradiol. 2012;33(11):2136–9.
23. Conforti R, Porto AM, Capasso R, Cirillo M, Fontanella G, Salzano A, Fabrazzo M, Cappabianca S. Magnetic resonance imaging of a transient splenial lesion of the corpus callosum resolved within a week. Radiography. 2016;22(1):97–9, ISSN 1078-8174. https://doi.org/10.1016/j.radi.2015.03.001.
24. Conforti R, Cirillo M, Sardaro A, Caiazzo G, Negro A, Paccone A, Sacco R, Sparaco M, Gallo A, Lavorgna L, Tedeschi G, Cirillo S. Dilated perivascular spaces and fatigue: is there a link? Magnetic resonance retrospective 3 tesla study. Neuroradiology. 2016;58(9):859–66. https://doi.org/10.1007/s00234-016-1711-0. Epub 2016 Jul 16.
25. Capalbo D, Elefante A, Spagnuolo MI, Mazza C, Betterle C, Pignata C, Salerno M. Posterior reversible encephalopathy syndrome in a child during an accelerated phase of a severe APECED phenotype due to an uncommon mutation of AIRE. Clin Endocrinol (Oxf). 2008;69(3):511–3. https://doi.org/10.1111/j.1365-2265.2008.03206.x.Epub.
26. Patkar D, Narang J, Yanamandala R, Lawande M, Shah GV. Central nervous system tuberculosis: pathophysiology and imaging findings. Neuroimaging Clin N Am. 2012;22(4):677–705.
27. Aggarwal A, Patra DP, Gupta K, Sodhi HB. Dural tuberculoma mimicking meningioma: a clinicoradiologic review of dural en-plaque lesions. World Neurosurg. 2016;88:686.e1–7.
28. Adachi K, Yoshida K, Tomita H, Niimi M, Kawase T. Tuberculoma mimicking falx meningioma—case report. Neurol Med Chir (Tokyo). 2004;44(9):489–92.
29. Boutarbouch M, Arkha Y, Gana R, El Maquili MR, Bellakhdar F. Tuberculoma of the cavernous sinus mimicking a meningioma: case report and review of the literature. J Neurol Sci. 2009;278(1–2):123–6.

30. Khanna PC, Godinho S, Patkar DP, Pungavkar SA, Lawande MA. MR spectroscopy-aided differentiation: "giant" extra-axial tuberculoma masquerading as meningioma. AJNR Am J Neuroradiol. 2006;27(7):1438–40.
31. Vogler R, Castillo M. Dural cavernous angioma: MR features. AJNR Am J Neuroradiol. 1995;16:773–5.
32. Rosso D, Lee DH, Ferguson GG, Tailor C, Iskander S, Hammond RR. Dural cavernous angioma: a preoperative diagnostic challenge. Can J Neurol Sci. 2003;30:272–7.
33. Makary MS, Kobalka P, Giglio P, Slone HW. Meningioangiomatosis: clinical, imaging, and histopathologic characteristics. J Clin Imaging Sci. 2020;12(10):36. https://doi.org/10.25259/JCIS_39_2020. PMID: 32637227; PMCID: PMC7332468.
34. Kashlan ON, LaBorde DV, Davison L, Saindane AM, Brat D, Hudgins PA, Gross RE. Meningioangiomatosis: A case report and literature review emphasizing diverse appearance on different imaging modalities. Case Rep Neurol Med. 2011;2011:361203, 8. https://doi.org/10.1155/2011/361203.
35. Ghosal N, Dadlani R, Gupta K, Furtado SV, Hegde AS. A clinicopathological study of diagnostically challenging meningioma mimics. J Neuro-Oncol. 2012;106(2):339–52. https://doi.org/10.1007/s11060-011-0669-3. Epub 2011 Aug 3

Radiomics in Meningiomas: Pathological and Biomolecular Correlation

Lorenzo Ugga, Teresa Perillo, Serena Capasso, Davide Negroni, and Renato Cuocolo

1 Introduction

Recently, there has been a growing interest in radiomics in all fields of medicine. Radiomics is based on the extraction and modeling of high-dimensional, quantitative features obtained from medical pictures. This strategy has gained importance as it can be used for precision medicine [1]. The association between previously identified biological variables of interest and measurable picture gray level heterogeneity inside a lesion serves as evidence for the theory supporting the use of radiomics in Radiology. Therefore, radiomics enables identification of non-invasive imaging biomarkers which could replace invasive pathological or genetic lesion profiling techniques. Furthermore, completely new radiomics biomarkers with diagnostic or prognostic significance (such as radiomics signatures) may be created [2]. This method may reveal tumoral characteristics that are missed by qualitative or conventional quantitative evaluations, and they may offer vital data for decision assistance.

Radiomics workflow typically begins with picture acquisition and preprocessing, then moves on to lesion segmentation and images characteristics extraction [3] (Fig. 1).

Preprocessing phase is crucial as it affects reproducibility. Variations in image acquisition parameters, different scanner vendors or technology bring noise into radiomics features, possibly causing biased models [4]. Therefore, image preprocessing techniques are used to reduce preprocessing variations, such as denoising techniques and gray level normalization [5]. Expert groups are creating common reference standards for features and workflow aimed to standardize radiomics. For instance, the Image Biomarker Standardization Initiative (IBSI) has created guidelines for boosting repeatability and enabling translation of radiomic investigations to the clinical setting [6].

Morphologic characteristics and first-, second-, and higher-order statistics are included in radiomics data. Morphologic features characterize the region of interest's geometric properties, including volume, surface area, compactness, and sphericity [7]. Gray level histogram-derived features are included in first-order statistics, whereas second-order characteristics (texture features) represent the spatial variation in pixel intensities.

L. Ugga (✉) · S. Capasso · D. Negroni
Department of Advanced Biomedical Sciences, University of Naples "Federico II", Naples, Italy
e-mail: lorenzo.ugga@unina.it

T. Perillo
Department of Radiology, CTO Hospital, Azienda Ospedaliera dei Colli, Naples, Italy

R. Cuocolo
Department of Medicine, Surgery and Dentistry, University of Salerno, Baronissi, Italy

F. Maiuri, M. Del Basso De Caro (eds.), *Meningiomas*,
https://doi.org/10.1007/978-3-031-76680-0_12

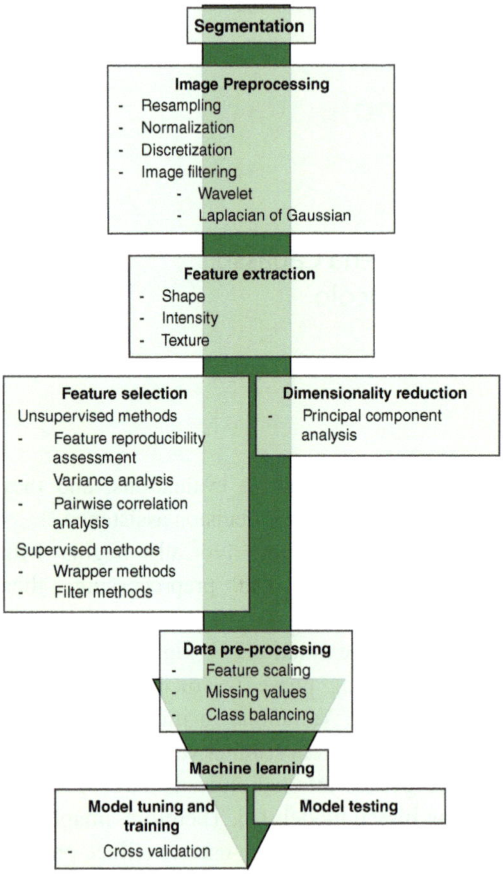

Fig. 1 Radiomics workflow

In the setting of radiomics, standard statistical procedures are often not preferred to perform the analysis due to the vast amount of data to be studied. Conversely, artificial intelligence (AI)-based data analysis methods show better performance for this kind of work. A common feature of radiomics pipelines is the integration of machine learning (ML) techniques and classical statistics. The former are frequently employed for feature selection and data preprocessing, while the latter is typically used for regression modeling or feature selection and classification.

ML tools are classified based on the type of learning [8]:

- Supervised learning is the most widely used ML method and it is based on labeled training dataset (ground truth) as algorithm input. Supervised learning encompasses regression and classification methods. Common classification algorithm types include ensemble approaches, decision trees, support vector machines, and linear classifiers.
- Unsupervised learning is an algorithm that identifies hidden patterns in data and makes predictions without needing a labeled database. Consequently, it requires a huge amount of training data and the post-hoc interpretation of the resulting clusters may be highly complicated.
- Reinforcement learning is a process that involves trial and error and is dependent on feedback loops (positive and/or negative reinforcement). This method has been widely used in the domains of game theory, robotics, and telecommunications.

The above-mentioned types of learning can be used separately or in combination (for instance semi-supervised learning).

A branch of ML known as deep learning (DL) uses several tiers of data processing nodes arranged in a hierarchy of increasing complexity and abstraction to gradually extract higher-level features from the raw input data. It is built on neural networks, whose circuits of linked artificial neurons (or nodes), are based on the structure of the human brain.

Any model derived from radiomics data should go through extensive validation, regardless of the kind of training and algorithm design. Better estimation of the model's performance on fresh data can be achieved by different techniques, such as resampling the data using k-fold cross-validation. This method divides the training data into k folds, which are then used repeatedly to train the model (k-1 folds of data), with the final subset being used for validation. It is also important that the final model is validated outside to evaluate its performance in the clinical setting.

Meningiomas are the most common primary tumors of the central nervous system, more common in females with the median age at diagnosis being 66 years old [9]. Arising from meningothelial arachnoid cap cells connected to the dura or choroid plexus, they are generally benign lesions.

The most current guidelines (WHO 2021) established 3 grades [10]:

- WHO grade I (70–75%): benign histology.
- WHO grade II (20–25%): locally aggressive meningiomas.
- WHO grade III (1–6%): meningiomas with malignant behavior.

Several predisposing variables, such as radiation exposure, hormonal therapy, genetic abnormalities, and family history, raise the risk of occurrence [11].

While some patients may be asymptomatic, others may suffer from headache and different neurological signs due to compression of adjacent structures [12]. Magnetic resonance imaging is the imaging technique of choice to diagnose and study meningioma [13]. Computed tomography (CT) can be used in the emergency setting and to better evaluate calcifications and bony alterations. The development of novel prognostic indicators and possible treatment targets has been possible thanks to understanding the genomic and epigenetic landscapes of Meningiomas [14].

In the past few years, radiomics has shown multiple promising applications in diagnosis and treatment of meningioma. In this chapter, we describe the application of ML in management and treatment of meningioma.

2 Image Acquisition, Segmentation, and Feature Extraction

Radiomics workflow begins with image acquisition and reconstruction [12]. This phase includes gathering of raw data and annotations of imaging parameters, such as field of view, echo time, and repetition time for MRI and tube voltage and current in CT [15]. Regarding reconstruction, standardization is important as different algorithms affect final analysis [16].

This phase is followed by segmentation, which is a crucial phase in radiomics workflow since features are extracted through segmented regions of interest (ROI). In addition, volumetric assessment may be useful for therapy planning and management [17]. Fully automated and semi-automatic segmentation techniques are faster and usually associated with reduced inter-observer variability. On the other hand, manual segmentation is more detailed, time-consuming, and susceptible to inter-observer variability [3].

Feature extraction consists of decoding high-dimensional visual data into quantitative outputs [16]. Feature extraction patterns are classified as either having human commands or not [18]. The latter are usually based on convolutional neural networks, and feature extraction, selection, and classification can take place across several layers.

There are two categories of features in radiomics, which are called semantic or agnostic. The first one includes form, size, and placement; the second category is composed of quantitative descriptors which are derived theoretically and can be first-second-, and higher-order [19].

AI has been extensively used to create models for automatic Meningioma segmentation. For instance, Laukamp et al. used a multiparametric DL model for fully automated detection and segmentation of Meningiomas, showing a significant correlation with manual segmentation [3]. They used FLAIR and contrast-enhanced T1-weighted images, and the average Dice coefficients for overall tumor volume was 0.81 ± 0.10 (range: 0.46–0.93). Similarly, Chen et al. produced Dice scores of 0.920 ± 0.009 using a modified U-Net convolutional neural network segmentation model which used contrast-enhanced T1-weighted images [20]. Consistent lesion segmentation is also crucial before radiation therapy and radiosurgery to establish the gross target volume. In this setting, Florenz et al. proposed the use of multiparametric MRI texture-based features to improve the differentiation between tumor and edema for gross target volume definition (AUC > 0.71) [21].

3 Grading and Molecular Status

Determination of meningioma grading through traditional methods can be challenging due to the need for invasive biopsy and potential sampling errors. Recently, radiomics has shown great promise in non-invasive grading of meningiomas as it can analyze tumor's texture, shape, and intensity, providing a detailed quantitative assessment that can reflect the underlying pathology. This approach potentially enhances the accuracy of preoperative grading, leading to more tailored treatment strategies. It can also help in monitoring tumor progression and response to therapy. Numerous studies have studied the applications of ML tools to predict tumor grading from presurgical images. For instance, Khanna et al. used a ML-based tool to extract features from preoperative MRI to predict Ki-67 indexes, which relates to cellular replication [22]. It showed high accuracy, with areas under the curve (AUC) of 0.83–0.84. Zhao et al. created and evaluated a nomogram that used clinical markers and multiparametric MRI to determine the grading of Meningioma [23]. They used a dataset of 568 pathologically confirmed meningiomas, a support vector technique and a nomogram including both clinical markers and radiomics properties. In terms of grading prediction, multiparametric MRI features combining contrast-enhanced T1 and T2 sequences showed a pooled AUC of 0.885 and 0.860 in the training set and test groups, respectively. When they combined radiomics score, peritumoral edema volume, and gender, the radiomic tool showed an AUC of 0.912 in predicting grading. Similarly, Park et al. assessed the predictive power of a tool based on radiomic and clinical imaging parameters for meningioma histologic grading [24]. The dataset was composed of 212 preoperative MRI, and they chose five radiomics features to create the model. The algorithm combining both clinical and imaging information showed high AUC and precision (0.86 and 0.73, respectively). Herrgott et al. used ML to identify DNA methylation markers which could be used to predict prognosis and recurrence in Meningiomas. They examined DNA methylation levels in the serum and plasma of 155 patients with Meningioma comparing them with those of other CNS tumors and non-tumor samples. They demonstrated that liquid biopsy is potentially a reliable and non-invasive method for diagnosis and prognosis prediction in patients with Meningiomas. Similarly, radiomics can be used to detect specific DNA methylation patterns which are associated with specific tumors, as shown [25].

Duan et al. compared the accuracy of 2D and 3D T1-weighted images radiomics characteristics to predict Meningioma grading [26]. The accuracy of the 2D and 3D characteristics in predicting meningioma grade was similar thus, given the labor and time constraints, 2D sequences could be chosen for radiomics research in Meningiomas. Yang et al. used a T1 weighted image-based DL model to distinguish between low-and high-grade meningiomas [27]. They discovered that the combination of a DL and handcrafted radiomics model performed better than either model alone. Similarly, Cai et al. used AI to create a tool for prediction of meningioma grading from preoperative MRI, which showed high accuracy (AUC = 0.78) [28]. Apparent diffusion coefficient (ADC) sequences are also important to predict grading, as demonstrated by the ML tool developed by Lu et al. [29].

Since TERT promotor mutations in meningiomas are found in anaplastic tumors, Akkurt et al. developed a ML tool which used preoperative MRI to predict this mutation (AUC 0.92–0.94) [30]. They discovered that low ADC values were linked to high-grade Meningioma and low sphericity to higher risk of local failure and lower overall survival. They also found that integrating clinical, radiologic, and radiomic data into prognostic models led to increased accuracy in predicting meningioma grading, local failure, and overall survival (AUC = 0.78, 0.75, and 0.78, respectively), in contrast to models that solely rely on clinical characteristics.

Regarding specific radiomic features, high-grade meningiomas are associated with more spherical disproportion than low-grade tumors [31]. Low sphericity is also linked higher rate of local recurrence and lower overall survival rate.

Apart from radiomic characteristics, another crucial element influencing prediction success is the modeling algorithm [32]. In particular, random forest, logistic regression, naïve Bayes, support vector machine, eXtreme gradient boosting, multilayer perceptron, and linear discriminant analysis are the classification methods currently used for grade prediction [33]. Among the above-mentioned modeling algorithm, Random Forest has several advantages such as accurate high-dimensional data analysis and low overfitting [34].

4 Differential Diagnosis

Radiomics can be used to distinguish meningioma from other conditions which have similar imaging findings, such as solitary fibrous tumors, craniopharyngiomas, and malignant dural-based tumors.

Solitary fibrous tumor/hemangiopericytoma are highly vascularized tumors that may easily and extensively bleed after resection. On MRI imaging, these lesions share similar characteristics with the angiomatous subtype of meningioma. Some studies explored the ability of radiomics in improving differentiations of these two conditions. For instance, Kanazawa et al. used a texture analysis approach to evaluate differences in ADC maps, finding that it was higher in case of meningiomas [35]. Lie et al. trained four different support vector machine classifiers in analyzing FLAIR, contrast-enhanced T1 and diffusion-weighted imaging (DWI) sequences, showing the highest AUC value (0.90) when using a contrast-enhanced T1-weighted sequence-based classifier [36]. In this setting, Wei et al. developed an integrated diagnostic approach and Meningioma Diagnostic Tool through combination of clinical and radiological data and radiomics signature obtained using a ML algorithm. The study showed a very high diagnostic accuracy with an AUC of 0.91 in distinguishing meningioma from hemangiopericytoma [37]. Similarly, Fan et al. developed a diagnostic model capable of distinguishing the two lesions with a remarkable accuracy in the validation set (AUC = 0.91) [38].

Craniopharyngiomas show similar imaging features to meningiomas of the sellar region. In this regard, the study of Tian et al. investigated the possibility of an easier differential diagnosis using both qualitative and quantitative MRI features, selecting four qualitative and three quantitative parameters that showed significant difference between the two neoplasms. However, only one qualitative parameter (cystic alteration) could be considered as a diagnostic independent predictor [39].

Regarding dural-based tumors of the anterior cranial fossa, Zhang et al. tested the accuracy of ML models in differentiating skull base lesions, using both clinical and radiomic features. With a linear discriminant analysis as the classification algorithm, the study reported an AUC of 0.81 in the testing group [40].

Up to 2% of malignant dural-based tumors share similar imaging features to meningiomas. In this setting, Panyaping et al. investigated different MRI features that can improve the process of differential diagnosis [41]. In particular, they discovered that rim enhancement, both complete or more than 50%, can be used to distinguish between meningioma and other malignant dural-based tumors with high accuracy (AUC of 0.95). In this study, exceptions were convexity and intracavernous meningiomas, which show little to none pial supply (key factor in the determination of rim enhancement). In addition, bone hyperostosis showed an excellent specificity (100%) and positive predictive value (100%), while the dural tail sign had high sensitivity and its absence yields high specificity (98%). The combination of cortical breakthrough and leptomeningeal enhancement also proved useful, with high accuracy (AUC of 0.91).

5 Tumor Consistency

Consistency of meningiomas plays a crucial role in surgical planning and influences the extent of resection [42]. In particular, softer tumors can be easily managed through cutting and suctioning and show lesser degree of recurrence and morbidity. On the other hand, firmer tumors, especially skull base meningiomas, require additional effort, technical skills, and devices, such as ultrasonic aspiration, intraoperative navigation, and electrophysiological monitoring). Therefore, prediction of meningioma consistency in the presurgical setting is crucial to plan neurosurgery.

Some recent studies have shown the utility of radiomics in this setting. For instance, Alkubeyer et al. combined supervised learning algorithms with typical descriptor reaching an AUC of 0.87 when coupling local binary patterns with k-nearest neighbor classifier [43]. Zhai et al. developed a radiomic nomogram using five different radiomic-based classifiers [44]. Logistic regression was the classifier which performed best, with an AUC of 0.96 in predicting tumor consistency. Through histogram analysis of tensor-valued diffusion MRI, Brabec et al. have reported promising results in prediction not only of meningiomas consistency, but also of grading and histological type [45]. Finally, coupling feature selecting filters with ML algorithm, Cepeda et al. reached an AUC of 0.96 and a classification accuracy of 0.94 in consistency prediction using an algorithm which used preoperative MRI (in particular T2, ADC and contrast-enhanced images) [46].

6 Prediction of Prognosis/Recurrence and Response to Radiotherapy

In 1957, Simpson et al. introduced a grading scale describing different extents of meningioma resection [47]. This classification yields a vital prognostic significance, considering parameters like residual tumor or infiltrated dura and has been extensively used in neurosurgery ever since. The presence of residual tumor tissue and/or infiltrated dura are risk factors for tumor regrowth. On conventional imaging, MRI findings correlated to progression and/or recurrence are tumor size, bone invasion and proximity to major dural venous sinuses, but they lack accuracy and consistency.

Multiple studies have explored the possible role of radiomics in predicting progression and/or recurrence rate of meningioma after surgical resection. Ko et al. developed a radiomic model which used features extracted from T2 and contrast-enhanced T1 sequences to predict progression and/or recurrence [48]. They used a support vector machine and achieved an AUC of 0.81 in predicting progression and/or recurrence in WHO I meningiomas which underwent gross total resection. According to their results, a higher support vector machine score was associated with a shorter progression-free survival.

Han et al. built a clinical-radiomics model (which performed best when compared to the radiomics model alone) that could theoretically predict the fibroblastic nature of meningiomas with an AUC of 0.84 in the training cohort and 0.76 in the validation set. Fibroblastic meningiomas tend to be more challenging during surgery and are associated with a greater risk of complications and overall worse prognosis [49]. Krahling et al. tested a radiomic model to predict the number of mitosis per 10 microscopic high-power fields of meningiomas, key factor in diagnosing and differentiating an atypical (WHO grade II) from a anaplastic meningioma (WHO grade III) [50]. They reached an AUC of 0.852 and an accuracy of 0.794 in the validation set and report the best performance for cases with zero mitoses or more than one. Similarly, Li et al. developed a radiomic model to predict biological behavior of meningiomas from the perspective of WHO grade, Ki-67 index and a combination of both using contrast-enhance T1 sequences [51]. They achieved an AUC of 0.878, 0.802, 0.884 for prediction of grading, Ki-67 index and the two of them combined respectively in the training set and an AUC of 0.807, 0.792, 0.840 in the validation cohort.

Invasion of a venous sinus is associated with a higher risk for morbidity and mortality. Predicting the risk of venous sinus invasion of meningiomas using a radiomic approach was the purpose of the study of Wang and colleagues [52]. It showed an AUC of 0.86 and 0.82 in the training and validation cohort, respectively, using a clinical-radiomics model (based on tumor size and several radiomics features).

Up to 30% of meningiomas originates from the skull base, where a complete surgical resection is unlikely to be achieved, due to the complex neurovascular anatomy of the region and the consequent risk of neurological complications [53]. In this subset, a subtotal resection is thus performed but it is associated with a higher rate of progression and recurrence. Zhang et al. examined a pool of skullbase meningiomas which underwent total or subtotal resection and developed a MRI radiomic model to predict progression and/or recurrence rate, achieving an accuracy of 0.90 and an AUC of 0.88 [54].

Radiotherapy (RT) is performed as adjuvant therapy (ART) to lower progression and/or recurrence rate for patients who suffer from a high-grade meningioma or underwent subtotal tumor resection [55]. RT represents the primary treatment for lesions lying in proximity to critical structures, such as nerves or vessels or because of poor performance status of the patient. An appropriate patient selection and allocation is required to identify lesions that respond to RT with minimal risk of side effects. Jo et al. evaluated the effect of RT in terms of radiological and radiomics features changes on routine brain MRI, showing not only reduction of maximum area and diameter of meningioma as radiological signs, but also alteration of several radiomics features [56]. Their algorithm displayed an AUC of 0.76 and 0.79 in the training and validation set respectively, thus demonstrating a positive effect of RT in the control of meningioma. Similarly, the study of Park et al. developed a combined model (built from the association of a clinicopathological and radiomics model) to select from a pool of WHO II meningiomas those with high risk of progression and/or recurrence [57]. The study showed that the cohort of patients considered as high risk by the combined model, performed best after ART in terms of survival and delayed recurrence when compared to the no-treatment cohort, also validating the role of ART in the after-surgery management.

Stereotactic radiosurgery (SRS) is an additional therapeutic option for patients considered unfit for surgery, as a complementary or definitive treatment modality for these selected patients [58]. As for conventional radiotherapy, a prediction of SRS response (in terms of volume shrinkage) is crucial. In 2017, Speckter et al. used a radiomics approach to select features which performed best in predicting positive volumetric responses in meningiomas after gamma knife radio surgery (GKRS) suing pre-surgical brain MRI and in particular T1 and T2 sequences and diffusion tensor imaging (DTI) [59]. In particular, the study displayed a significative correlation between meningioma volume reduction and DTI-derived features. In 2022, the same team developed a combined radiomics model (coupling radiomic features and Karnofsky performance status) from T1 images, reporting an AUC of 0.88 in predicting volume reduction [59].

7 Limitation and Future Perspective

While radiomics algorithms are extremely promising, they come with some drawbacks. Firstly, the majority of radiomic studies are unicentric and retrospective, leading to potential patient selection biases, with samples showing limited representativeness of the general population [60]. Secondly, lack of high-quality raw data, due to heterogeneity of patient cohorts or imaging data and small sample size, could potentially lead to misinterpretation of changes as solely due to biological effects, error rates and overfitting of radiomics models [61]. Also, the wide variety in the staging process of patients further undermines the generalizability of radiomics analyses.

Considering the aforementioned drawbacks, future efforts need to be put on standardization of the radiomic pipeline in all processes of the workflow. Studies should be prospectively designed and be comprehensive not only of imaging data but also of other parameters to predict patients' response to treatment [3]. Another step that could potentially aid a clinical translation of radiomics analyses would be the improvement of data sharing, possibly creating high-quality, public datasets capable of integrating different data, like radiomics, genomics, clinical-prognostic, neuropathological data in a view of an even more fitting precision medicine [62]. It is vital to raise awareness on technical issues affecting current radiomics research and effective clinical relevance in readers, editors and reviewers and any potential buyers of commercial solutions based on radiomics.

8 Conclusion

Radiomics is a developing and promising field, theoretically capable of providing valuable information for the best management of patients affected by meningioma and several studies have explored the feasibility of such technology in a real clinical setting. Some implementations are already commercially available (such as segmentation), but others, especially those that purposely address grading and prognostic prediction, are far from capillary utilization in the near future. In effect, these models lack reproducibility and generalizability to be included in the clinical decision support software or to be positively involved in meningioma patient management.

References

1. Lambin P, Rios-Velazquez E, Leijenaar R, et al. Radiomics: extracting more information from medical images using advanced feature analysis. Eur J Cancer. 2012;48:441–6. https://doi.org/10.1016/j.ejca.2011.11.036.
2. Huang Y, Liu Z, He L, et al. Radiomics signature: a potential biomarker for the prediction of disease-free survival in early-stage (I or II) non—small cell lung cancer. Radiology. 2016;281:947–57. https://doi.org/10.1148/radiol.2016152234.
3. Van Timmeren JE, Cester D, Tanadini-Lang S, et al. Radiomics in medical imaging—"how-to" guide and critical reflection. Insights Imaging. 2020;11:91. https://doi.org/10.1186/s13244-020-00887-2.
4. Lennartz S, O'Shea A, Parakh A, et al. Robustness of dual-energy CT-derived radiomic features across three different scanner types. Eur Radiol. 2022;32:1959–70. https://doi.org/10.1007/s00330-021-08249-2.
5. Demircioğlu A. The effect of preprocessing filters on predictive performance in radiomics. Eur Radiol Exp. 2022;6:40. https://doi.org/10.1186/s41747-022-00294-w.
6. Zwanenburg A, Vallières M, Abdalah MA, et al. The image biomarker standardization initiative: standardized quantitative radiomics for high-throughput image-based phenotyping. Radiology. 2020;295:328–38. https://doi.org/10.1148/radiol.2020191145.
7. Chuanji Z, Zheng W, Shaolv L, et al. Comparative study of radiomics, tumor morphology, and clinicopathological factors in predicting overall survival of patients with rectal cancer before surgery. Transl Oncol. 2022;18:101352. https://doi.org/10.1016/j.tranon.2022.101352.
8. Cuocolo R, Caruso M, Perillo T, et al. Machine learning in oncology: a clinical appraisal. Cancer Lett. 2020;481:55–62. https://doi.org/10.1016/j.canlet.2020.03.032.
9. Hortobágyi T, Bencze J, Varkoly G, et al. Meningioma recurrence. Open Medicine. 2016;11:168–73. https://doi.org/10.1515/med-2016-0032.
10. Louis DN, Perry A, Wesseling P, et al. The 2021 WHO classification of tumors of the central nervous system: a summary. Neuro-Oncology. 2021;23:1231–51. https://doi.org/10.1093/neuonc/noab106.
11. Barnholtz-Sloan JS, Kruchko C. Meningiomas: causes and risk factors. FOC. 2007;23:E2. https://doi.org/10.3171/FOC-07/10/E2.
12. Ogasawara C, Philbrick BD, Adamson DC. Meningioma: a review of epidemiology, pathology, diagnosis, treatment, and future directions. Biomedicines. 2021;9:319. https://doi.org/10.3390/biomedicines9030319.
13. Galldiks N, Hattingen E, Langen K-J, Tonn JC. Imaging characteristics of Meningiomas. In: Zadeh G, Goldbrunner R, Krischek B, Nassiri F, editors. Biological and clinical landscape of meningiomas. Cham: Springer; 2023. p. 21–33.
14. Bi WL, Mei Y, Agarwalla PK, et al. Genomic and Epigenomic landscape in meningioma. Neurosurg Clin N Am. 2016;27:167–79. https://doi.org/10.1016/j.nec.2015.11.009.
15. Court LE, Fave X, Mackin D, et al. Computational resources for radiomics. Transl Cancer Res. 2016;5:340–8. https://doi.org/10.21037/tcr.2016.06.17.
16. Gillies RJ, Kinahan PE, Hricak H. Radiomics: images are more than pictures, they are data. Radiology. 2016;278:563–77. https://doi.org/10.1148/radiol.2015151169.
17. Funingana I-G, Piyatissa P, Reinius M, et al. Radiomic and volumetric measurements as clinical trial endpoints—a comprehensive review. Cancers. 2022;14:5076. https://doi.org/10.3390/cancers14205076.
18. Bodalal Z, Trebeschi S, Nguyen-Kim TDL, et al. Radiogenomics: bridging imaging and genomics. Abdom Radiol. 2019;44:1960–84. https://doi.org/10.1007/s00261-019-02028-w.
19. Park JE, Kim HS. Radiomics as a quantitative imaging biomarker: practical considerations and the current standpoint in neuro-oncologic studies. Nucl Med Mol Imaging. 2018;52:99–108. https://doi.org/10.1007/s13139-017-0512-7.
20. Chen C, Cheng Y, Xu J, et al. Automatic meningioma segmentation and grading prediction: a hybrid deep-learning method. JPM. 2021;11:786. https://doi.org/10.3390/jpm11080786.
21. Florez E, Nichols T, Parker E, et al. Multiparametric magnetic resonance imaging in the assessment of primary brain tumors through radiomic features: a metric for guided radiation treatment planning. Cureus. 2018;10:e3426. https://doi.org/10.7759/cureus.3426.
22. Khanna O, Fathi Kazerooni A, Arif S, et al. Radiomic signatures of meningiomas using the Ki-67 proliferation index as a prognostic marker of clinical outcomes.

Neurosurg Focus. 2023;54:E17. https://doi.org/10.3171/2023.3.FOCUS2337.
23. Zhao Z, Nie C, Zhao L, et al. Multi-parametric MRI-based machine learning model for prediction of WHO grading in patients with meningiomas. Eur Radiol. 2023;34:2468. https://doi.org/10.1007/s00330-023-10252-8.
24. Park J, Quang L, Yoon W, et al. Predicting histologic grade of Meningiomas using a combined model of Radiomic and clinical imaging features from preoperative MRI. Biomedicines. 2023;11:3268. https://doi.org/10.3390/biomedicines11123268.
25. Herrgott GA, Snyder JM, She R, et al. Detection of diagnostic and prognostic methylation-based signatures in liquid biopsy specimens from patients with meningiomas. Nat Commun. 2023;14:5669. https://doi.org/10.1038/s41467-023-41434-z.
26. Duan C, Li N, Liu X, et al. Performance comparison of 2D and 3D MRI radiomics features in meningioma grade prediction: a preliminary study. Front Oncol. 2023;13:1157379. https://doi.org/10.3389/fonc.2023.1157379.
27. Yang L, Xu P, Zhang Y, et al. A deep learning radiomics model may help to improve the prediction performance of preoperative grading in meningioma. Neuroradiology. 2022;64:1373–82. https://doi.org/10.1007/s00234-022-02894-0.
28. Cai Z, Wong L, Wong Y, et al. Dual-level augmentation Radiomics analysis for multisequence MRI meningioma grading. Cancers. 2023;15:5459. https://doi.org/10.3390/cancers15225459.
29. Lu Y, Liu L, Luan S, et al. The diagnostic value of texture analysis in predicting WHO grades of meningiomas based on ADC maps: an attempt using decision tree and decision forest. Eur Radiol. 2019;29:1318–28. https://doi.org/10.1007/s00330-018-5632-7.
30. Akkurt BH, Spille DC, Peetz-Dienhart S, et al. Radiomics-based prediction of TERT promotor mutations in intracranial high-grade meningiomas. Cancers. 2023;15:4415. https://doi.org/10.3390/cancers15174415.
31. Morin O, Chen WC, Nassiri F, et al. Integrated models incorporating radiologic and radiomic features predict meningioma grade, local failure, and overall survival. Neuro-Oncol Adv. 2019;1:vdz011. https://doi.org/10.1093/noajnl/vdz011.
32. Parmar C, Grossmann P, Bussink J, et al. Machine learning methods for quantitative Radiomic biomarkers. Sci Rep. 2015;5:13087. https://doi.org/10.1038/srep13087.
33. Zhu Y, Man C, Gong L, et al. A deep learning radiomics model for preoperative grading in meningioma. Eur J Radiol. 2019;116:128–34. https://doi.org/10.1016/j.ejrad.2019.04.022.
34. Jain R, Lui YW. How far are we from using Radiomics assessment of gliomas in clinical practice? Radiology. 2018;289:807–8. https://doi.org/10.1148/radiol.2018182033.
35. Kanazawa T, Minami Y, Jinzaki M, et al. Preoperative prediction of solitary fibrous tumor/Hemangiopericytoma and angiomatous meningioma using magnetic resonance imaging texture analysis. World Neurosurg. 2018;120:e1208–16. https://doi.org/10.1016/j.wneu.2018.09.044.
36. Li X, Lu Y, Xiong J, et al. Presurgical differentiation between malignant haemangiopericytoma and angiomatous meningioma by a radiomics approach based on texture analysis. J Neuroradiol. 2019;46:281–7. https://doi.org/10.1016/j.neurad.2019.05.013.
37. Wei J, Li L, Han Y, et al. Accurate preoperative distinction of intracranial Hemangiopericytoma from meningioma using a multihabitat and multisequence-based Radiomics diagnostic technique. Front Oncol. 2020;10:534. https://doi.org/10.3389/fonc.2020.00534.
38. Fan Y, Liu P, Li Y, et al. Non-invasive preoperative imaging differential diagnosis of intracranial Hemangiopericytoma and angiomatous meningioma: a novel developed and validated multiparametric MRI-based Clini-Radiomic model. Front Oncol. 2022;11:792521. https://doi.org/10.3389/fonc.2021.792521.
39. Tian Z, Chen C, Zhang Y, et al. Radiomic analysis of Craniopharyngioma and meningioma in the Sellar/Parasellar area with MR images features and texture features: a feasible study. Contrast Media Mol Imaging. 2020;2020:1–9. https://doi.org/10.1155/2020/4837156.
40. Zhang Y, Shang L, Chen C, et al. Machine-learning classifiers in discrimination of lesions located in the anterior Skull Base. Front Oncol. 2020;10:752. https://doi.org/10.3389/fonc.2020.00752.
41. Panyaping T, Punpichet M, Tunlayadechanont P, Tritanon O. Usefulness of a rim-enhancing pattern on the contrast-enhanced 3D-FLAIR sequence and MRI characteristics for distinguishing meningioma and malignant Dural-based tumor. AJNR Am J Neuroradiol. 2023;44:247–53. https://doi.org/10.3174/ajnr.A7780.
42. Sauvigny T, Ricklefs FL, Hoffmann L, et al. Features of tumor texture influence surgery and outcome in intracranial meningioma. Neuro-Oncol Adv. 2020;2:vdaa113. https://doi.org/10.1093/noajnl/vdaa113.
43. AlKubeyyer A, Ben Ismail MM, Bchir O, Alkubeyyer M. Automatic detection of the meningioma tumor firmness in MRI images. XST. 2020;28:659–82. https://doi.org/10.3233/XST-200644.
44. Zhai Y, Song D, Yang F, et al. Preoperative prediction of meningioma consistency via machine learning-based Radiomics. Front Oncol. 2021;11:657288. https://doi.org/10.3389/fonc.2021.657288.
45. Brabec J, Szczepankiewicz F, Lennartsson F, et al. Histogram analysis of tensor-valued diffusion MRI in meningiomas: relation to consistency, histological grade and type. Neuro Image: Clinical. 2022;33:102912. https://doi.org/10.1016/j.nicl.2021.102912.
46. Cepeda S, Arrese I, García-García S, et al. Meningioma consistency can be defined by combining

the Radiomic features of magnetic resonance imaging and ultrasound Elastography. A pilot study using machine learning classifiers. World Neurosurg. 2021;146:e1147–59. https://doi.org/10.1016/j.wneu.2020.11.113.
47. Simpson D. The recurrence of intracranial MENINGIOMAS after surgical treatment. J Neurol Neurosurg Psychiatry. 1957;20:22–39. https://doi.org/10.1136/jnnp.20.1.22.
48. Ko C-C, Zhang Y, Chen J-H, et al. Pre-operative MRI Radiomics for the prediction of progression and recurrence in Meningiomas. Front Neurol. 2021;12:636235. https://doi.org/10.3389/fneur.2021.636235.
49. Han T, Xu Z, Geng Y, et al. Meningioma typing model construction using radiomics-based multi-parameter magnetic resonance imaging. 2022.
50. Krähling H, Musigmann M, Akkurt BH, et al. A magnetic resonance imaging based radiomics model to predict mitosis cycles in intracranial meningioma. Sci Rep. 2023;13:969. https://doi.org/10.1038/s41598-023-28089-y.
51. Li M, Liu L, Qi J, et al. MRI-based machine learning models predict the malignant biological behavior of meningioma. BMC Med Imaging. 2023;23:141.
52. Wang L, Cao Y, Zhang G, et al. A radiomics model enables prediction venous sinus invasion in meningioma. Ann Clin Transl Neurol. 2023;10:1284–95. https://doi.org/10.1002/acn3.51797.
53. Zada G, Başkaya MK, Shah MV. Introduction: surgical management of skull base meningiomas. Neurosurgical Focus. 2017;43:intro. https://doi.org/10.3171/2017.10.FocusVid.Intro.
54. Zhang Y, Chen J-H, Chen T-Y, et al. Radiomics approach for prediction of recurrence in skull base meningiomas. Neuroradiology. 2019;61:1355–64. https://doi.org/10.1007/s00234-019-02259-0.
55. Caccese M, Busato F, Guerriero A, et al. The role of radiation therapy and systemic treatments in meningioma: the present and the future. Cancer Med. 2023;12:16041–53. https://doi.org/10.1002/cam4.6254.
56. Jo SW, Kim ES, Yoon DY, Kwon MJ. Changes in radiomic and radiologic features in meningiomas after radiation therapy. BMC Med Imaging. 2023;23:164. https://doi.org/10.1186/s12880-023-01116-0.
57. Park CJ, Choi SH, Eom J, et al. An interpretable radiomics model to select patients for radiotherapy after surgery for WHO grade 2 meningiomas. Radiat Oncol. 2022;17:147. https://doi.org/10.1186/s13014-022-02090-7.
58. Jensen RL, Minshew L, Shrieve AF, et al. Stereotactic radiosurgery and radiotherapy for meningiomas: biomarker predictors of patient outcome and response to therapy. J Radiosurg SBRT. 2012;2:41–50.
59. Speckter H, Bido J, Hernandez G, et al. Pretreatment texture analysis of routine MR images and shape analysis of the diffusion tensor for prediction of volumetric response after radiosurgery for meningioma. J Neurosurg. 2018;129:31–7. https://doi.org/10.3171/2018.7.GKS181327.
60. Hatt M, Lucia F, Schick U, Visvikis D. Multicentric validation of radiomics findings: challenges and opportunities. EBioMedicine. 2019;47:20–1. https://doi.org/10.1016/j.ebiom.2019.08.054.
61. Mali SA, Ibrahim A, Woodruff HC, et al. Making Radiomics more reproducible across scanner and imaging protocol variations: a review of harmonization methods. JPM. 2021;11:842. https://doi.org/10.3390/jpm11090842.
62. Gu H, Zhang X, Di Russo P, et al. The current state of Radiomics for Meningiomas: promises and challenges. Front Oncol. 2020;10:567736. https://doi.org/10.3389/fonc.2020.567736.

PET and SPECT in Meningiomas: Pathological Correlation

Evangelos Kalogirou, Sotiria Founta, Paraskevi Papaspirou, and George A. Alexiou

Abbreviations

CNS	Central Nervous System
CT	Computational Tomography
FDG	Fluorodeoxyglucose
FDOPA	18F-fluorodopa
FET	Fluoroethyl tyrosine
FLT	Fluorothymidine
HPF	High-Power Field
L/N	Ratio Lesion-to-Normal uptake ratio
MET	C-methionine
MRI	Magnetic Resonance Imaging
N:C ratio	Nuclear-Cytoplasmic ratio
PET	Positron-Emitting Tomography
ROI	Region Of Interest
SPECT	Single-photon emission computed tomography
SUV	Standardized Uptake Value
T/NT	Tumor-Normal-Tissue Ratio
Tc-TF	Technetium-Tetrofosmin
TGR	Tumor-To-Gray matter Ratio
TUR	Thallium Uptake Ratio
TWBR	Tumor-to-Whole-Blood Ratio
WHO	World Health Organization

E. Kalogirou · S. Founta · P. Papaspirou
Department of Neurosurgery, University of Ioannina, Ioannina, Greece

G. A. Alexiou (✉)
Department of Neurosurgery, University of Ioannina, Ioannina, Greece

University of Ioannina, School of Medicine, Ioannina, Greece
e-mail: galexiou@uoi.gr

1 Introduction

Meningiomas are dural-based tumors. Classically, they are thought to originate from the arachnoid cap cells, due to their cytological similarity. The origin cell is not yet proven though [1]. The CBTRUS study found out that 36.7% of primary brain and other central nervous system (CNS) tumors are meningiomas, constituting the most frequent CNS tumor [2]. Meningioma incidence is on average 8.3 cases per 100,000 persons per year. Nevertheless, in children, meningiomas are exceedingly rare [3]. Meningiomas occurrence is age dependent, with the incidence being proportional to increasing age (from 0.14 per 100,000 in the 0–19 age group to 37.75 per 100,000 in the 75–84 age group) [2]. In many cases, meningiomas can also be asymptomatic and recognized incidentally during imaging for other reasons [4]. Incidental meningiomas can be found on brain MRI of up to 1.0% of the general population [5]. They are grossly categorized in three grades of malignancy, namely grade 1 (meningioma), grade 2 (atypical meningioma), and grade 3 (malignant meningioma) [6]. Epidemiological data suggests that approximately 80.5% are grade I, 18% are grade II, and 1.5% are grade III [2]. Meningiomas can be further subdivided into 15 subtypes, based on histological features [7]. See Table 1 for more information. The diagnosis of meningiomas is mainly made by MRI and for the patients that cannot undergo MRI, contrast-enhanced CT can also be used. Positron-emitting

F. Maiuri, M. Del Basso De Caro (eds.), *Meningiomas*,
https://doi.org/10.1007/978-3-031-76680-0_13

Table 1 Meningioma histological subtypes

Meningothelial meningioma
Fibrous meningioma
Transitional meningioma
Psammomatous meningioma
Angiomatous meningioma
Microcystic meningioma
Secretory meningioma
Lymphoplasmacyte-rich meningioma
Metaplastic meningioma
Chordoid meningioma
Clear cell meningioma
Rhabdoid meningioma
Papillary meningioma
Atypical meningioma
Anaplastic (malignant) meningioma

tomography (PET) and single-photon emission computed tomography (SPECT) have been also utilized for the diagnosis and grading of these tumors [8, 9].

2 PET and SPECT

SPECT is a nuclear medicine imaging technique which makes use of gamma-emitting radioisotopes. These radioisotopes, after being injected into the patient, emit gamma rays, which are picked up by single or multiple gamma cameras. Reconstruction algorithms produce a 3-dimensional image from the data acquired [10].

Positron emission tomography (PET) is an imaging technique based on the beta plus decay phenomenon of certain radioisotopes. The positron emitted by this process annihilates almost instantaneously, producing a pair of gamma rays in counter directions. These gamma rays are then picked up by sensors and a highly-detailed three-dimensional image is being produced [11, 12].

Both techniques are clinical useful, because they essentially mark the biodistribution of a specific substance throughout the body. The radioisotopes are conjugated with several molecules (producing radioligands), which can be used to track the metabolism of different pathologies. The spatial resolution of a PET scan, though, exceeds that of a SPECT scan [13], due to the fact that the pair of gamma rays produced by each positron is coincidentally detected, permitting a more accurate localization of the event. It was theorized that PET scan would completely replace other radioisotope techniques in due time [14]. However, latest studies [15–17] have shown that there is a role for both techniques, since SPECT is less expensive and more accessible option in many countries. In addition, newer radiopharmaceuticals are being developed in conjunction for use in PET and SPECT scans [16]. Their parallel evolving direction prompts us, therefore, to study both of those two techniques and their usefulness in neuro-oncology.

3 PET and SPECT in Neuro-Oncology

There are several roles of neuroimaging in neuro-oncology such as anatomical localization of intracranial tumors and nearby structures, assessment of tumor grade and aggressiveness, delineate optimal stereotactic biopsy site, assessment of early treatment response, distinguishing between tumor recurrence and post-therapy changes, and aiding in prognosis evaluation [18].

3.1 Radiopharmaceuticals Used in Neuro-Oncology

As previously explained, both PET and SPECT need specific radioisotopes to work. These radioisotopes can be used on their own or combined with biological compounds to form radioligands.

3.1.1 SPECT Radiopharmaceuticals

The radiotracers mostly used are seen on Table 2, along with their half-life. The main radiotracers that have been used in neuro-oncology are ^{201}Thallium, (99 m)Tc-tetrofosmin, (99 m) Tc-sestamibi, 99mTc-DTPA-bis(Met), 99mTc-HYNIC-FAPI-04 [20–24]. As seen, technetium labeled compounds are mainly used due to their superior imaging properties.

Table 2 SPECT radioisotopes [19]

Radioisotope	Half-life
^{67}Ga	3.26 d
^{111}In	2.8 d
^{123}I	13.2 h
^{133}Xe	5.25 d
^{201}Tl	3.05 d
^{99m}Tc	6 h

Table 3 PET radioisotopes [19]

Radioisotope	Half-life
^{11}C	20 min
^{13}N	10 min
^{15}O	2 min
^{18}F	110 min
^{68}Ga	68.3 min
^{82}Rb	1.27 min

3.1.2 PET Radiopharmaceuticals

The radiotracers mostly used in PET can be seen in Table 3, along with their half-life.

As with SPECT, there are numerous radiopharmaceutical compounds containing these isotopes which are being produced and tested. Some of the compounds currently being used or tested in neuro-oncology are (18)F-fluorodeoxyglucose (FDG), C-methionine (MET), O-(2-[18F] fluoroethyl)-L-tyrosine (FET), 18F-fluorodopa (FDOPA), anti-3-[18F]FACBC, 68Ga-DOT ATOC [25–28].

3.2 PET and SPECT in Meningiomas

Nuclear medicine techniques, even if not yet routinely used in the characterization of meningiomas, can provide numerous information that cannot be presented by classical MRI or CT. In the context of meningioma diagnosis, PET and SPECT offer enhanced capabilities in demarcating tumors situated in proximity to the cranial base or those exhibiting substantial osseous expansion. Furthermore, these imaging modalities can provide clues on the histological grade of these tumors (Figs. 1 and 2) [29]. Additionally, PET and SPECT find utility in apprising the outcome of tumor excision procedures and in the long-term follow-up of patients, as they provide superior discrimination between fibrous tissue and residual or recurrent neoplastic tissue when compared to other radiology techniques.

Herewith, we present data on predicting tumor grading using PET and SPECT, since meningioma grades serve as crucial indicators of biological behavior [30]. Other imaging modalities, such as classic MRI and CT, fall short in providing sufficient information for tumor grading [31]. Through the utilization of diverse radiopharmaceuticals, we can adequately correlate imaging findings with potential tumor grade and risk of recurrence, thereby possibly facilitating the

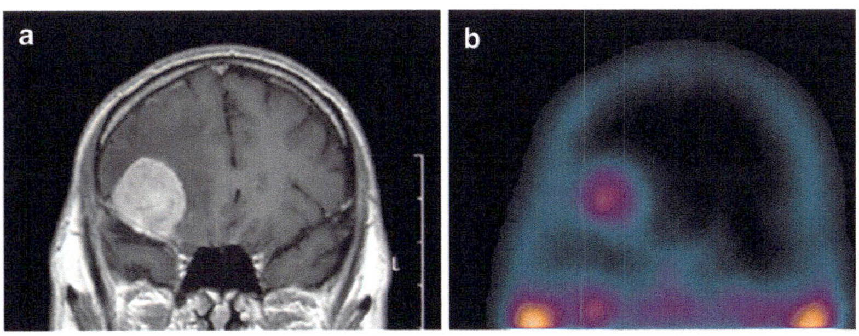

Fig. 1 (**a**) T1-weighed coronal gadolinium-enhanced MRI demonstrating a meningioma. (**b**) 99mTc-Tetrofosmin brain SPECT showing radiotracer accumulation (L/N = 3.5) in the same region as that on MRI. Histopathology verified a typical meningioma. The Ki-67 index was approximately 2%. (Reproduced from Alexiou et al. [29])

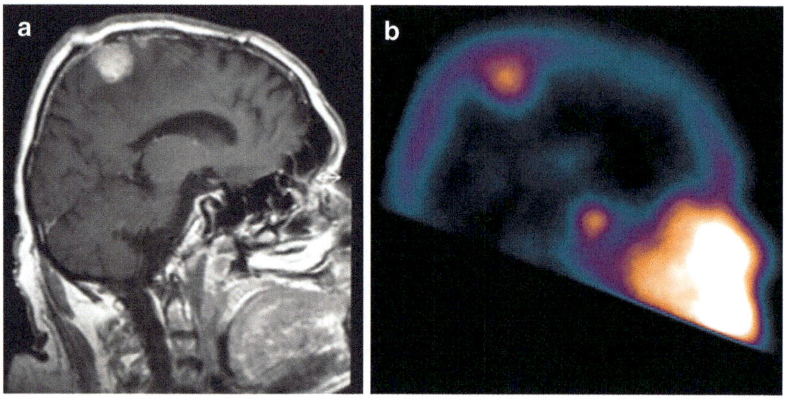

Fig. 2 (**a**) Gadolinium-enhanced T1-weighed sagittal MRI demonstrating a histopathologically verified anaplastic meningioma. (**b**) Profound focal 99mTc-Tetrofosmin uptake (L/N = 9.9) in the reconstructed sagittal SPECT slice. The Ki-67 expression was approximately 15%. (Reproduced from Alexiou et al. [29])

design of more effective therapeutic interventions and follow-up strategies. This could be especially useful in the context of incidental meningiomas, as a way to recognize and treat potentially aggressive tumors earlier.

3.2.1 PET in Meningiomas

Overall, PET can detect meningioma grade with a pooled sensitivity of 76% and specificity of 89% [32]. The results vary between different radiopharmaceuticals.

In the context of PET scan grading, it is important to clarify the following terms, which are the ones mostly used as cutoff values in the following studies.

Standardized uptake value (SUV) is a normalized value describing the concentration of a radioisotope in an area of interest, while taking into account the administered dose of the radiotracer and the patient's body weight [33, 34]. The aim of SUV is to provide a standardized metric for comparing uptake across different individuals and scans.

The mathematical expression of SUV is [34]:

$$\mathrm{SUV} = \frac{\mathrm{Tissueactivity}\left(\mathrm{millicurie} / \mathrm{milliliter}\right)}{\mathrm{Injecteddose}\left(\mathrm{millicurie}\right) / \mathrm{weight}\left(\mathrm{grams}\right)} \tag{1}$$

SUV is the most used metric in analyzing PET scans [35], as it can easily indicate variations in metabolic activity, aiding in the identification of areas with abnormal cellular function, such as tumors or regions of inflammation [36].

SUVmax is another commonly used metric and it refers to the maximum SUV value reported in a region of interest (ROI) [37].

SUVmean is the median value of SUV in a ROI [37] and it is mostly being used in calculating other parameters.

Tumor-to-gray matter ratio (TGR) is a subterm of the more generalized term tumor-to-background ratio (TBR) and it is defined as the SUV of the region of interest (ROI) in relation to the SUVmean of a background region, usually in the contralateral hemisphere [35].

The mathematical expression of TGR is

$$\mathrm{TGR} = \frac{\mathrm{SUV}\left(\mathrm{ROI}\right)}{\mathrm{SUVmean}\left(\mathrm{Background}\right)} \tag{2}$$

TGR is a useful metric because it normalizes the SUV value in relation to background.

Lastly, more specialized metrics used in the following studies are TGRmax and TGRmean, which refers to SUVmax and SUVmean values, respectively, normalized by the SUVmean of reference areas [38].

$$\text{TGRmax} = \frac{\text{SUVmax}\left(\text{ROI}\right)}{\text{SUVmean}\left(\text{Background}\right)} \quad (3)$$

$$\text{TGRmean} = \frac{\text{SUVmean}\left(\text{ROI}\right)}{\text{SUVmean}\left(\text{Background}\right)} \quad (4)$$

Fluorine-18 Fluorodeoxyglucose (18F-FDG)

18F-FDG is perhaps the earliest pet-tracer used to differentiate between meningioma grades; U. Cremerius et al. used 18F-FDG in their 1997 study, in which they reported a maximum sensitivity of 89% and specificity of 88% in differentiating high from low-grade meningiomas (using a TGR of 0.85 and 1.05 as cutoff values for primary and recurrent meningiomas, respectively) [39]. However, the study was limited due to the low amount of high-grade meningiomas in their patient group (9 out of 75 patients, i.e., 12% instead of the expected 19.5% [2]).

J. Won Lee et al. in their 2009 study agreed that high-grade meningiomas have significantly higher TGR than low-grade meningiomas. They used a TGR of 1.0 as the best cutoff value and achieved a sensitivity of 43% and specificity of 95%. Regrettably, this implies a notable prevalence of false negative outcomes, as merely 7 out of the 16 high-grade meningiomas were identified. However, they also found out that patients with a TGR > 1.0 had an overall recurrence rate of 33% (3 of 9 patients), which was significantly higher than that of patients with a TGR $\leq$ 1.0 (4%, 2 of 50 patients) ($p = 0.02$) [40].

In their 2015 study, S. Okuchi et al. perpetuated the pattern by demonstrating a positive correlation between the grade of meningiomas and the uptake of FDG. Their study yielded a sensitivity of 72.7% and specificity of 87.5% using SUVmax with a cutoff threshold of 7.2. Furthermore, the researchers established a noteworthy association between SUVmax and TGRmax with the Ki67 index, albeit with coefficients of relatively modest magnitude [38].

Overall, we can see that the existing studies most likely support the use of PET scan with FDG as a way to assess high-grade meningiomas preoperatively.

3′-Deoxy-3′-[18F]Fluorothymidine (18F-FLT)

18F-FLT was recently studied as a PET radiopharmaceutical by Bashir et al. in their 2020 study with promising results [41]. 18F-FLT is a thymidine analogue and its accumulation is a marker of cellular proliferation [42]. Utilizing different metrics, the study ascertained that the SUVmax normalized to the average radioactivity concentrations within whole-blood, denoted as tumor-to-whole-blood ratio (TWBR), emerged as the most efficacious metric for distinguishing between meningioma grades 1 and 2. Employing a TWBR threshold of 2.7 yielded a sensitivity of 100% and specificity of 95%. Furthermore, the study demonstrated a notable and robust correlation with various proliferative biomarkers, thereby enabling the prognostication of tumor aggressiveness and recurrence rates [41]. It is important to note, however, that the study is limited due to the small study size, comprising merely 17 participants.

O-(2-[18F]Fluoroethyl)-L-Tyrosine (18F-FET)

O-(2-[18F]Fluoroethyl)-l-tyrosine (18F-FET), which is one of the most long-standing radiopharmaceuticals, was used in the 2014 study of Cornelius et al. to evaluate the tracer's uptake from different histological subtypes of meningiomas. The authors showed that a TBR cutoff of 2.3 can be used to differentiate low from high-grade meningiomas with both sensitivity and specificity of 83%. 18F-FET was also studied to identify meningiomas located in the skull base but demonstrated poor outcomes in detecting tumors around the cavernous sinus [43].

3.2.2 SPECT in Meningiomas

As before, it is necessary to define some metrics used in the following studies.

Lesion-to-Normal (L/N) uptake ratio refers to the average counts in a ROI around the lesion on the transverse slice with the highest tumoral tracer uptake divided by the average counts of a second identical mirror region on the contralateral normal side of the brain [44]. It is analogous to the TGR metric used in PET. It is also referred to by the following studies as tumor-normal-tissue ratio (T/NT) and thallium uptake ratio (TUR) [38, 45].

(99 m)Tc-Tetrofosmin

In their 2008 investigation, Fotopoulos et al. demonstrated a robust linear positive association between the intensity of 99 m Tc-TF uptake and the expression of MIB-1 ($r = 0.930$, $P < 0.001$), as well as meningioma grading ($r = 0.722$, $P = 0.001$) (Figs. 3 and 4). Notably, meningio-

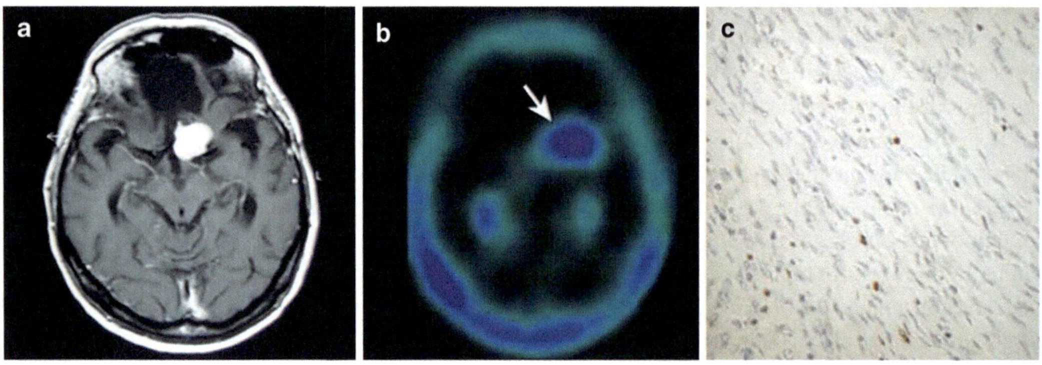

Fig. 3 (**a**) Gadolinium-enchanced gradient echo T1-weighed MRI slice in a 49-year-old female patient with a benign meningioma; (**b**), Low tracer uptake in the corresponding reconstructed transverse SPECT slice (arrow); (c), Ki-67 expression about 1.5% (immunohistochemical stain MIB-1). (Reproduced from Fotopoulos et al. [44])

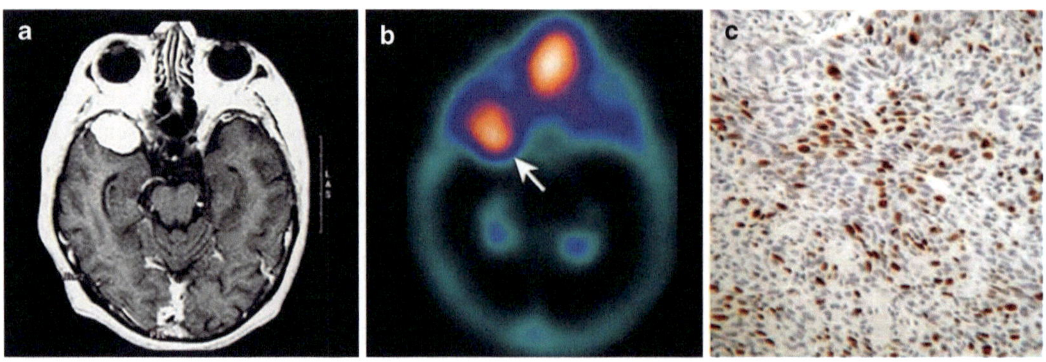

Fig. 4 (**a**). Gadolinium-enhanced gradient echo T1-weighed MRI slice in a 76-year-old male with; (**b**) Profound focal 99mTcTetrofosmin uptake in the reconstructed transverse SPECT slice (arrow); (**c**) Ki-67 expression about 20% (immunohistochemical stain MIB-1). (Reproduced from Fotopoulos et al. [44])

mas of higher grades consistently exhibited a L/N ratio exceeding 9, whereas those of lower grades manifested a ratio below 6 with one exception of L/N ratio 8.9. Furthermore, a significant correlation between the intensity of tracer uptake and tumor recurrence at 1 year postoperative was reported [44].

In a latter 2011 study, it was determined that typical and anaplastic meningiomas can be differentiated with 96% sensitivity and 100% specificity, using a L/N ratio of 9.6 as the optimum cutoff value [46]. Meningioma aggressiveness, as assessed by flow cytometry, was also found to correlate with meningioma grade. S-phase fraction of the cell cycle was positive correlated with 99 m Tc-TF uptake [47].

Thallium-201

Okuchi et al. studied the applicability of Thallium-201 SPECT scan in their 2015 study, alongside 18F-FDG PET scan [38].

Thallium is a potassium analogue which is up taken by the sodium- and potassium-activated adenosine triphosphatase (Na + -K+ ATPase) pump [48].

According to their results, using a threshold value of 7.26 for the L/N ratio, a sensitivity of 72.7% and specificity of 82.%1 could be achieved for differentiating between low-grade and high-grade meningiomas [38].

(99 m)Tc-HYNIC-Octreotide

Shengjun Wang et al. in their 2013 study explored (99 m)Tc-HYNIC-octreotide as a radiopharmaceutical in meningioma imaging [45]. Octreotide is a somatostatin analogue and somatostatin receptor 2 is expressed highly in meningiomas [49]. The study showed that the L/N ratio of a meningioma was positively correlated with its pathological grade ($r = 0.78$, $P < 0.01$) [45]. A threshold value was not recommended by the authors. The L/N ratio was also positively correlated with SSTR2 expression ($r = 0.81$, $P < 0.01$) [45], which in turn has been used as a biomarker for tumor behavior by other studies [19].

4 Critical Evaluation and Clinical Implications

In summary, PET emerges as a valuable tool for preoperative meningioma grading, offering significant advantages, particularly in the context of incidentally discovered tumors. This capability could contribute substantively to the guidance favoring observation over surgery for low-grade asymptomatic tumors. Furthermore, the preoperative grading facilitated by PET holds promise for advancing innovative treatment modalities, as a prerequisite for the effective modeling of non-invasive interventions. SPECT offers a noteworthy alternative, since it provides similar results while characterized by broader accessibility, potentially benefiting a larger demographic.

However, it is essential to critically assess the limitations inherent in the mentioned studies. Notably, these studies span different timeframes, each employing different criteria for meningioma grading. Discrepancies arise, for instance, in the application of 18F-fluorodeoxyglucose (FDG); Cremerius et al. used the 1997 WHO grading system [39], while J. Won Lee et al. and S. Okuchi et al. used the 2007 WHO grading system [38, 40]. No studies across different radiopharmaceuticals had yet to account for the newer 2021 WHO grading system. The 2021 grading system permits grading decoupled from histological characteristics [6]. In previous studies, there was an absence of documentation elucidating the rationale behind the assigned grades, which prevents accurate retrospective grading [38–41, 43, 45, 46]. The modifications introduced in the newer WHO system have certainly influenced the classification of particular meningiomas with distinct histological subtypes, such as the papillary and rhabdoid subtypes, which are no longer automatically characterized as high-grade [6]. Nevertheless, the likelihood of misclassified tumors is presumed to be relatively low, suggesting limited impact on the outcomes of preceding studies.

The absence of standardized grading protocols however and the lack of explicit justification for grading decisions in prior studies underscore the need for a consensus on meningioma grading. To ascertain the efficacy of PET and SPECT in distinguishing low-grade from high-grade meningiomas, it is imperative to conduct updated investigations that align with a universally accepted grading framework. To bypass this problem, an alternative approach would be correlating PET and SPECT scans with biological markers predictive of meningioma aggressiveness and recurrence, a strategy already explored in several of the studies, with overall good results [17].

Additionally, the heterogeneous use of various radiopharmaceuticals poses a challenge [50]. Perhaps it would be better to concentrate research efforts initially on a single most promising radiopharmaceutical, proving its efficacy through comprehensive, long-term studies and adopting its use in clinical settings before considering comparisons with alternative agents. This focused strategy can contribute to a more methodical and impactful advancement in the field of meningioma imaging.

Funding None.

Conflict of Interest None.

References

1. Riemenschneider MJ, Perry A, Reifenberger G. Histological classification and molecular genetics of meningiomas. Lancet Neurol. 2006;5(12):1045–54. https://doi.org/10.1016/S1474-4422(06)70625-1.
2. Ostrom QT, Gittleman H, Xu J, et al. CBTRUS statistical report: primary brain and other central nervous system tumors diagnosed in the United States in 2009–2013. Neuro-Oncology. 2016;18(Suppl 5):v1–v75. https://doi.org/10.1093/neuonc/now207.
3. Alexiou GA, Mpairamidis E, Psarros A, Sfakianos G, Prodromou N. Intracranial meningiomas in children: report of 8 cases. Pediatr Neurosurg. 2008;44(5):373–5. https://doi.org/10.1159/000149903.
4. Alexiou GA, Gogou P, Markoula S, Kyritsis AP. Management of meningiomas. Clin Neurol Neurosurg. 2010;112(3):177–82. https://doi.org/10.1016/j.clineuro.2009.12.011.
5. Håberg AK, Hammer TA, Kvistad KA, et al. Incidental intracranial findings and their clinical impact; the HUNT MRI study in a general population of 1006 participants between 50-66 years. PLoS One. 2016;11(3):e0151080. https://doi.org/10.1371/journal.pone.0151080.
6. Louis DN, Perry A, Wesseling P, et al. The 2021 WHO classification of tumors of the central nervous system: a summary. Neuro-Oncology. 2021;23(8):1231–51. https://doi.org/10.1093/neuonc/noab106.
7. WHO Classification of Tumours EditorialBoard. Central nervous system tumours, vol. 6. 5th ed. World Health Organization; 2021. https://publications.iarc.fr/Book-And-Report-Series/Who-Classification-Of-Tumours/Central-Nervous-System-Tumours-2021. Accessed 1 Nov 2023.
8. Siempis T, Tsakiris C, Alexiou GA, Xydis VG, Voulgaris S, Argyropoulou MI. Diagnostic performance of diffusion and perfusion MRI in differentiating high from low-grade meningiomas: a systematic review and meta-analysis. Clin Neurol Neurosurg. 2020;190:105643. https://doi.org/10.1016/j.clineuro.2019.105643.
9. Goldbrunner R, Stavrinou P, Jenkinson MD, et al. EANO guideline on the diagnosis and management of meningiomas. Neuro-Oncology. 2021;23(11):1821–34. https://doi.org/10.1093/neuonc/noab150.
10. Warwick JM. Imaging of brain function using SPECT. Metab Brain Dis. 2004;19(1):113–23. https://doi.org/10.1023/B:MEBR.0000027422.48744.a3.
11. Berger A. Positron emission tomography. BMJ. 2003;326(7404):1449.
12. Maisey MN. Positron emission tomography in clinical medicine. In: Bailey DL, Townsend DW, Valk PE, Maisey MN, editors. Positron emission tomography: basic sciences. Cham: Springer; 2005. p. 1–12. https://doi.org/10.1007/1-84628-007-9_1.
13. Lodge MA, Braess H, Mahmoud F, et al. Developments in nuclear cardiology: transition from single photon emission computed tomography to positron emission tomography-computed tomography. J Invasive Cardiol. 2005;17(9):491–6.
14. Alavi A, Basu S. Planar and SPECT imaging in the era of PET and PET-CT: can it survive the test of time? Eur J Nucl Med Mol Imaging. 2008;35(8):1554–9. https://doi.org/10.1007/s00259-008-0813-2.
15. Eilsberger F, Lapa C, Kläsner B, Schäfers MA, Luster M. Hand in hand -the future of SPECT and PET imaging. Nuklearmedizin. 2023;62(2):52–3. https://doi.org/10.1055/a-1998-8402.
16. Mariani G, Strauss HW. Positron emission and single-photon emission imaging: synergy rather than competition. Eur J Nucl Med Mol Imaging. 2011;38(7):1189–90. https://doi.org/10.1007/s00259-011-1767-3.
17. Fotopoulos AD, Alexiou GA. Is there still a place for SPET in the era of PET brain imaging? Hell J Nucl Med. 2012;15(2):89–91. https://doi.org/10.1967/s002449910026.

18. Kas A, Rozenblum L, Pyatigorskaya N. Clinical value of hybrid PET/MR imaging: brain imaging using PET/MR imaging. Magn Reson Imaging Clin N Am. 2023;31(4):591–604. https://doi.org/10.1016/j.mric.2023.06.004.
19. Roytman M, Kim S, Glynn S, et al. PET/MR Imaging of somatostatin receptor expression and tumor vascularity in meningioma: implications for pathophysiology and tumor outcomes. Front Oncol. 2021;11:820287. https://doi.org/10.3389/fonc.2021.820287.
20. Alexiou GA, Xourgia X, Gerogianni P, et al. 99mTc-Tetrofosmin uptake correlates with the sensitivity of glioblastoma cell lines to Temozolomide. World J Nucl Med. 2017;16(1):45–50. https://doi.org/10.4103/1450-1147.181155.
21. Alexiou GA, Xourgia X, Vartholomatos E, et al. Comparison of (99m)tc-tetrofosmin and (99m)tc-sestamibi uptake in glioma cell lines: the role of p-glycoprotein expression. Int J Mol Imaging. 2014;2014:471032. https://doi.org/10.1155/2014/471032.
22. Perek N, Denoyer D, Dubois F, Koumanov F. Malignant gliomas display altered plasma membrane potential and pH regulation—interaction with Tc-99m-MIBI and Tc-99m-Tetrofosmin uptakes. Gen Physiol Biophys. 2002;21(4):381–404.
23. Hazari PP, Yadav SK, Kumar PK, et al. Preclinical and clinical use of indigenously developed 99mTc-Diethylenetriaminepentaacetic acid-Bis-methionine: l-type amino acid transporter 1-targeted single photon emission computed tomography radiotracer for glioma management. ACS Pharmacol Transl Sci. 2023;6(9):1233–47. https://doi.org/10.1021/acsptsci.3c00091.
24. Luo X, Zhang Z, Cheng C, et al. SPECT Imaging with Tc-99m-labeled HYNIC-FAPI-04 to extend the differential time window in evaluating tumor fibrosis. Pharmaceuticals (Basel). 2023;16(3):423. https://doi.org/10.3390/ph16030423.
25. Katsanos AH, Alexiou GA, Fotopoulos AD, Jabbour P, Kyritsis AP, Sioka C. Performance of 18F-FDG, 11C-methionine, and 18F-FET PET for glioma grading: a meta-analysis. Clin Nucl Med. 2019;44(11):864–9. https://doi.org/10.1097/RLU.0000000000002654.
26. Cobes N, Tran S, Mathon B, et al. Exploring the mechanism of 18F-fluorodopa uptake in recurrent high-grade gliomas: a comprehensive histomolecular-positron emission tomography analysis. Eur J Neurol. 2023;12:e16093. https://doi.org/10.1111/ene.16093.
27. Milosevic A, Styczen H, Haubold J, et al. Correlation of the apparent diffusion coefficient with the standardized uptake value in meningioma of the skull plane using ^{68}Ga-DOTATOC PET/MRI. Nucl Med Commun. 2023;12:1106. https://doi.org/10.1097/MNM.0000000000001774.
28. Karlberg A, Pedersen LK, Vindstad BE, et al. Diagnostic accuracy of anti-3-[18F]-FACBC PET/MRI in gliomas. Eur J Nucl Med Mol Imaging. 2023;30:496. https://doi.org/10.1007/s00259-023-06437-4.
29. Alexiou GA, Tsiouris S, Goussia A, Polyzoidis KS, Fotopoulos AD, Kyritsis AP. (99m)Tc-Tetrofosmin SPECT for the assessment of meningioma malignancy. J Clin Neurosci. 2008;15(6):721–2. https://doi.org/10.1016/j.jocn.2007.08.031. author reply 722-3. Epub 2008 Apr 10. PMID: 18406141.
30. Yang SY, Park CK, Park SH, Kim DG, Chung YS, Jung HW. Atypical and anaplastic meningiomas: prognostic implications of clinicopathological features. J Neurol Neurosurg Psychiatry. 2008;79(5):574–80. https://doi.org/10.1136/jnnp.2007.121582.
31. Saloner D, Uzelac A, Hetts S, Martin A, Dillon W. Modern meningioma imaging techniques. J Neuro-Oncol. 2010;99(3):333–40. https://doi.org/10.1007/s11060-010-0367-6.
32. Filis P, Alexiou GA, Zigouris A, Sioka C, Filis N, Voulgaris S. Meningioma grading based on positron emission tomography: a systematic review and meta-analysis. World Neurosurg X. 2023;18:100167. https://doi.org/10.1016/j.wnsx.2023.100167.
33. Bennett P, Mintz A, Perry B, Trout AT, Vergara-Wentland P. Approach to interpretation. In: Specialty imaging: PET specialty imaging. Amsterdam: Elsevier; 2018. p. 8–13. https://doi.org/10.1016/B978-0-323-52484-1.50010-8.
34. Ziessman HA, O'Malley JP, Thrall JH. Oncology: positron emission tomography. In: Nuclear medicine. 4th ed. Philadelphia: W.B. Saunders; 2014. p. 227–64. https://doi.org/10.1016/B978-0-323-08299-0.00011-0.
35. Kamson D, Tsien C. Novel magnetic resonance imaging and positron emission tomography in the RT planning and assessment of response of malignant gliomas. In: Ross BD, Gambhir SS, editors. Molecular imaging. 2nd ed. Boca Raton: Academic Press; 2021. p. 1031–48. https://doi.org/10.1016/B978-0-12-816386-3.00078-8.
36. Mah K, Caldwell CB. Biological target. In: Paulino AC, Teh BS, editors. PET-CT in radiotherapy treatment planning. Amsterdam: Elsevier; 2008. p. 52–89. https://doi.org/10.1016/B978-1-4160-3224-3.50007-4.
37. Gallamini A, Zwarthoed C, Borra A. Positron emission tomography (PET) in oncology. Cancers (Basel). 2014;6(4):1821–89. https://doi.org/10.3390/cancers6041821.
38. Okuchi S, Okada T, Yamamoto A, et al. Grading Meningioma. Medicine (Baltimore). 2015;94(6):e549. https://doi.org/10.1097/MD.0000000000000549.
39. Cremerius U, Bares R, Weis J, et al. Fasting improves discrimination of grade 1 and atypical or malignant meningioma in FDG-PET. J Nucl Med. 1997;38(1):26–30.
40. Lee JW, Kang KW, Park SH, et al. 18F-FDG PET in the assessment of tumor grade and prediction of tumor recurrence in intracranial meningioma. Eur J Nucl Med Mol Imaging. 2009;36(10):1574–82. https://doi.org/10.1007/s00259-009-1133-x.

41. Bashir A, Binderup T, Vestergaard MB, et al. In vivo imaging of cell proliferation in meningioma using 3′-deoxy-3′-[18F]fluorothymidine PET/MRI. Eur J Nucl Med Mol Imaging. 2020;47(6):1496–509. https://doi.org/10.1007/s00259-020-04704-2.
42. Chalkidou A, Landau DB, Odell EW, Cornelius VR, O'Doherty MJ, Marsden PK. Correlation between Ki-67 immunohistochemistry and 18F-fluorothymidine uptake in patients with cancer: a systematic review and meta-analysis. Eur J Cancer. 2012;48(18):3499–513. https://doi.org/10.1016/j.ejca.2012.05.001.
43. Cornelius JF, Stoffels G, Filß C, et al. Uptake and tracer kinetics of O-(2-18F-fluoroethyl)-l-tyrosine in meningiomas: preliminary results. Eur J Nucl Med Mol Imaging. 2015;42(3):459–67. https://doi.org/10.1007/s00259-014-2934-0.
44. Fotopoulos AD, Alexiou GA, Goussia A, et al. (99m)Tc-Tetrofosmin brain SPECT in the assessment of meningiomas-correlation with histological grade and proliferation index. J Neuro-Oncol. 2008;89(2):225. https://doi.org/10.1007/s11060-008-9611-8.
45. Wang S, Yang W, Deng J, Zhang J, Ma F, Wang J. Correlation between 99mTc-HYNIC-octreotide SPECT/CT somatostatin receptor scintigraphy and pathological grading of meningioma. J Neuro-Oncol. 2013;113(3):519–26. https://doi.org/10.1007/s11060-013-1146-y.
46. Fotopoulos AD, Kyritsis AP, Tsiouris S, et al. Characterization of intracranial space-occupying lesions by 99(m)Tc-Tetrofosmin SPECT. J Neuro-Oncol. 2011;101(1):83–9. https://doi.org/10.1007/s11060-010-0230-9.
47. Alexiou GA, Vartholomatos G, Tsiouris S, Papadopoulos A, Kyritsis AP, Polyzoidis KS, Voulgaris S, Fotopoulos AD. Evaluation of meningioma aggressiveness by (99m)Tc-Tetrofosmin SPECT. Clin Neurol Neurosurg. 2008;110(7):645–8.
48. Sato T, Indo H, Kawabata Y, et al. Thallium-201 chloride (Tl-201) accumulation and Na+/K+-ATPase expression in tumours of the head and neck. Dentomaxillofac Radiol. 2005;34(4):212–7. https://doi.org/10.1259/dmfr/50773431.
49. Durand A, Champier J, Jouvet A, et al. Expression of c-Myc, neurofibromatosis type 2, somatostatin receptor 2 and erb-B2 in human meningiomas: relation to grades or histotypes. Clin Neuropathol. 2008;27(5):334–45. https://doi.org/10.5414/npp27334.
50. Alqahtani FF. SPECT/CT and PET/CT, related radiopharmaceuticals, and areas of application and comparison. Saudi Pharm J. 2023;31(2):312–28. https://doi.org/10.1016/j.jsps.2022.12.013.

Part IV

Meningioma Pathology and Management

The Simpson Grading: Is It Still Valid in the Imaging Era?

Rupen Desai, Gülce Gel,
Victor Ramzes Chavez-Herrera,
and Theodore H. Schwartz

1 Introduction

Meningiomas are the most common primary tumors in the central nervous system, representing more than one-third of all brain tumors [1]. The first historical characterization of meningiomas by Bailey and Cushing in 1928 focused on histologic features [2–4]. Today, we categorize meningiomas according to the World Health Organization (WHO) classification, which distributes the tumors into three grades meant to mirror increasing recurrence rates: WHO grade 1 (low-grade), WHO grade 2 (atypical), and WHO grade 3 (anaplastic).

The WHO classification was initially designed based on the histopathological features of tumors to predict clinical course, recurrence rates, and guide appropriate postoperative management options. However, histopathological features alone are not sufficient to accurately predict outcomes in all patients [5–7]. Importantly, the heavy reliance on histology to distinguish meningioma subtypes remained until the most recent WHO classification. Thus, for nearly 90 years, histology dominated the clinical decision-making for meningiomas, and although nearly 80% of meningiomas were characterized as WHO grade 1, recurrence rates were incongruent with the low pathologic grade [8]. In prior WHO schema, meningiomas recurrence risk varied significantly, ranging from 7% to 25% for grade 1, 29% to 52% for grade 2, and 50% to 94% for grade 3 [9].

Historically, gross total resection of meningioma was considered curative independent of meningioma grade, and in 1957, Donald Simpson published his findings analyzing extent of surgical resection, describing the Simpson grading system to predict recurrence independently of histology [10]. In his overwhelming belief that a gross total resection is curative, Simpson stated "the development of a recurrence must mean either that the initial resection was inadequate, or that the neoplastic change was multifocal, or that viable tumor was accidentally implanted during an otherwise successful resection". Notably, the Simpson grade was designed on the intraoperative interpretation of extent of resection which, prior to the advent of MRI and CT imaging, was the most accurate indication.

A significant limitation of the Simpson scale is that it only focuses on extent of resection. Although maximum surgical resection plays a crucial role in predicting recurrence rates, numerous other prognostic factors, including

R. Desai (✉) · G. Gel · V. R. Chavez-Herrera · T. H. Schwartz
Department of Neurological Surgery, Weill Cornell Medical College, New York Presbyterian Hospital, New York, NY, USA

F. Maiuri, M. Del Basso De Caro (eds.), *Meningiomas*,
https://doi.org/10.1007/978-3-031-76680-0_14

tumor location, histopathology, and molecular profiles, influence patient outcomes [11, 12]. As we will describe in this chapter, the 2021 WHO meningioma classification attempts to assimilate these features to better characterize the recurrence risk and guide treatment strategies of meningiomas [13, 14]. Finally, we will describe the role of advanced imaging in the modern era as a more sensitive marker of extent of resection that may be used to better delineate clinical outcomes in the context of WHO grade.

2 Histopathology

Since the first published schema nearly 100 years ago, and despite significant advances in molecular genetics, tumor histology has served as the underlying basis of meningioma classification [1]. Both the 2016 and 2021 WHO classification defined 15 histologic subtypes of meningioma; in the most recent iteration, there are nine grade 1 subtypes, three grade 2 subtypes, and three grade 3 subtypes. The nine WHO Grade 1 histologic subtypes are: meningothelial, fibrous, transitional, psammomatous, angiomatous, microcystic, secretory, lymphoplasmacyte-rich, and metaplastic. The WHO Grade 2 subtypes include chordoid, clear cell, atypical. The WHO Grade 3 subtypes include rhabdoid, papillary, and anaplastic. Importantly in the 2021 WHO classification, additional genetic criteria may be applied regardless of histologic subtype to characterize tumor grade [4, 15].

Histologically, meningiomas are characterized by spindled cells and microscopic features that include whorls, nuclear pseudoinclusions, and psammomatous calcifications [15]. Generally, WHO Grade 1 meningiomas have well-demarcated borders, WHO Grades 2 and 3 subtypes have higher cellularity, and WHO Grade 3 meningiomas also frequently infiltrate the surrounding tissue [13, 16, 17]. Immunohistochemical (IHC) staining may demonstrate reduction or loss of epithelial membrane antigen (EMA) in higher grade lesions, and meningiomas predominantly demonstrate somatostatin receptor 2A (SSTR2A) expression [15, 18]. Ki-67, an immunohistochemical marker of cell proliferation, correlates with disease prognosis when greater than 4% [19, 20].

3 Genomic and Epigenomic Profiles

In the twentieth century, more than 60% of sporadic meningiomas were found to have loss of heterozygosity (LOH) on chromosome 22; the first description of *NF2* as a causal gene was published in 1994 [21]. Merlin (moesin-ezrin-radixin-like protein) is an important tumor suppressor protein encoded by the *NF2* gene that serves as an intracellular scaffolding protein that engages during direct cell–cell contact. When functional, merlin helps regulate cell proliferation, cell growth, and cytoskeleton structure to prevent further cell growth [22–24]. Inactivating *NF2* mutations, therefore, result in dysregulated cell cycle and tumorigenesis. Studies have indicated a correlation between *NF2* mutations and higher grade histological changes as well as a diminished recurrence-free survival in meningiomas [25–29]. Meningiomas with *NF2* mutations are commonly found along the cerebral and cerebellar convexities, lateral skull base, or spinal cord [30].

More recent analyses reveal additional genetic correlates of meningioma aggressiveness. In 2013, Clark et al. described key genetic mutations found in a sample size of 300 meningiomas specifically analyzing the 40% that do not bear *NF2* mutations, finding this meningioma subset is clinically more benign and more likely to originate from the medial skull base [30–32] The group found *TRAF7* is found in 25% of meningiomas and is a mutation exclusive of *NF2* mutations. *KLF4* frequently co-occurs with *TRAF7* mutations, representing approximately 10% of meningiomas; *KLF4* belongs to the DNA-binding transcription factor family, playing an important role in the regulation of proliferation, differentiation, migration, inflammation, and pluripotency [32–34]. Tumors with *KLF4/TRAF7* mutations exhibit secretory histology, a characteristic known to induce peritumoral edema [34, 35]. Methylation in the promoters of *TRAF7* and *KLF4* has been identified in association with meningioma recurrence and elevated tumor grade [36]. Finally, the group found meningiomas exclusively with the *SMO* mutation are commonly found along the midline anterior skull base, and these meningiomas are frequently associated with the meningothelial histopathologic class of WHO Grade 1 [5, 30].

Other groups have described key meningioma mutations with increasing clinical relevance. *TERT* encodes the telomerase reverse transcriptase protein that extends cell telomeres; the promoter region is commonly found to have mutations at two sites (C228T and C250T) in cancer, enabling indefinite cellular division by maintaining telomere length in proliferating tumor cells [37, 38]. *TERT* promoter mutations have increasingly been associated with malignant disease, with Sahm et al. describing time to progression in patients with *TERT* mutation of just 10.1 months compared to 179.0 months in wild-type cases [39–41]. Thus, *TERT* promoter mutations were integrated into the 2021 WHO classification; regardless of histopathologic grade, tumors bearing *TERT* promotor mutations are classified as WHO Grade 3 [1].

Homozygous deletions of *CDKN2A/B* (genes encoding proteins regulating cell cycle) are also associated with an aggressive clinical course in meningioma [42–44]. In 2020, an analysis of 528 discrete meningioma samples demonstrated nearly 5% had *CDKN2A/B* deletions, 27% of which were histologically graded as WHO grade 2 and 73% as WHO grade 3 according to the 2016 WHO classification [42]. A closer analysis of these patients demonstrated meningiomas with *CDKN2A/B* deletions had significantly worse outcomes (median time to progression of 8 versus 101 months) even when adjusting for WHO grade or *TERT* promoter status. Thus, *CDKN2A* and *CDKN2B* deletions were also added to the 2021 WHO classification to identify WHO grade 3 meningiomas regardless of underling histopathology [1].

Chromosomal abnormalities are additional valuable biomarkers in cancer diagnosis, serving as predictors for recurrence and treatment response. In 2013, Brastianos et al. published data focusing on meningioma cytogenetics, or chromosomal structural abnormalities and copy number alterations [45]. While WHO Grade 1 meningiomas were marked by losses on chromosomes 1p, 7p, 14q, and 19, higher grade meningiomas also had losses on chromosomes 10q and 14q. The group found mutations of the *PI3K/AKT/mTOR* pathway occurred exclusively of either *NF2* or *SMO* mutations, and similar to Clark et al., found meningiomas with either *SMO* and *AKT* are found at the skull base and are of the meningothelial subtype, while also finding *NF2*-mutated grade 1 meningiomas more likely of the fibroblastic and transitional subtypes. Bi et al. correlated increasing tumor mutational burden and copy number alterations with higher tumor grade, further demonstrating the role of genetic instability in aggressive meningioma recurrence [46].

Recent years have borne rapid expansion of the study of how meningioma epigenetics (alterations in DNA methylation patterns, histone modifications, and changes in chromatin structure) collectively impact gene expression and contribute to disease progression [47–49]. The loss of H3K27me3 correlates with unfavorable prognosis and recurrence, emphasizing the vital role of methylation profiling in meningioma classification [50, 51]. The seminal methylation classification schema for meningioma was described in 2017, performing a genome-wide methylation analysis of 497 meningiomas and 309 extra-axial tumors, defining 3 methylation classes of benign meningiomas, 2 classes of intermediate disease, and 1 class representing malignant disease. More contemporary publications describe methylation profiles of meningiomas that not only better predict disease course than the 2016 WHO classification, but also better correlated with signs of aggressive clinical course such as genetic instability as marked by copy number variation [52, 53].

Understanding the molecular biology of meningiomas provides valuable insights into potential targeted therapies, predicting recurrence rates. But before such biomarkers can be fully integrated into the WHO classification, multi-institutional studies are needed to validate their significance.

4 Contemporary Meningioma Grading

In its most recent iterations, the WHO classification has significantly changed, acknowledging the sampling error, interobserver variation, and subjectivity of pathologic features such as brain invasion [1, 54]. Given the vast progress in the field of molecular genetics since its last iteration, the 2021 WHO classification begins to incorporate key biomarkers of aggressive clinical course, including *TERT*, *CDK2NA/B*, *NF2*, and loss of H3K27me3 while suggesting DNA methylation may help to further stratify borderline lesions [1]. Notably, copy number variation, tumor mutational burden, and detailed descriptions of the role of methylation profiles are not yet included in the most recent WHO classification of meningiomas.

Other groups have generated classification schema attempting to integrate this data. Driver et al. divides patients into three grades based on mitotic count, loss of chromosomes, and *CDK2NA* status, finding their classification schema better predicts recurrence than the 2016 WHO classification in both discovery and validation cohorts [55]. Similarly, Maas et al. incorporates copy number variation, histology, and methylation to better predict meningioma disease progression [56]. Importantly, while detailed, multimodal assimilation of data appears to vastly improve meningioma classification compared to historical histopathologic-centric schema, the resources for such analyses are not yet widely accessible and is therefore not feasible on a routine basis. Thus, WHO classification, by necessity, will continue to include the most essential genetic and epigenetic information while suggesting additional stratification when needed for more nuanced distinction.

5 The Simpson Grade and Its Relevance in the Modern Era

The basis of the Simpson Grade is, in its essence, achieving a gross total resection of all meningioma cells to improve postoperative outcomes. While recurrence rates tend to increase with higher pathologic grade, a clear distinction can be made between subtotal resection (STR) and gross total resection (GTR) within a tumor grade [57–60]. The original Simpson Grade ranges from grade 1 (macroscopic complete tumor resection with removal of affected dura and bone) to grade 5 (decompression with/without biopsy); grades 1–3 are generally categorized as GTR, while grade 4–5 are classified as subtotal resection (STR) [10, 12].

Importantly, in the years following the Simpson Grade's publication, significant variability was encountered in recurrence rates even within a single Simpson Grade [61, 62]. Other recent publications have found Simpson Grade to have clinical significance only in convexity locations [63]. Presumably, this variation could, in part, be explained by (1) the subjectivity of intraoperative assessments of extent of resection, (2) differences in technique in dealing with the dural tail of the meningioma, and (3) a reliance on clinical symptoms to diagnose recurrence as the scale was introduced in a period without access to advanced imaging studies. However, a discovery of both microscopic and macroscopic meningioma clusters as far as 4 cm from the tumor edge led to the description of a grade 0 resection, that includes a margin of dural resection, with vastly improved recurrence rates [64, 65]. Notably, given the lack of advanced imaging at the time of its conception, the Simpson Grade does not take into account the volume of residual tumor, and we have since found that volume of residual is an independent predictor of recurrence [66, 67].

Thus, in the modern era, CT scans and contrast-enhanced MRI and have become the clinical standards for the evaluation of tumor resection. A 2016 publication of 901 patients by Gousias et al. utilized such advanced imaging in conjunction with intraoperative assessment of gross total resection to correlate aggressive surgical resection with improved disease control; the correlation of STR with increased recurrence rates was correlated in another cohort of 1571 patients in 2021 [68, 69].

Therefore, the Simpson Grade alone does not serve as a predictor of recurrence-free survival. Instead, prognostication should include factors such as the residual tumor volume, meningioma location, histopathological and molecular subtypes [47, 70–72].

6 DOTATATE Pet

Contrast-enhanced MRI is the current clinical standard for evaluating extent of meningioma resection. However, in the postoperative and post-radiotherapy periods, MRI has limited utility in distinguishing meningiomas from scarring and inflammation, detecting small size lesions, infiltrative or "en plaque" lesions, distinguishing hyperostosis from bony tumor invasion, and assessing meningiomas in challenging locations such as the skull base and cavernous sinus [73–75]. In such contexts, functional imaging modalities such as positron emission tomography (PET) are increasingly being utilized to provide additional diagnostic information [76]. Meningiomas express the surface somatostatin receptor, SSTR2, and therefore may be imaged with the nuclear imaging study ^{68}Ga-DOTATATE PET/MRI [77–83]. Compared to standard contrast-enhanced MRI, ^{68}Ga-DOTATATE PET/MRI exhibits superior precision in delineating the extent of the tumor and can be particularly valuable for assessing the extent of disease in anatomically complex locations, such as the skull base, orbit, parasagittal sinus, and the cavernous sinus [74, 84, 85]. In a study of 112 patients, ^{68}Ga-DOTATATE PET/MRI was found to have excellent sensitivity overall, with reduced detection of very small (volume < 0.5 cm^3) or petroclival tumor location, suggesting ^{68}Ga-DOTATATE PET/MRI may serve as an excellent tool to detect clinically significant primary or recurrent meningioma [83]. Thus, such ^{68}Ga-DOTATATE PET/MRI has increasing utility in clinical practice, with class 3 evidence for its utilization to diagnose unclear recurrence or tumor extension in the 2021 EANO guidelines [86].

A controversial area of meningioma resection is the pathologic role of the dural tail and, therefore, the need to treat a potentially wide dural margin [87]. In a retrospective series of 179 convexity meningiomas, tumor invasion was exclusively found <1.5 cm from the tumor edge in cases with a smooth dural tail and < 2.5 cm from the tumor edge with a nodular dural tail; the group thereby recommended a 2.5 cm margin of dural resection to ensure total resection [88]. However, such a dural margin is not only infeasible in certain locations, but the dural tail has also been described as a purely reactive change without tumor invasion, potentially obviating the need for such a wide margin [89–91]. ^{68}Ga-DOTATATE PET/MRI needs further evaluation to differentiate tumor invasion from reactive changes to potentially guide surgical and radiation planning [85, 92, 93].

7 Expert Opinion

In the surgical management of intracranial meningiomas, maximal safe resection is undeniably integral to minimizing tumor growth or recurrence. Over time, the boundary of maximal safe resection has vastly expanded with the advent of less morbid, minimally invasive approaches, such as the endoscopic endonasal approach for midline anterior skull base lesion. While the Simpson Grade initially tried to categorize surgical resection, its reliance on subjective measures and its inability to fully incorporate the nuances of dural margins and bony infiltration have made the scoring system less relevant in the modern era. Extent of resection should now rely on advanced imaging, bolstered by burgeoning technologies like ^{68}Ga-DOTATATE PET/MRI when readily available imaging is equivocal.

Occasionally, gross total resection may be impossible due to tumor location or anatomic constraints, though the tumor is genetically benign. In other cases, gross total resection may be possible, though the tumor may represent an aggressive phenotype by its genetic or epigenetic profile. Neither the Simpson Grade nor gross total resection distinguish such clinical cases. Meningiomas can clearly have differing underlying biological behavior, therefore aggressive surgical resection and postoperative radiation should be tailored to disease specifics in each of these instances. The 2021 WHO classification has made significant strides into incorporating genetic markers of disease aggression, but there is not yet widespread availability or a deep understanding of its suggested methylation profiling.

At the current time, the senior author's institution utilizes the 2021 WHO classification for meningioma grading. Maximal safe resection is universally attempted when possible. Postoperatively, WHO grade 1 meningiomas are imaged 3 months postoperatively, and every 6 months subsequently. Patients with WHO grade

2 meningiomas undergo ^{68}Ga-DOTATATE PET/MRI at 3 months postoperatively; patients with residual undergo radiation therapy, while those without residual are followed serially with contrast-enhanced MRI. Finally, patients with WHO grade 3 meningioma universally undergo radiation therapy and are closely followed with serial contrast-enhanced MRI.

8 Conclusion

In summary, the integration of traditional prognostic factors with emerging molecular and radiological advancements provides a comprehensive approach to enhance post-meningioma resection prognostication. Predicting the recurrence of meningiomas requires the examination of various parameters including the assessment of molecular alterations, histopathological features, and surgical resection grades. This multidimensional assessment and imaging considerations show potential for customizing patient management strategies, ultimately contributing to improved outcomes in the treatment of meningiomas.

References

1. Louis DN, Perry A, Wesseling P, Brat DJ, Cree IA, Figarella-Branger D, et al. The 2021 WHO classification of tumors of the central nervous system: a summary. Neuro-Oncology. 2021;23(8):1231–51. https://doi.org/10.1093/neuonc/noab106.
2. Nowosielski M, Galldiks N, Iglseder S, Kickingereder P, von Deimling A, Bendszus M, et al. Diagnostic challenges in meningioma. Neuro-Oncology. 2017;19(12):1588–98. https://doi.org/10.1093/neuonc/nox101.
3. Patel B, Desai R, Pugazenthi S, Butt OH, Huang J, Kim AH. Identification and Management of Aggressive Meningiomas. Front Oncol. 2022;12:851758. https://doi.org/10.3389/fonc.2022.851758.
4. Gritsch S, Batchelor TT, Gonzalez Castro LN. Diagnostic, therapeutic, and prognostic implications of the 2021 World Health Organization classification of tumors of the central nervous system. Cancer. 2022;128(1):47–58. https://doi.org/10.1002/cncr.33918.
5. Clark VE, Harmancı AS, Bai H, Youngblood MW, Lee TI, Baranoski JF, et al. Recurrent somatic mutations in POLR2A define a distinct subset of meningiomas. Nat Genet. 2016;48(10):1253–9.
6. Goldbrunner R, Stavrinou P, Jenkinson MD, Sahm F, Mawrin C, Weber DC, et al. EANO guideline on the diagnosis and management of meningiomas. Neuro-Oncology. 2021;23(11):1821–34.
7. Patel B, Desai R, Pugazenthi S, Butt OH, Huang J, Kim AH. Identification and management of aggressive meningiomas. Front Oncol. 2022;12:851758.
8. Perry A, Stafford SL, Scheithauer BW, Suman VJ, Lohse CM. Meningioma grading: an analysis of histologic parameters. Am J Surg Pathol. 1997;21(12):1455–65. https://doi.org/10.1097/00000478-199712000-00008.
9. Ostrom QT, Cioffi G, Waite K, Kruchko C, Barnholtz-Sloan JS. CBTRUS statistical report: primary brain and other central nervous system tumors diagnosed in the United States in 2014–2018. Neuro-Oncology. 2021;23(Supplement_3):iii1–iii105.
10. Simpson D. The recurrence of intracranial meningiomas after surgical treatment. J Neurol Neurosurg Psychiatry. 1957;20(1):22–39. https://doi.org/10.1136/jnnp.20.1.22.
11. Oya S, Kawai K, Nakatomi H, Saito N. Significance of Simpson grading system in modern meningioma surgery: integration of the grade with MIB-1 labeling index as a key to predict the recurrence of WHO grade I meningiomas. J Neurosurg. 2012;117(1):121–8.
12. Sughrue ME, Kane AJ, Shangari G, Rutkowski MJ, McDermott MW, Berger MS, et al. The relevance of Simpson grade I and II resection in modern neurosurgical treatment of World Health Organization grade I meningiomas. J Neurosurg. 2010;113(5):1029–35.
13. Louis DN, Perry A, Wesseling P, Brat DJ, Cree IA, Figarella-Branger D, et al. The 2021 WHO classification of tumors of the central nervous system: a summary. Neuro-Oncology. 2021;23(8):1231–51.
14. Buerki RA, Horbinski CM, Kruser T, Horowitz PM, James CD, Lukas RV. An overview of meningiomas. Future Oncol. 2018;14(21):2161–77.
15. Davidson C. Histopathologic and molecular evaluation of meningioma. Neurosurg Clin N Am. 2023;34(3):311–8. https://doi.org/10.1016/j.nec.2023.02.001.
16. Louis DN, Perry A, Reifenberger G, Von Deimling A, Figarella-Branger D, Cavenee WK, et al. The 2016 World Health Organization classification of tumors of the central nervous system: a summary. Acta Neuropathol. 2016;131:803–20.
17. Louis DN. WHO classification of tumours of the central nervous system. (No Title). 2016.
18. Boulagnon-Rombi C, Fleury C, Fichel C, Lefour S, Marchal Bressenot A, Gauchotte G. Immunohistochemical approach to the differential diagnosis of Meningiomas and their mimics. J Neuropathol Exp Neurol. 2017;76(4):289–98. https://doi.org/10.1093/jnen/nlx008.
19. Menger R, Connor DE Jr, Chan AY, Jain G, Nanda A, Menger RP. Degree of resection and Ki-67 labeling index for recurring meningiomas. Cureus. 2017;9(11):e1820.

20. Liu N, Song S-Y, Jiang J-B, Wang T-J, Yan C-X. The prognostic role of Ki-67/MIB-1 in meningioma: a systematic review with meta-analysis. Medicine. 2020;99(9):e18644.
21. Ruttledge MH, Sarrazin J, Rangaratnam S, Phelan CM, Twist E, Merel P, et al. Evidence for the complete inactivation of the NF2 gene in the majority of sporadic meningiomas. Nat Genet. 1994;6(2):180–4. https://doi.org/10.1038/ng0294-180.
22. Rouleau GA, Merel P, Lutchman M, Sanson M, Zucman J, Marineau C, et al. Alteration in a new gene encoding a putative membrane-organizing protein causes neuro-fibromatosis type 2. Nature. 1993;363(6429):515–21.
23. Evans DR. Neurofibromatosis type 2 (NF2): a clinical and molecular review. Orphanet J Rare Dis. 2009;4(1):1–11.
24. McClatchey AI, Giovannini M. Membrane organization and tumorigenesis—the NF2 tumor suppressor. Merlin Genes Dev. 2005;19(19):2265–77.
25. Ueki K, Wen-Bin C, Narita Y, Asai A, Kirino T. Tight association of loss of merlin expression with loss of heterozygosity at chromosome 22q in sporadic meningiomas. Cancer Res. 1999;59(23):5995–8.
26. Ruttledge MH, Sarrazin J, Rangaratnam S, Phelan CM, Twist E, Merel P, et al. Evidence for the complete inactivation of the NF2 gene in the majority of sporadic meningiomas. Nat Genet. 1994;6(2):180–4.
27. Harmancı AS, Youngblood MW, Clark VE, Coşkun S, Henegariu O, Duran D, et al. Integrated genomic analyses of de novo pathways underlying atypical meningiomas. Nat Commun. 2017;8(1):14433.
28. Loewenstern J, Rutland J, Gill C, Arib H, Pain M, Umphlett M, et al. Comparative genomic analysis of driver mutations in matched primary and recurrent meningiomas. Oncotarget. 2019;10(37):3506.
29. Pemov A, Dewan R, Hansen NF, Chandrasekharappa SC, Ray-Chaudhury A, Jones K, et al. Comparative clinical and genomic analysis of neurofibromatosis type 2-associated cranial and spinal meningiomas. Sci Rep. 2020;10(1):12563.
30. Clark VE, Erson-Omay EZ, Serin A, Yin J, Cotney J, Ozduman K, et al. Genomic analysis of non-NF2 meningiomas reveals mutations in TRAF7, KLF4, AKT1, and SMO. Science. 2013;339(6123):1077–80. https://doi.org/10.1126/science.1233009.
31. Brastianos PK, Horowitz PM, Santagata S, Jones RT, McKenna A, Getz G, et al. Genomic sequencing of meningiomas identifies oncogenic SMO and AKT1 mutations. Nat Genet. 2013;45(3):285–9.
32. Clark VE, Erson-Omay EZ, Serin A, Yin J, Cotney J, Özduman K, et al. Genomic analysis of non-NF2 meningiomas reveals mutations in TRAF7, KLF4, AKT1, and SMO. Science. 2013;339(6123):1077–80.
33. Yuzawa S, Nishihara H, Tanaka S. Genetic landscape of meningioma. Brain Tumor Pathol. 2016;33:237–47.
34. Youngblood MW, Duran D, Montejo JD, Li C, Omay SB, Özduman K, et al. Correlations between genomic subgroup and clinical features in a cohort of more than 3000 meningiomas. J Neurosurg. 2019;133(5):1345–54.
35. Reuss DE, Piro RM, Jones DT, Simon M, Ketter R, Kool M, et al. Secretory meningiomas are defined by combined KLF4 K409Q and TRAF7 mutations. Acta Neuropathol. 2013;125:351–8.
36. Berghoff AS, Hielscher T, Ricken G, Furtner J, Schrimpf D, Widhalm G, et al. Prognostic impact of genetic alterations and methylation classes in meningioma. Brain Pathol. 2022;32(2):e12970.
37. Huang FW, Hodis E, Xu MJ, Kryukov GV, Chin L, Garraway LA. Highly recurrent TERT promoter mutations in human melanoma. Science. 2013;339(6122):957–9. https://doi.org/10.1126/science.1229259.
38. Horn S, Figl A, Rachakonda PS, Fischer C, Sucker A, Gast A, et al. TERT promoter mutations in familial and sporadic melanoma. Science. 2013;339(6122):959–61. https://doi.org/10.1126/science.1230062.
39. Koelsche C, Sahm F, Capper D, Reuss D, Sturm D, Jones DT, et al. Distribution of TERT promoter mutations in pediatric and adult tumors of the nervous system. Acta Neuropathol. 2013;126(6):907–15. https://doi.org/10.1007/s00401-013-1195-5.
40. Goutagny S, Nault JC, Mallet M, Henin D, Rossi JZ, Kalamarides M. High incidence of activating TERT promoter mutations in meningiomas undergoing malignant progression. Brain Pathol. 2014;24(2):184–9. https://doi.org/10.1111/bpa.12110.
41. Sahm F, Schrimpf D, Olar A, Koelsche C, Reuss D, Bissel J, et al. TERT promoter mutations and risk of recurrence in meningioma. J Natl Cancer Inst. 2016;108(5):djv377. https://doi.org/10.1093/jnci/djv377.
42. Sievers P, Hielscher T, Schrimpf D, Stichel D, Reuss DE, Berghoff AS, et al. CDKN2A/B homozygous deletion is associated with early recurrence in meningiomas. Acta Neuropathol. 2020;140(3):409–13. https://doi.org/10.1007/s00401-020-02188-w.
43. Bostrom J, Meyer-Puttlitz B, Wolter M, Blaschke B, Weber RG, Lichter P, et al. Alterations of the tumor suppressor genes CDKN2A (p16(INK4a)), p14(ARF), CDKN2B (p15(INK4b)), and CDKN2C (p18(INK4c)) in atypical and anaplastic meningiomas. Am J Pathol. 2001;159(2):661–9. https://doi.org/10.1016/S0002-9440(10)61737-3.
44. Guyot A, Duchesne M, Robert S, Lia AS, Derouault P, Scaon E, et al. Analysis of CDKN2A gene alterations in recurrent and non-recurrent meningioma. J Neuro-Oncol. 2019;145(3):449–59. https://doi.org/10.1007/s11060-019-03333-6.
45. Brastianos PK, Horowitz PM, Santagata S, Jones RT, McKenna A, Getz G, et al. Genomic sequencing of meningiomas identifies oncogenic SMO and AKT1 mutations. Nat Genet. 2013;45(3):285–9. https://doi.org/10.1038/ng.2526.
46. Bi WL, Greenwald NF, Abedalthagafi M, Wala J, Gibson WJ, Agarwalla PK, et al. Genomic landscape of

high-grade meningiomas. NPJ Genom Med. 2017;2:2. https://doi.org/10.1038/s41525-017-0014-7.
47. Sahm F, Schrimpf D, Stichel D, Jones DT, Hielscher T, Schefzyk S, et al. DNA methylation-based classification and grading system for meningioma: a multicentre, retrospective analysis. Lancet Oncol. 2017;18(5):682–94.
48. Nassiri F, Mamatjan Y, Suppiah S, Badhiwala JH, Mansouri S, Karimi S, et al. DNA methylation profiling to predict recurrence risk in meningioma: development and validation of a nomogram to optimize clinical management. Neuro-Oncology. 2019;21(7):901–10.
49. Olar A, Wani KM, Wilson CD, Zadeh G, DeMonte F, Jones DT, et al. Global epigenetic profiling identifies methylation subgroups associated with recurrence-free survival in meningioma. Acta Neuropathol. 2017;133:431–44.
50. Behling F, Fodi C, Gepfner-Tuma I, Kaltenbach K, Renovanz M, Paulsen F, et al. H3K27me3 loss indicates an increased risk of recurrence in the Tübingen meningioma cohort. Neuro-Oncology. 2021;23(8):1273–81.
51. Nassiri F, Wang JZ, Singh O, Karimi S, Dalcourt T, Ijad N, et al. Loss of H3K27me3 in meningiomas. Neuro-Oncology. 2021;23(8):1282–91. https://doi.org/10.1093/neuonc/noab036.
52. Patel AJ, Wan YW, Al-Ouran R, Revelli JP, Cardenas MF, Oneissi M, et al. Molecular profiling predicts meningioma recurrence and reveals loss of DREAM complex repression in aggressive tumors. Proc Natl Acad Sci USA. 2019;116(43):21715–26. https://doi.org/10.1073/pnas.1912858116.
53. Nassiri F, Liu J, Patil V, Mamatjan Y, Wang JZ, Hugh-White R, et al. A clinically applicable integrative molecular classification of meningiomas. Nature. 2021;597(7874):119–25. https://doi.org/10.1038/s41586-021-03850-3.
54. Rogers CL, Perry A, Pugh S, Vogelbaum MA, Brachman D, McMillan W, et al. Pathology concordance levels for meningioma classification and grading in NRG oncology RTOG trial 0539. Neuro-Oncology. 2016;18(4):565–74. https://doi.org/10.1093/neuonc/nov247.
55. Driver J, Hoffman SE, Tavakol S, Woodward E, Maury EA, Bhave V, et al. A molecularly integrated grade for meningioma. Neuro-Oncology. 2022;24(5):796–808. https://doi.org/10.1093/neuonc/noab213.
56. Maas SLN, Stichel D, Hielscher T, Sievers P, Berghoff AS, Schrimpf D, et al. Integrated molecular-morphologic meningioma classification: a multicenter retrospective analysis, retrospectively and prospectively validated. J Clin Oncol. 2021;39(34):3839–52. https://doi.org/10.1200/JCO.21.00784.
57. Ostrom QT, Gittleman H, Xu J, Kromer C, Wolinsky Y, Kruchko C, et al. CBTRUS statistical report: primary brain and other central nervous system tumors diagnosed in the United States in 2009–2013. Neuro-Oncology. 2016;18(suppl_5):v1–v75.
58. Rogers L, Barani I, Chamberlain M, Kaley TJ, McDermott M, Raizer J, et al. Meningiomas: knowledge base, treatment outcomes, and uncertainties. A RANO review. J Neurosurg. 2015;122(1):4–23.
59. Sun SQ, Hawasli AH, Huang J, Chicoine MR, Kim AH. An evidence-based treatment algorithm for the management of WHO grade II and III meningiomas. Neurosurg Focus. 2015;38(3):E3.
60. Schwartz TH, McDermott MW. The Simpson grade: abandon the scale but preserve the message. J Neurosurg. 2020;135:488. https://doi.org/10.3171/2020.6.JNS201904.
61. Adegbite AB, Khan MI, Paine KW, Tan LK. The recurrence of intracranial meningiomas after surgical treatment. J Neurosurg. 1983;58(1):51–6. https://doi.org/10.3171/jns.1983.58.1.0051.
62. Mirimanoff RO, Dosoretz DE, Linggood RM, Ojemann RG, Martuza RL. Meningioma: analysis of recurrence and progression following neurosurgical resection. J Neurosurg. 1985;62(1):18–24. https://doi.org/10.3171/jns.1985.62.1.0018.
63. Hasseleid BF, Meling TR, Ronning P, Scheie D, Helseth E. Surgery for convexity meningioma: Simpson grade I resection as the goal: clinical article. J Neurosurg. 2012;117(6):999–1006. https://doi.org/10.3171/2012.9.JNS12294.
64. Borovich B, Doron Y. Recurrence of intracranial meningiomas: the role played by regional multicentricity. J Neurosurg. 1986;64(1):58–63. https://doi.org/10.3171/jns.1986.64.1.0058.
65. Kinjo T, al-Mefty O, Kanaan I. Grade zero removal of supratentorial convexity meningiomas. Neurosurgery. 1993;33(3):394–9; discussion 9. https://doi.org/10.1227/00006123-199309000-00007.
66. Fukushima Y, Oya S, Nakatomi H, Shibahara J, Hanakita S, Tanaka S, et al. Effect of dural detachment on long-term tumor control for meningiomas treated using Simpson grade IV resection. J Neurosurg. 2013;119(6):1373–9. https://doi.org/10.3171/2013.8.JNS13832.
67. Materi J, Mampre D, Ehresman J, Rincon-Torroella J, Chaichana KL. Predictors of recurrence and high growth rate of residual meningiomas after subtotal resection. J Neurosurg. 2020;134(2):410–6. https://doi.org/10.3171/2019.10.JNS192466.
68. Gousias K, Schramm J, Simon M. The Simpson grading revisited: aggressive surgery and its place in modern meningioma management. J Neurosurg. 2016;125(3):551–60. https://doi.org/10.3171/2015.9.JNS15754.
69. Behling F, Fodi C, Hoffmann E, Renovanz M, Skardelly M, Tabatabai G, et al. The role of Simpson grading in meningiomas after integration of the updated WHO classification and adjuvant radiotherapy. Neurosurg Rev. 2021;44(4):2329–36. https://doi.org/10.1007/s10143-020-01428-7.
70. Brokinkel B, Spille DC, Brokinkel C, Hess K, Paulus W, Bormann E, et al. The Simpson grading: defining the optimal threshold for gross total

resection in meningioma surgery. Neurosurg Rev. 2021;44:1713–20.
71. Spille DC, Hess K, Bormann E, Sauerland C, Brokinkel C, Warneke N, et al. Risk of tumor recurrence in intracranial meningiomas: comparative analyses of the predictive value of the postoperative tumor volume and the Simpson classification. J Neurosurg. 2020;134(6):1764–71.
72. Chotai S, Schwartz TH. The Simpson grading: is it still valid? Cancers. 2022;14(8):2007.
73. Ivanidze J, Roytman M, Lin E, Magge RS, Pisapia DJ, Liechty B, et al. Gallium-68 DOTATATE PET in the evaluation of intracranial meningiomas. J Neuroimaging. 2019;29(5):650–6.
74. Kunz WG, Jungblut LM, Kazmierczak PM, Vettermann FJ, Bollenbacher A, Tonn JC, et al. Improved detection of transosseous meningiomas using 68Ga-DOTATATE PET/CT compared with contrast-enhanced MRI. J Nucl Med. 2017;58(10):1580–7.
75. Nyuyki F, Plotkin M, Graf R, Michel R, Steffen I, Denecke T, et al. Potential impact of 68 Ga-DOTATOC PET/CT on stereotactic radiotherapy planning of meningiomas. Eur J Nucl Med Mol Imaging. 2010;37:310–8.
76. Soffietti R, Wen P, Weller M, Tonn J, Galldiks N, Langen K-J, et al. Investigational PET tracers in neuro-oncology-Whats on the horizon? A report of the PET/RANO group. Neuro-Oncology. 2022;24(11):1815.
77. Wu W, Zhou Y, Wang Y, Liu L, Lou J, Deng Y, et al. Clinical significance of somatostatin receptor (SSTR) 2 in meningioma. Front Oncol. 2020;10:1633. https://doi.org/10.3389/fonc.2020.01633.
78. Reubi JC, Kappeler A, Waser B, Laissue J, Hipkin RW, Schonbrunn A. Immunohistochemical localization of somatostatin receptors sst2A in human tumors. Am J Pathol. 1998;153(1):233–45. https://doi.org/10.1016/S0002-9440(10)65564-2.
79. Schulz S, Pauli SU, Schulz S, Handel M, Dietzmann K, Firsching R, et al. Immunohistochemical determination of five somatostatin receptors in meningioma reveals frequent overexpression of somatostatin receptor subtype sst2A. Clin Cancer Res. 2000;6(5):1865–74.
80. Mahase SS, Roth O'Brien DA, No D, Roytman M, Skafida ME, Lin E, et al. [(68)Ga]-DOTATATE PET/MRI as an adjunct imaging modality for radiation treatment planning of meningiomas. Neuro Oncol Adv. 2021;3(1):vdab012. https://doi.org/10.1093/noajnl/vdab012.
81. Roytman M, Kim S, Glynn S, Thomas C, Lin E, Feltus W, et al. PET/MR imaging of somatostatin receptor expression and tumor vascularity in meningioma: implications for pathophysiology and tumor outcomes. Front Oncol. 2021;11:820287. https://doi.org/10.3389/fonc.2021.820287.
82. Kim SH, Roytman M, Madera G, Magge RS, Liechty B, Ramakrishna R, et al. Evaluating diagnostic accuracy and determining optimal diagnostic thresholds of different approaches to [(68)Ga]-DOTATATE PET/MRI analysis in patients with meningioma. Sci Rep. 2022;12(1):9256. https://doi.org/10.1038/s41598-022-13467-9.
83. Einhellig HC, Siebert E, Bauknecht HC, Tietze A, Graef J, Furth C, et al. Comparison of diagnostic value of 68 Ga-DOTATOC PET/MRI and standalone MRI for the detection of intracranial meningiomas. Sci Rep. 2021;11(1):9064. https://doi.org/10.1038/s41598-021-87866-9.
84. Rachinger W, Stoecklein VM, Terpolilli NA, Haug AR, Ertl L, Pöschl J, et al. Increased 68Ga-DOTATATE uptake in PET imaging discriminates meningioma and tumor-free tissue. J Nucl Med. 2015;56(3):347–53.
85. Galldiks N, Albert NL, Sommerauer M, Grosu AL, Ganswindt U, Law I, et al. PET imaging in patients with meningioma—report of the RANO/PET group. Neuro-Oncology. 2017;19(12):1576–87.
86. Goldbrunner R, Stavrinou P, Jenkinson MD, Sahm F, Mawrin C, Weber DC, et al. EANO guideline on the diagnosis and management of meningiomas. Neuro-Oncology. 2021;23(11):1821–34. https://doi.org/10.1093/neuonc/noab150.
87. Piper K, Yu S, Taghvaei M, Fernandez C, Mouchtouris N, Smit RD, et al. Radiation of meningioma dural tail may not improve tumor control rates. Front Surg. 2022;9:908745. https://doi.org/10.3389/fsurg.2022.908745.
88. Qi ST, Liu Y, Pan J, Chotai S, Fang LX. A radiopathological classification of dural tail sign of meningiomas. J Neurosurg. 2012;117(4):645–53. https://doi.org/10.3171/2012.6.JNS111987.
89. Tokumaru A, O'Uchi T, Eguchi T, Kawamoto S, Kokubo T, Suzuki M, et al. Prominent meningeal enhancement adjacent to meningioma on Gd-DTPA-enhanced MR images: histopathologic correlation. Radiology. 1990;175(2):431–3. https://doi.org/10.1148/radiology.175.2.2326470.
90. Wilms G, Lammens M, Marchal G, Demaerel P, Verplancke J, Van Calenbergh F, et al. Prominent dural enhancement adjacent to nonmeningiomatous malignant lesions on contrast-enhanced MR images. AJNR Am J Neuroradiol. 1991;12(4):761–4.
91. Aoki S, Sasaki Y, Machida T, Tanioka H. Contrast-enhanced MR images in patients with meningioma: importance of enhancement of the dura adjacent to the tumor. AJNR Am J Neuroradiol. 1990;11(5):935–8.
92. Perlow HK, Siedow M, Gokun Y, McElroy J, Matsui J, Zoller W, et al. 68Ga-DOTATATE PET-based radiation contouring creates more precise radiation volumes for patients with meningioma. Int J Radiation Oncol Biol Physics. 2022;113(4):859–65.
93. Ueberschaer M, Vettermann FJ, Forbrig R, Unterrainer M, Siller S, Biczok A-M, et al. Simpson grade revisited–intraoperative estimation of the extent of resection in meningiomas versus postoperative somatostatin receptor positron emission tomography/computed tomography and magnetic resonance imaging. Neurosurgery. 2021;88(1):140–6.

Aggressive Versus More Conservative Surgery for Invasive Meningiomas

Francesco Maiuri
and Marialaura Del Basso De Caro

1 Introduction

Meningiomas are more often benign tumors but with variable histologic and clinical behavior and tendency to recur. Based on the histopathologic features and brain invasion, three tumor grades are considered in the WHO classification: grade I (benign), grade II (atypical), and grade III (anaplastic) [1, 2]. Although WHO grade II and III are often defined together high grade meningiomas, they show different clinical behaviors and outcomes. A major distinguish factor that separates grade II from grade I is the presence of brain invasion. Some biomolecular markers introduced in the last classification [2], such as TERT promoter mutations and CDKN2 A and/or CDKN 2B homogeneous deletion [1], have been found to be correlates with the outcome. According to these parameters, grades I and II meningiomas were reclassified and account for up to 30% [3]. These data also reflect the high number of cases with local invasiveness.

Surgery is the primary treatment option for symptomatic meningiomas at the first diagnosis. Maximal safe resection with preservation of the neurologic function is the main aim. However, in some cases, the invasion of the contiguous structures may hind the complete resection. The definition of invasive meningiomas is referred to tumors invading the brain, bone and surrounding dura. For these meningiomas, the neurosurgeon must decide a more aggressive versus a more conservative surgery.

The chapter discusses the best surgical options for invasive meningiomas and the risk of neurosurgical complications and recurrence.

2 Definition of Complete Resection

The concept of complete resection of meningiomas was significantly modified since the introduction of the Simpson classification [4]. Simpson considered as complete resection of the intradural tumor with removal of the affected dura and bone. The grades of resection were associated to different outcomes. However, this classification is referred to the pre-microsurgical era. Sughrue et al. [5] found no statistically significant difference of progress free survive (PFS) between Simpson grades I–IV resection; this results from more extensive dural coagulation and the small tumor remnant in grade IV resections. However, it should be more suitable to consider the complete resection of the intradural

F. Maiuri (✉)
Department of Neurosciences, Reproductive and Odontostomatological Sciences, University of Naples "Federico II", Naples, Italy
e-mail: frmaiuri@unina.it

M. Del Basso De Caro
Department of Advanced Biomedical Sciences, University of Naples "Federico II", Naples, Italy

F. Maiuri, M. Del Basso De Caro (eds.), *Meningiomas*,
https://doi.org/10.1007/978-3-031-76680-0_15

tumor with resection or extensive coagulation of the dural attachment; significant residual tumor in the venous sinuses and bone is at risk of recurrence, mainly for atypical meningiomas.

The meningiomas location is also relevant. A Simpson grade I resection of a convexity meningioma is probably more radical than the grade I resection of a tumor in the skull base, where the extensive dural removal is less easy [6].

Several studies have compared the rates of complete resection in the different surgical periods, but controversial results are reported. Three studies [7–9] have found significant improvement in Simpson grade of resection through the surgical periods, likely due to improved intraoperative tools, mainly the neuronavigation [10]. On the other hand, low gross-total resection rates are reported also in recent studies and excellent results in case series before 2000 [11–12]. These differences may be explained by the patient selection and different distribution of meningiomas locations.

3 Invasive Meningiomas: Pathological Imaging and Prognostic Relevance

3.1 Brain Invasion

The term "invasive meningiomas" mainly refers to central nervous system (CNS) invasion. This has been identified as unfavorable prognostic factor [13]. Thus, the brain invasion has been included in the WHO classification of CNS tumors as a standalone criterion for assessing a meningioma to the WHO grade II [2].

The concept of brain tumor interface and brain invasion is discussed in chapter "Brain–Tumor Interface, Brain Invasion, and Peritumoral Edema in Meningiomas: Pathology: Clinical Implications" of the book. The last WHO classification defines brain invasion as "irregular tonger like protrusions of the tumor cells into underlying GFPA positive parenchyma without intervening leptomeninges" [2, 14].

However, the diagnosis of brain invasion is often differently made. The brain tumor interface is not sufficiently sampled during surgery [14–16]. Besides, focal brain invasion in a very limited area of the meningioma surface is less significant. Brain invasion meningiomas most often show additional histological features of atypia or anaplasia [17]; a positive correlation between invasion and higher Ki67 proliferation rate has been demonstrated [18]. However, the prognostic relevance of the isolated brain invasion without further characteristics of atypia or malignancy (the so-called brain invasive but otherwise benign meningiomas or BIOBM) is controversial [17].

Two recent literature reviews [17, 19] include seven studies focusing on the outcome of BIOBM [20–26]. Four of them with compared BIOBM with other meningiomas have shown higher recurrence rates of invasive tumors [21], significant better PFS in BIOBM versus atypical meningiomas [20], brain invasion as predictor of recurrence in WHO grade II meningiomas with necrosis [22] or no association between brain invasion and PFS [23]. Three other studies which compared BIOBM versus the other meningiomas WHO grade I have shown shorter PFS [24], significantly higher rates of recurrence [25], and significant increase of recurrence risk [26] in BIOBM. These data seem to confirm the prognostic role of the brain invasion. However, in the reviewed studies, the criteria for diagnosis of the brain invasion are not uniform and ill-defined and consequently the rates of brain invasion are different between the different neurosurgical centers [14, 26–28]. Thus, a prospective study including only cases with complete exploration of the whole meningiomas surface aim to clarify the problem.

Basic MR imaging may help to identifying brain invasion. Suggested features include significant peritumoral edema, irregular tumor margins, and morphology and irregular contrast enhancement [29–31]. Luo et al. [32] have shown that MR imaging features of BIOBM are more similar to WHO grade II than I. Brain invasion and meningioma grade may be better predicted preoperatively by radiomics [33–36].

3.2 Bone Invasion

Meningiomas may often invade the bone and may grow extracranially. The most common locations of brain invasion include the cranial convexity, the sphenoid wing, and the anterior cranial fossa. Meningiomas of the cranial convexity may extend in the extracranial tissues as a subcutaneous mass (Fig. 1).

Sphenoid wing meningioma may extensively invade the bone and extend in the orbit and infratemporal fossa; these of the anterior cranial fossa may grow in the ethmoidal sinus and nasal cavities (Fig. 2).

The correlation between the bone invasion and the recurrence risk is variably reported in the literature. In surgical series of WHO grade II meningiomas bone invasion was significantly associated with tumor progression, multiple recurrences and poor outcome [37–39]. On the other hand, in a series of non-NF2 type meningiomas, bone invasion and reactive hyperostosis were not associated with WHO grade, progression and recurrence [41]; similar results have been also reported in two surgical series including all meningioma types [41, 42]. Lemee et al. [43], in a large series of 1469 meningiomas, have found that bone invasion was significantly associated to lower rates of GTR (Simpson grade I-II-III). Thus, bone invasion is an indirect predictor of recurrence [6, 44, 45].

Invasive imaging features, including bone invasion, are associated to WHO high grade histology [46–48]; another study demonstrated negative association between bone invasion and WHO grade [49].

The bone invasion, both osteolysis and reactive hyperostosis, are already evidenced on the basic

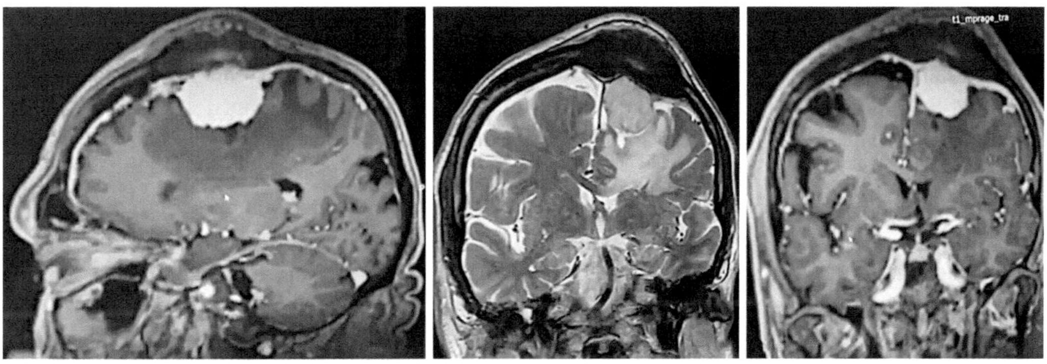

Fig. 1 Contrast-enhanced cranial MRI in sagittal (A) and coronal (B,C) sequences: invasive left parasagittal WHO II with infiltration of the bone cranial vault and marked perilesional brain edema

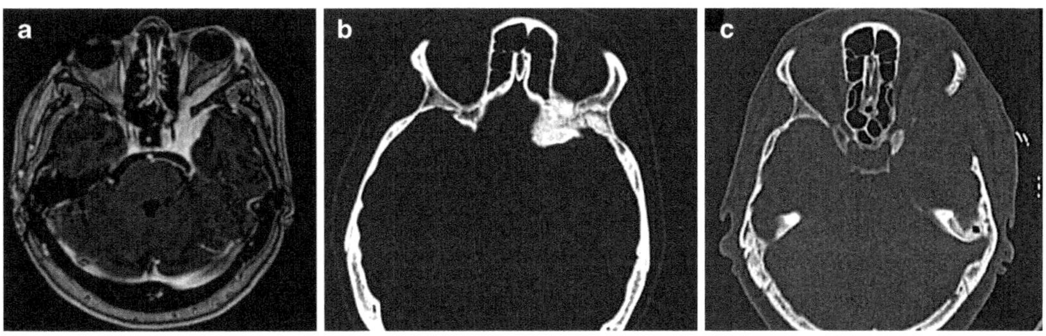

Fig. 2 Preoperative contrast-enhanced cranial MRI (**a**) and CT scan (**b**): left parasellar meningioma extending to the lateral orbital wall and sphenoid wing; significant dura and periorbita infiltration and reactive hyperostosis. (**c**) Postoperative CT scan showing large resection of the involved bone

MR study of meningiomas and better defined on CT scan with a bone window. A photo dynamic diagnosis has also been suggested; a reactive fluorescence signal from the diploe to the inner table has been demonstrated in cases of bone invasion [50]. Radiomics, a new technique imaging providing data on the biological proprieties of the tumor, has also the potential for predict bone invasion by meningiomas [33]. Scintigraphy and PET are suitable techniques for meningiomas detection. The scintigraphy with radiomarked somestatin (In DTPA octeotride) may help to identify to detect bone invasion and may help guide the surgical removal of bone invasive meningiomas [1, 51, 52]. Whole body, 18 F-fluoride PET-TC may also allow to detect the bone invasion [53], but it often provides false positive results [45].

3.3 Extensive Dural Infiltration

Some meningiomas may present at MRI variable contrast enhancement of the dura mater outside and surrounding the tumor attachment. Two mainly invasion may be observed.

(a) The dural tail, first described in 1989 [54], is a contrast-enhanced thickening of the dura contiguous to the meningioma dural attachment; it may extend for 1–2 cm and shows regular surface. It most probably arises through reactive changes as hypervascularity and dilatation of blood vessels; variable tumor infiltration has been reported in 35% to 88%of the histological specimens [55]. The dural tails does not represent a finding of invasive meningiomas.
(b) some invasive meningiomas may present variable infiltration of the surrounding dura mater; it appears to be thickened, with irregular margins and nodular aspect (differently from the dural tail) and often with extradural growth. The extension of the dural thickness may be of some centimeters.

At the cranial convexity the infiltration may involve the superior sagittal and transvers sinuses, according to tumor location. At the skull base the sphenoid wing is the most frequent site of dural infiltration, mainly in spheno-orbital meningiomas [56].

The mechanism of extensive dural infiltration outside the dural attachment has been discussed for recurrent meningiomas, but it may be considered also for tumors at initial diagnosis [57, 58]. First, localized dural multifocal nodules may grow and join to form a tumor plaque. Another theory suggests that distant nodules arise from the primary meningioma from dissemination by subarachnoid spread via CSF [59].

4 Surgery of Invasive Meningiomas

4.1 Brain Invasion

The aspect of brain invasion from a meningioma at microsurgical exploration may be variable. Most cases of BIOBM and atypical meningioma show slight invasion of the pia and superficial cortex in one or more points, whereas significant and extensive invasion of the brain parenchyma is evident in malignant meningiomas [13, 17].

The surgical management mainly depends on the tumor location and extension of the invasion along the tumor surface. Meningiomas of no-critical areas and no extensive invasion may completely be resected. For meningiomas in eloquent areas (mainly motor and speech), the extent of resection must be balanced with the risk of neurological deficits. Thus, several intraoperative techniques are used with the aim to maximize the tumor resection while reducing the risk of neurological complications. These include awake surgery [60, 61], intraoperative neurophysiological monitoring (IONM) [62–70], navigated transcranial magnetic stimulation (nTMS) [71], intraoperative 3D tractography [72], fluorescence-guided microsurgery [73–75].

Although IONM is routinely used during surgery of intracerebral tumors of the motor area, only four studies report its use in surgery of rolandic area meningiomas [63, 67–69]. A recent review [70] comparing surgical series that include patients operated with IONM versus those with no monitoring has shown that IONM use does not reduce the rate of postoperative neurological deficits. Several reason why IONM does not protect against postoperative motor deficits may be suggests. First, when the motor cortex is displaced and located below the meningioma, it is inaccessible to the direct stimulation [70, 71]. Besides, the direct cortical damage probably is not the main mechanism. Other involved factors include damage of small cortical veins, minor local ischemia, local edema, which should be evidenced by the rarely used early postoperative MRi; IONM does not protect against these anatomical changes.

The nTMS in a useful technique which allows to well localizing the motor cortex and predicting the lack of arachnoid cleavage plane, thus, selecting the cases which need for IOM. This seems the most promising tool for neurophysiological monitoring of meningioma of the rolandic region [71].

The microsurgical management of the brain tumor interface is very important. The tumor mass must progressively be reduced as much as possible up to obtain a thin tumor layer in front of the cortex. If only lack of arachnoidal layer with tumor pial adhesion is evidenced, gentle dissection and tumor resection may easily be accomplished. If cortical invasion is evidenced in some points of the residual tumor, the adherence is harder and the dissection is difficult; nevertheless, it may be realized in not eloquent region. On the other hand, in cortical motor area, the IOM is useful in deciding the management of the more adherent tumor fragments (subpial resection versus maximal reductions) [70]. The bipolar coagulation at the brain tumor interface should be avoided because of the risk of vascular injury to the cortex.

4.2 Bone and Dural Invasion

The bone invasion is a preoperative factor related to incomplete resection, in addition to tumor location at the skull base [40, 42, 43] and thus is a risk factor for recurrence. Thus, maximal safe resection of the involved bone would be preferable [9]. However, it mainly depends on the extent of invasion and the meningioma location. The bone invasion of the cranial vault may occur in convexity and parasagittal meningiomas. If the involved bone area is small, a limited resection is sufficient. Larger area of bone invasion is associated with extradural growth and hypervascularization of the dural and scalp feeders. In these cases, a large craniectomy is required, followed by cranial reconstruction [76–78]. Preoperative embolization of larger feeders may help to reduce the intraoperative bleeding [79]. Several alloplastic materials may be used for the reconstruction, both prefabricated (bPMMA, titanium plates) and intraoperatively fashioned (h PMMA, titanium mesh) [78]. Prefabricated material on the CT scan is more functional, improve cosmesis, and reduce the risk of infection. However, because the invasive bone component is often not too aggressive, the risk and benefit of the extensive resection should be balanced.

The dural infiltration is more often observed in association with bone invasion in patients with invasive meningiomas of the cranial vault or base.

At the brain convexity, the dural resection should include at least 2 cm outside the involved area. However, this may need for a too large craniotomy with consequent higher rate of complications. Thus, the opportunity of completely resecting the infiltrated dura with reconstruction must be evaluated in single cases.

At the skull base, the bone invasion is more frequent at the sphenoid wing and lesser at the anterior cranial fossa.

The sphenoid wing "en plaque" meningiomas are low grade slowly growing tumors with extensive and hard bone hyperostosis and no or small intradural component. The complete bone resection is often not possible. The resection must obtain decompression of the optic canal and superior orbital fissure (to prevent visual and oculomotor dysfunction) and good cosmetic result. On the other hand, sphenoid wing meningiomas with not "en plaque" growth show a more limited bone invasion which may be resected more easily.

Sphenoorbital meningiomas are defined as tumors with both intracranial and intraorbital component. They often invade extensively the sphenoid bone, dura mater, superior, and lateral bone of the orbit and periorbita [56]. Cophignon et al. [80] stated that to cure a patient with sphenorbital meningioma, one has to remove the entire intradural tumor, all involved dura and periorbita and invaded hyperostotic bone. Thus, complete resection can be obtained only in selected cases. The cases with more extensive invasion of the bone and dura are a surgical challenge. The invaded sphenoid wing bone must be resected as much as possible, also with the aim to reduce the proptosis [56, 81, 82]; however, residual bone fragments may remain stable even for many years. The involved dura and periorbita must also be resected by preserving the optic canal and superior orbital fissure [82].

Some patients with meningiomas of the anterior cranial fossa (both of the ethmoidal plane and olfactory groove) may rarely show invasion of the basal bone and dura. In such cases, the invaded dura must be dissected from the bone, extensively coagulated to reduce the bleeding, and then resected. This results in complete loss of sense of smell and increases the risk of cerebrospinal fluid leak. The entity of bone invasion may be variable. Cases with scarce tumor nodules may be treated conservatively with radiological follow-up, also because of the usually low-grade histology and low recurrence rate of these meningiomas [83]. Cases with significant bone invasion, mainly with extension into the ethmoidal sinus and nasal cavities need of a surgical resection [84–86]. An aggressive resection of the basal bone may result difficult and insufficient through transcranial approach; therefore, an endoscopic endonasal transsphenoidal or transethmoidal approach, alone or combined with a transcranial approach, is more indicated [84, 87–89]. Thus allows to obtain greater resection and better repair of the bone deficit.

5 Adjuvant Treatments for Invasive Meningiomas

The indication of adjuvant treatments after resection of invasive meningiomas depends on several factors, including the site of invasion, the entity of the residual tumor, and the histological grade. Radiosurgery is an alternative for small or medium sized residual meningiomas; larger remnants are also eligible for fractioned radiosurgery [1, 90]. However, more invasive meningiomas require external radiotherapy (see chapter “Meningioma Pathology and Radiotherapy at Initial Diagnosis”).

Radiosurgery is a useful treatment for residual tumor nodules at the tumor-brain surface. There is increased evidence that adjuvant radiosurgery improves local control of WHO grade II meningiomas irrespective of the extent of the initial resection [39, 91]. On the other hand, radiosurgery for bone invading component and dural infiltration of meningiomas has been less studied and remains controversial [45]. The main problem is the target delineation. Hyperostosis or direct bone invasion of WHO grade II and III meningiomas should be included in the gross tumor volume with an additional margin of 3–5 mm [92].

The role of the medical therapies in the management of invasive meningiomas is not well defined. The European Association of neuro-oncology does not provide specific recommendation [1]. Somestatin receptor agonists and VEGF inhibitors are suggested [93], but their efficacy is limited. The role of the biomolecular related medical therapies is discussed in chapter “Biomolecular-Related Medical Therapy for Meningiomas” of this book.

6 Conclusion

Meningiomas may sometimes diffusively invade the brain, surrounding dura, and bone. Invasive meningiomas are mainly atypical or malignant but sometimes also benign (WHO grade I). The maximal safe resection should be obtained particularly in WHO grade II and III tumors, where the extent of resection, is much strongly correlated with the recurrences. However, most pertinent studies provide evidence in favor of radical resection also for benign WHO grade I tumors, with major caution for cases with brain invasion at level of critical cortical areas.

References

1. Goldbrunner R, Stavrinou P, Jenkinson MD, Sahm F, Mawrin C, Weber DC, et al. EANO guideline on the diagnosis and management of meningiomas. Neuro-Oncology. 2021;23:1821–34. https://doi.org/10.1093/neuonc/noab150.

2. Louis DN, Perry A, Wesseling P, Brat D, Cree IA, Figarella-Branger D, et al. The 2021 WHO classification of tumors of the central nervous system: a summary. Neuro-Oncology. 2021;23:1231–51. https://doi.org/10.1093/neuonc/noab106.
3. Rogers L, Barani I, Chamberlain M, Kaley IJ, McDermott M, Raizer J, Schiff D, Weber DC, Wen PY, Vogelbaum MA. Meningiomas: knowledge base, treatment outcomes, and uncertainties. A RANO review. J Neurosurg. 2015;122:4–23. [CrossRef].
4. Simpson D. The recurrence of intracranial meningiomas after surgical treatment. J Neurol Neurosurg Psychiatry. 1957;20:22–39.
5. Berger MS, et al. The relevance of Simpson Grade I and II resection in modern neurosurgical treatment of World Health Organization Grade I meningiomas. J Neurosurg. 2010;113:1029–35. https://doi.org/10.3171/2010.3.JN5091971.
6. Gousias K, Schramm J, Simon M. The Simpson grading revisited: aggressive surgery and its place in modern meningioma management. J Neurosurg. 2016;125:551–60. https://doi.org/10.3171/2015.9.
7. Heald JB, Canoll TA, Mair RJ. Simpson grade: an opportunity to reassess the need for complete resection of meningiomas. Acta Neurochir (Wirn). 2013;156:383–8.
8. Bir SC, Konar SK, Maiti TK, Thakur ID, Guthikonda B, Nanda A. Utility of neuro-navigation in intracranial meningioma resection: a single-center retrospective study. World Neurosurg. 2016;90:546–555-eT.
9. Meling TR, Da Broi M, Scheie D, Helseth E, Smoll NR. Meningioma surgery-are we making progress? World Neurosurg. 2019;125:e205–13. https://doi.org/10.1016/j.wneu.2019.01.042.
10. Colen-Inbar O, Soustiel JE, Zaaroor M. Meningiomas in the elderly, the surgical benefit and a new scoring system. Acta Neurochir. 2009;152:87–97. discussion: 97
11. Coru P, Chatellier G, Dagreou E, et al. Intracranial meningiomas patients. Post-factors predictive of outcome. Acta Neurochir (Wien). 1990;102:98–102. https://doi.org/10.1007/BF01405421.
12. Maurice-Williams RS, Kitchen ND. Intractanial tumours in the elderly: the effect of age on the outcome of first time surgery for meningionias. Br J Neurosurg. 1992;6:131–7.
13. Perry A, Scheithauer BW, Stafford SL, Lohse CM, Wollan PC. "Malignancy" in meningiomas: a clinicopathologic study of 116 patients, with grading implications. Cancer. 1999;85:2046–56.
14. Timme M, et al. Brain invasion in meningiomas: does surgical sampling impact specimen characteristics and histology? Neurosurg Rev. 2020;43:793–800. https://doi.org/10.1007/510143-019-01125-0.
15. Jenkinson MD, Santarius T, Zadeh G, Aldape KD. Atypical meningioma-is it time to standardize surgical sampling techniques? Neuro-Oncology. 2017;19:453–4. https://doi.org/10.1093/neuonc/now245.
16. Picart T, Dumot C, Guyotat J, Pavlov V, Streichenberger N, Vasiljevic A, et al. Clinical and pathological impact of an optimal assessment of brain invasion for grade 2 meningioma diagnosis: lessons from a series of 291 cases. Neurosurg Rev. 2022;45:2797–809. https://doi.org/10.1007/810143-022-01792-6.
17. Gousias K, Trakolis L, Simon M. Meningiomas with CNS invasion. Front Neurosci. 2023; https://doi.org/10.3389/fins.2023.1189606.
18. Behling F, Fodi C, Wang S, Hempel JM, Hoffmann E, Tabatabai G, et al. Increased proliferation is associated with CNS invasion in meningiomas. Neuro Oncol. 2021;155:247–54. https://doi.org/10.1007/11060-021-03892-7.
19. Nakasu S, Nakasu Y. Prognostic significance of brain invasion in meningiomas: systematic review and meta-analysis. Brain Tumor Pathol. 2021;38:81–95. https://doi.org/10.1007/10014-020-00390-y.
20. Baumgarten P, Gessler E, Schittenhelm J, Skardelly M, Tews DS, Senft C, et al. Brain invasion in otherwise benign meningiomas does not predict tumor recurrence. Acta Neuropathol. 2016;132:479–81. https://doi.org/10.1007/500401-016-1598-1.
21. Spille DC, Hess K, Sauerland C, Sanai N, Stummer W, Paulus W, et al. Brain invasion in meningiomas: incidence and correlations with clinical variables and prognosis. World Neurosurg. 2016;93:346–54. https://doi.org/10.1016/j.wneu.2016.06.055.
22. Garcia-Segura ME, Erickson AW, Jairath R, Munoz DG, Das S. Necrosis and brain invasion predict radioresistance and tumor recurrence in atypical meningiomas: a retrospective cohort study. Neurosurgery. 2020;88:E42–8. https://doi.org/10.1093/neuro/nyaa348.
23. Kim MS, Chun SW, Dho YS, Seo Y, Lee JH, Won K, et al. Histopathological predictors of progression-free survival in atypical meningioma: a single-center retrospective cohort and meta-analysis. Brain Tumor Pathol. 2022;39:99–110. https://doi.org/10.1007/s10014-021-00419-w.
24. Biczok A, Jungk C, Egensperger R, von Deimling A, Suchorska B, Tonn JC, et al. Microscopic brain invasion in meningiomas previously classified as WHO grade 1 is not associated with patient outcome. J Neuro Oncol. 2019;145:469–77. https://doi.org/10.1007/11060-019-03312-x.
25. Banan R, Abbetmeier-Basse M, Hong B, Dumitru CA, Sahm E, Nakamura M, et al. The prognostic significance of clinicopathological features in meningiomas: microscopic brain invasion can predict patient outcome in otherwise benign meningiomas. Neuropathol Appl Neurobiol. 2021;47:724–35. https://doi.org/10.1111/nan.12700.
26. Traylor JL, Plitt AR, Hicks WH, Mian TM, Mickey BE, Barnett SL. Evaluating risk of recurrence in patients with meningioma. J Neurosurg. 2023;138:621–8. https://doi.org/10.3171/2022.6.JNS221162.

27. Perry A. The definition and role of brain invasion in meningioma grading: still controversial after all these years. Free Neuropathol. 2021;2:8. https://doi.org/10.17879/freeneuropathology-2021-3276.
28. Von Spreckelsen N, Kesseler C, Brokinkel B, Goldbrunner R, Perry A, Mawrin C. Molecular neuropathology of brain-invasive meningiomas. Brain Pathol. 2022;32:e13048. https://doi.org/10.1111/bpa.13048.
29. Adeli A, Hess K, Mawrin C, Streckert EMS, Stummer W, Paulus W, et al. Prediction of brain invasion in patients with meningiomas using preoperative magnetic resonance imaging. Oncotarget. 2018;9:35974–82. https://doi.org/10.18632/oncotarget.26313.
30. Joo L, Park JE, Park SY, Nam SJ, Kim YH, Kim JH, et al. Extensive peritumoral edema and brain-to-tumor interface MRI features enable prediction of brain invasion in meningioma: development and validation. Neuro-Oncology. 2021;23:324–33. https://doi.org/10.1093/neuonc/noaa190.
31. Ong T, Bharatha A, Alsufayan R, Das S, Lin AW. MRI predictors for brain invasion meningiomas. Neuroradiology. 2021; https://doi.org/10.1177/1971400920953417.
32. Luo X, Jiang H, Liu XJ, Zhang Z, Deng K, Lin F, et al. Base MRI imaging characteristics of meningioma patients to discuss the WHO classification of brain invasion otherwise benign meningiomas. Technol Cancer Res Treat. 2023;22 https://doi.org/10.1177/15330338231171470.
33. Zhang J, Yao K, Liu P, Liu Z, Han T, Zhao Z, et al. A radiomics model for preoperative prediction of brain invasion in meningioma non-invasively based on MRI: a multicentre study. EBioMedicine. 2020;58:102933. https://doi.org/10.1016/j.ebiom.2020.102933.
34. Peng S, Cheng Z, Guo Z. Diagnostic nomogram model for predicting preoperative pathological grade of meningioma. Transl Cancer Res. 2021;10:4057–61. https://doi.org/10.21037/tcr-21-798.
35. Xiao D, Zhao Z, Liu J, Wang X, Fu P, Le Grange M, et al. Diagnosis of invasive meningioma based on brain-tumor interface radiomics features on brain MR images: a multicenter study. Front Oncol. 2021;11:708040. https://doi.org/10.3389/fonc.2021.708040.
36. Ugga L, Spadarella G, Pinto L, Cuocolo R, Brunetti A. Meningioma radiomics: at the nexus of imaging, pathology and biomolecular characterization. Cancers (Basel). 2022;14:2605. https://doi.org/10.3390/cancers14112605.
37. Gabeau-Lacet D, Aghi M, Betensky RA, Barker PG, Loefller IS, Louis DN. Bone involvement predicts poor outcome in atypical meningioma. J Neurosurg. 2009;111:464–71. https://doi.org/10.3171/2009.2.JNS08877.
38. Shakir SI, Souhami L, Petrecca K, Mansure J, Singh K, Panet-Raymond V, et al. Prognostic factors for progression in atypical meningioma. J Neurosurg. 2018;129:1240–8. https://doi.org/10.3171/2017.6.JNS17120.
39. Zima L, Baine MJ, Sleightholm R, Wang B, Punsoni M, Aizenberg M, et al. Pathologic characteristics associated with local recurrence of atypical meningiomas following surgical resection. J Clin Med Res. 2021;13:143–50. https://doi.org/10.14740/jocmr+444.
40. Hwang WL, Marciscano AE, Niemierko A, Kim DW, Stemmer-Rachamimov AO, Curry WT, et al. Imaging and extent of surgical resection predict risk of meningioma recurrence better than WHO histopathological grade. Neuro-Oncology. 2016;18:863–72. https://doi.org/10.1093/neuonc/nov285.
41. Corniola MV, Lemee JM, Meling TR. Histological transformation in recurrent WHO grade I meningiomas. Sci Rep. 2020;10:11220. https://doi.org/10.1038/541598-020-68177-x.
42. Haddad AF, Young IS, Kanungo I, Sudhir S, Chen JS, Raleigh DR, et al. WHO grade I meningioma recurrence: identifying high risk patients using histopathological features and the MIB-1 index. Front Oncol. 2020;10:1522. https://doi.org/10.3389/fonc.2020.01522.
43. Lemee IM, Corniola MV, Da Broi M, Joswig H, Scheie D, Schaller K, et al. Extent of resection in meningioma: predictive factors and clinical implications. Sci Rep. 2019;9:5944. https://doi.org/10.1038/s41598-019-42451-2.
44. Voss KM, Spille DC, Sauerland C, Suero Molina E, Brokinkel C, Paulus W, et al. The Simpson grading in meningioma surgery: does the tumor location influence the prognostic value? Neuro-Oncology. 2017;133:641–51. https://doi.org/10.1007/s11060-017-2481-1.
45. Takase H, Yamamoto T. Bone invasive meningioma: recent advances and therapeutic perspectives. Front Oncol. 2022;12:895374.
46. Kawahara Y, Nakada M, Hayashi Y, Kai Y, Hayashi Y, Uchiyama N, et al. Prediction of high-grade meningioma by preoperative MRI assessment. Neuro-Oncology. 2012;108:147–52. https://doi.org/10.1007/s11060-012-0809-4.
47. Bj L, Chou KN, Kao HW, Lin C, Tsai WC, Feng SW, et al. Correlation between magnetic resonance imaging grading and pathological grading in meningioma. Neurosurgery. 2014;121:1201–8. https://doi.org/10.3171/2014.7.JNS132359.
48. Liu Y, Chotai S, Chen M, Jin S, Qi ST, Pan J. Preoperative radiologic classification of convexity meningioma to predict the survival and aggressive meningioma behavior. PLoS One. 2015;10:e0118908. https://doi.org/10.1371/journal.pone.0118908.
49. Saygin L, Çakir E. Retrospective analysis of meningioma and alternative method of grading. Erciyes Med J. 2020;42:380–5. https://doi.org/10.14744/etd.2020.69851.
50. Morofuji Y, Matsuo T, Hayashi Y, Suyama K, Nagata I. Usefulness of intraoperative photodynamic diagnosis using 5-aminolevulinic acid for meningiomas with

cranial invasion: technical case report. Neurosurgery. 2008;62:102–3.;discussion 103–104. https://doi.org/10.12271/01.neu.0000317378.22820.46.
51. Gay E, Vuillez JP, Palombi O, Brard PY, Bessou P, Passagia JG. Intraoperative and postoperative gamma detection of somatostatin receptors in bone-invasive en plaque meningiomas. Neurosurgery. 2005;57:107–13.; discussion 107–113. https://doi.org/10.1227/01.NEU.0000163490.15578.FF.
52. Galldiks N, Lohmann P, Albert NL, Tonn JC, Langen KJ. Current status of PET imaging in neuro-oncology. Neurooncol Adv. 2019;1:vdz010. https://doi.org/10.1093/noajnl/vdz010.
53. Tateishi U, Tateishi K, Shizukuishi K, Shishikura A, Murata H, Inoue T, et al. 18f-fluoride PET/CT allows detection of hyperostosis and osseous involvement in meningioma: initial experience. Clin Nucl Med. 2013;38:el 25-131. https://doi.org/10.1097/RLU.06013e318279fd79.
54. Wilms G, Lammens M, Marchal G, Van Calenbergh F, Van Plets C, Fraeyenhoven L, Baert AL. Thickening of dura surrounding meningiomas. J Comput Assist Tomogr. 1989;13:763–8. [CrossRef].
55. Kalasauskas D, et al. Evaluation of resection margin after image-guided dural tail resection in convexity meningiomas. J Clin Med. 2021; https://doi.org/10.3390/jcm10061177.
56. Mariniello G, de Divitiis O, Corvino S, Strianese D, Iuliano A, Bonavolontà G, Maiuri F. Recurrences of spheno-orbital meningiomas: risk factors and management. World Neurosurg. 2022;161:e514–22. https://doi.org/10.1016/j.wneu.2022.02.048. Epub 2022 Feb 26. PMID: 35231620.
57. Borovich B, et al. Recurrence of intracranial meningiomas: the role played by regional multicentricity. J Neurosurg. 1986; https://doi.org/10.3171/jns.1986.64.1.0058.
58. Qi ST, et al. A radiopathological classification of dural tail sign of meningiomas. J Neurosurg. 2012;117:645–53.
59. von Deimling A, et al. Clonal origin of recurrent meningiomas. Brain Pathol. 2006;9(4):645–50. https://doi.org/10.1111/j.1750-3639.1999.tb00546.x.
60. Shinoura N, Midorikawa A, Hiromitsu K, Saito S, Yamada R. Preservation of cranial nerve function following awake surgery for benign brain tumors in 22 consecutive patients. J Clin Neurosci. 2019;61:189–95. https://doi.org/10.1016/j.jocn.2018.10.037.
61. Bakhshi SK, Jawed N, Shafiq F, Enam SA. Awake craniotomy for resection of intracranial meningioma: first case series from a low-and middle-income country. Cureus. 2021;13:18716. https://doi.org/10.7759/cureus.18716.
62. Bi N, Xu RX, Liu RY, Wu CM, Wang J, Chen WD, Liu J, Xu YS, Wei ZQ, Li T, Zhang J, Bai JY, Dong B, Fan SJ, Xu YH. Microsurgical treatment for parasagittal meningioma in the central gyrus region. Oncol Lett. 2013;6:781–4. https://doi.org/10.3892/ol.2013.1429.
63. Ostry S, Netuka D, Benes V. Rolandic area meningioma resection controlled and guided by intraoperative cortical map-ping. Acta Neurochir. 2012;154:843–53. https://doi.org/10.1007/00701-012-1279-3.
64. Lee SJ, Hwang SC, Im SB, Kim BT. Surgical resection of non-glial tumors in the motor cortex. Brain Tumor Res Treat. 2016;4:70–6. https://doi.org/10.14791/btrt.2016.4.2.70.
65. Elzarief AA, Ibrahim MF. Long-term follow-up of motor function deterioration following microsurgical resection of middle third parasagittal and falx meningioma Egypt. J Neurol Psychiatr Neurosurg. 2018;54(1):9. https://doi.org/10.1186/s41983-018-0013-3.
66. Ottenhausen M, Rumalla K, Younus I, Minkowitz S, Tsiouris AJ, Schwartz TH. Predictors of postoperative motor function in rolandic meningiomas. J Neurosurg. 2018;1-6 https://doi.org/10.3171/2017.12.JNS172423.
67. Tang H, et al. Intra-operative motor function preservation for resection of primary motor cortex meningioma. Transl Cancer Res. 2018;7(6):1666–74. https://doi.org/10.21037/tcr.2018.12.11.
68. Cosman M, Pantiru IM, Cucu A, Atomei AL, Dumitrecu GF, Poeată I. Surgical management of Rolandic area meningioma in the era of intraoperative neurophysiological monitoring. Rom Neurosurg. 2020;XXXIV(4):488–94. https://doi.org/10.33962/roneuro-2020-082.
69. Paldor L, Doron O, Peso D, Jubran M, Sviri GE. Intraoperative neuromonitoring during resection of cranial meningiomas and its effect on the surgical workflow. Neurosurg Rev. 2022;45:1481–90. https://doi.org/10.1007/s10143-021-01667-2.
70. Maiuri F, Corvino S. Meningiomas of the rolandic region: risk factors for motor deficit and role of intra-operative monitoring. Acta Neurochir. 2023;165:1707–16. https://doi.org/10.1007/s00701-023-05630-6.
71. Raffa G, Picht T, Scibilia A, Rösler J, Rein J, Conti A, Ric-ciardo G, Cardali SM, Vajkoczy P, Germanò A. Surgical treatment of meningiomas located in the rolandic area: the role of navigated transcranial magnetic stimulation for preoperative planning, surgical strategy, and prediction of arachnoidal cleavage and motor outcome. J Neurosurg. 2019; https://doi.org/10.3171/2019.3.JNS183411.
72. Kumar A, Chandra PS, Sharma BS, Garg A, Rath GK, Bithal PK, et al. The role of neuronavigation guided functional MRI and diffusion tensor tractography along with cortical stimulation in patients with eloquent cortex lesions. Br J Neurosurg. 2014;28:226233. https://doi.org/10.3109/02688697.2013.835370.
73. Linsler S, Ketter R, Oertel J, Urbschat S. Fluorescence imaging of meningioma cells with somatostatin receptor ligands: an in vitro study. Acta Neurochir. 2019;161:1017–24. https://doi.org/10.1007/s00701-019-03872-x.
74. Chotai S, Schwartz TH. The Simpson grading: is it still valid? Cancers (Basel). 2022;14:2007. https://doi.org/10.3390/cancers14082007.

75. Reichert D, et al. Flavin fluorescence lifetime and autofluorescence optical redox ratio for improved visualization and classification of brain tumors front. Oncology. 2023; https://doi.org/10.3389/fonc.2023.1105648.
76. Marbacher S, Coluccia D, Fathi AR, Andereggen L, Beck J, Fan-dino J. Intraoperative patient-specific reconstruction of partial bone flap defects after convexity meningioma resection. World Neurosurg. 2013;79(1):124–30.
77. Lindner D, Schlothofer-Schumann K, Kern BC, Marx O, Múns A, Meixensberger J. Cranioplasty using custom-made hydroxyapatite versus titanium: a randomized clinical trial. J Neurosurg. 2017;126(1):175–83. https://doi.org/10.3171/2015.10.JNS151245.
78. Clynch AL, Norrington M, Mustafa MA, Richardson GE, Doherty JA, Humphries TJ, Gillespie CS, Keshwara SM, McMahon CJ, Islim AI, Jenkinson MD, Millward CP, Brodbelt AR. Cranial meningioma with bone involvement: surgical strategies and clinical considerations. Acta Neurochir (Wien). 2023;165(5):1355–63. https://doi.org/10.1007/s00701-023-05535-4. Epub 2023 Mar 6. Erratum in: Acta Neurochir (Wien). 2023 Oct;165(10):2923. doi:10.1007/s00701-023-05710-7. PMID: 36877330; PMCID: PMC10140130.
79. Manaka H, Sakata K, Tatezuki J, Shinohara T, Shimohigoshi W, Yamamoto T. Safety and efficacy of preoperative embolization in patients with meningioma. J Neurol Surg B Skull Base. 2018;79:5328–33. https://doi.org/10.1055/s-0038-1667043.
80. Cophignon J, Lucena J, Clay C, Marchac D. Limits to radical treatment of spheno-orbital meningiomas. Acta Neurochir Suppl (Wien). 1979;28:375–80.
81. Terrier LM, Bernard F, Fournier HD, et al. Spheno-orbital meningiomas surgery: multicenter management study for complex extensive tumors. World Neurosurg. 2018;112:145–015б.
82. Mariniello G, Maiuri F, Strianese D, et al. Spheno-orbital meningiomas: surgical approaches and outcome according to the intraorbital tumor extent. Zentralbl Neurochir. 2008;69:175–81.
83. Maiuri F, Mariniello G, Guadagno E, Barbato M, Corvino S, Del Basso De Caro M. WHO grade, proliferation index, and progesterone receptor expression are different according to the location of meningioma. Acta Neurochir. 2019;161:2553–61. https://doi.org/10.1007/s00701-019-04084-Z.
84. Spektor S, Valarezo J, Fliss DM, et al. Olfactory groove meningiomas from neurosurgical and ear, nose, and throat perspectives: approaches, techniques, and outcomes. Neurosurgery. 2005;57(Suppl 4):268–80; discussion: 268–80.
85. Pallini R, Fernandez E, Lauretti L, Doglietto F, D'Alessandris QG, Montano N, Capo G, Meglio M, Maira G. Olfactory groove meningioma: report of 99 cases surgically treated at the Catholic University School of Medicine, Rome. World Neurosurg. 2015;83(2):219–31.1-3.
86. Zhang J, Sai K, Zhu ZQ, Lin FH, Wang ZF, Chen YM, Huang CY, Ye YL, Wang XL, Li YP, Sun SX, Zhong WY, Chen JB, Yang YQ. Prognostic factors for olfactory groove meningioma with nasal cavity extension. Oncotarget. 2017;9(4):4607–13.
87. Maiuri F, Salzano FA, Motta S, Colella G, Sardo L. Olfactory groove meningioma with paranasal sinus and nasal cavity extension: removal by combined subfrontal and nasal approach. J Craniomaxillofac Surg. 1998;26(5):314–7. https://doi.org/10.1016/s1010-5182(98)80060-7. PMID: 9819682.
88. Schwartz TH. Should endoscopic endonasal surgery be used in the treatment of olfactory groove meningiomas? Neurosurg Focus. 2014;37(4):E9.
89. Purohit A, Jha R, Khalafallah AM, Price C, Rowan NR, Mukherjee D. Endoscopic endonasal versus transcranial approach to resection of olfactory groove meningiomas: a systematic review. Neurosurg Rev. 2019; https://doi.org/10.1007/s10143-019-01193-2.
90. Kondziolka D, Mathieu D, Lunsford LD, Martin JI, Madhok R, Niranjan A, et al. Radiosurgery as definitive management of intracranial meningiomas. Neurosurgery. 2008;62:53–8; discussion 58–60. https://doi.org/10.1227/01.NEU.0000311061.72626.00.
91. Lee G, Lamba N, Niemierko A, Kim DW, Chapman PH, Loeffler JS, Curry WT, Martuza RL, Oh KS, Barker FG 2nd, Shih HA. Adjuvant radiation therapy versus surveillance after surgical resection of atypical meningiomas. Int J Radiat Oncol Biol Phys. 2021;109:252–66. https://doi.org/10.1016/jijrobp.2020.08.015.
92. Brastianos PK, Galanis E, Butowski N, Chan IW, Dunn IF, Goldbrunner R, et al. Advances in multidisciplinary therapy for meningiomas. Neuro-Oncology. 2019;21:118–31. https://doi.org/10.1093/neuonc/noy136.
93. Karsy M, et al. Medical management of meningiomas: current status, failed treatments, and promising horizons. Neurosurg Clin. 2016; https://doi.org/10.1016/j.nec.2015.11.002.

Meningioma Pathology and Radiotherapy at Initial Diagnosis

Giuseppe Minniti, L. Zaccaro, R. De Pietro, and C. Cattaneo

1 Introduction

The primary treatment for the majority of symptomatic or enlarging meningiomas is surgery; to date, there are no randomized trials comparing surgery to other therapies for meningioma [1].

The principal aim of meningioma surgery is maximum safe resection with preservation of neurological function. Gross total resection (GTR) is determined by tumor location, consistency, size, and proximity or involvement of critical neurovascular structures. Although extent of resection (EOR) is the only modifiable risk factor for recurrence, GTR should not be performed at the expense of neurological or cognitive function. When gross total resection is not possible, a planned subtotal resection (STR) should be considered to preserve neurological function. Residual meningioma can be monitored with observation through magnetic resonance imaging (MRI) or treated with postoperative radiotherapy (RT).

Simpson described a grading system for the extent of meningioma resection [2]. According to it, an optimal resection (grade 1) is a complete tumor removal along with its "dural attachment" and any "abnormal bone" or involved venous sinus. Simpson grade 2 and 3 resection required, respectively, coagulation of the dural attachment and complete resection of the intradural tumor with no coagulation or resection of the dura or other involved sites such as the "invaded dural sinus." Subtotal resections and biopsies were considered Simpson grade 4 or 5, respectively. With long-term follow-up in the pre-imaging era, Simpson observed the following recurrence rates: 9% Simpson grade 1, 19% Simpson grade 2, 29% Simpson grade 3, and 44% Simpson grade 4.

Behling et al. analyzed the clinical and histopathological data of 1571 meningiomas surgically resected between 2003 and 2017 [3]. After the integration of all prognostic variables including adjuvant radiation therapy and the updated WHO classification, the authors highlighted the strong prognostic effect of complete tumor resection (<Simpson grade IV) and the role of adjuvant RT in lowering tumor recurrence rates. This confirms how the status of resection influences the prognosis of the patients and their therapeutic journey.

2 RT Techniques and Target Delineation

In the last years' advances in radiological imaging, their application to radiation planning and new delivery techniques have led to more accu-

G. Minniti (✉) · L. Zaccaro · R. De Pietro · C. Cattaneo
Department of Radiological Sciences, Oncology and Anatomical Pathology, Sapienza University of Rome, Policlinico Umberto I, Rome, Italy
e-mail: giuseppe.minniti@uniroma1.it

F. Maiuri, M. Del Basso De Caro (eds.), *Meningiomas*,
https://doi.org/10.1007/978-3-031-76680-0_16

rate and focused treatment. Application of conventional RT to meningiomas has evolved with the development of conformal and stereotactic techniques. Fractionated stereotactic radiotherapy (FSRT) and stereotactic radiosurgery (SRS) are two available advanced radiation techniques, that allow a steeper dose gradient at the target and a lower dose to the surrounding normal tissue compared with conventional RT.

Current stereotactic techniques include Gamma Knife (Elekta Instruments AB, Stockholm, Sweden) and linear accelerator (LINAC)-based SRS systems, such as CyberKnife (Accuray, Sunnyvale, CA, USA) or Novalis (NTx) (BrainLAB AG, Feldkirchen, Germany). To ensure a correct immobilization of the patient, Gamma Knife SRS have a rigid stereotactic frame, while LINAC-based frameless SRS systems use thermoplastic mask. The modern image guided radiation therapy (IGRT) technologies, such as orthogonal X-rays (ExacTrac®Xray 6D system) or cone beam CT (CBCT), permit to control the position prior to the treatment [4]. Among the recent technologies, MRI guided radiotherapy (MRgRT) consent to control the target position before and during treatment. The MR-LINAC (Elekta Instruments AB, Stockholm, Sweden) has also the ability to adapt the radiotherapy plan while the patient is on the treatment couch, even though its use in brain tumors is mainly limited to gliomas [5].

To targets smaller than 3 cm in size, the radiation dose is usually delivered in a single daily fraction, while conventionally fractionated (1.8 or 2 Gy/fr) or hypofractioned (>2 Gy/fr) schedules are frequently used for treating larger tumors. A better dose distribution to the target can be achieved with IMRT (Intensity-Modulated Radiation Therapy) and VMAT (Volumetric Modulated Arc Therapy) techniques. VMAT is the delivery of IMRT with the gantry always in motion using dynamic leaf motion, allowing a reduction in treatment time [6] IMRT and VMAT result in better dose delivery to complex shaped targets, and in a reduction in radiation doses to surrounding sensitive structures, e.g., optic nerves, chiasm, brainstem. Currently, no comparative studies have demonstrated the clinical superiority of a technique over another in patients with brain tumors in terms of local control and treatment-related toxicity.

An accurate delineation of the tumor and the surrounding organs at risk (OARs) is essential to achieve good local control without acute or late side effects to normal tissue nearby. The current ESTRO-ACROP guidelines [7] recommend making an imaging fusion between MRI scans and thin-slice non-contrast-enhanced CT scans for planning purposes. For resected tumors, the treatment planning is based on postoperative MRI, although preoperative MRI may provide useful information on the initial extent of disease or persistent postoperative brain infiltration. The gross tumor volume (GTV) delineation is based on the resection cavity plus any residual tumor using pre- and post-contrast T1-weighted and T2-weighted images. In some cases, PET imaging with DOTATOC or DOTANOC-tracers improve target volume definition, e.g. large tumors infiltrating the parapharyngeal soft tissues or in the bony structures which are difficult to be distinguished on MRI and CT [8, 9]. The clinical target volume (CTV) is the volume of tissue that contains any microscopic disease and potential paths of microscopic spread. In general, additional margin expansion from GTV to CTV is unnecessary for benign skull base tumors; however, a small margin of 1–3 mm may be added to encompass potential areas of microscopic tumor infiltration, e.g., the intracavernous portion of rapidly growing lesions. For atypical and malignant meningiomas, large CTV margins in the range of 10 mm are frequently utilized to adequately cover the microscopic extent of disease, but that may be reduced to 5 mm around anatomic barriers, such as non-infiltrated bone or non-infiltrated brain. The CTV can be extended along the dura up to 20 mm to encompass thickened dural tail or clearly involved bone. The planning target volume (PTV) is defined as the margin accounting for both the internal and external uncertainties in treatment planning, delivery and patient setup. No GTV-to-PTV margin is usually applied to invasive frame-based SRS, while with frameless SRS systems (CyberKnife or LINAC-based) a margin of

1–2 mm is generally created for GTV-to-PTV expansion. For patients treated with IMRT, a larger PTV margin of 3–5 mm is required, depending on the institution protocol.

3 Radiation Therapy Indications in Meningiomas

Benign meningioma, WHO grade 1, are usually treated with surgery only, whereas radiotherapy is considered after incomplete resection or following recurrence.

In a phase II trial [10], 60 patients with WHO grade 1 meningioma were observed after GTR or STR. After a median follow-up of 9.1 years, the 3-, 5-, and 10-year progression free survival (PFS) after GTR and STR were 94.3/94.3/87.6% and 83.1/72.7/72.7%, respectively. These results raise questions over the optimal management following STR, a subgroup that could benefit from adjuvant therapy.

Atypical WHO grade 2 meningiomas account for 15–25% of all meningioma lesions. Radiotherapy treatment has a role both after gross total and subtotal resection, even though observation only can be considered as an option after GTR. Chun et al. [11] reviewed 25 studies reporting outcomes of WHO grade 2 meningiomas following GTR: the crude local recurrence (LC) rate in the adjuvant RT cohort was 18.23% (70/384), which was statistically lower than the 24.68% of the observation cohort (304/1232). Adjuvant RT improved LC and PFS in 384 patients after surgery.

Palmieri et al. [12] analyzed 94 WHO type II meningioma patients: 65.9% of patients underwent adjuvant RT, while 25.5% had only close radiologic follow-up. Patients who did not undergo adjuvant RT experienced 50% more recurrence than patients who did ($P = 0.024$), showing a significant effectiveness of radiotherapy in controlling recurrence in this meningioma group. Two prospective non-randomized trials support the current evidence of adjuvant RT for WHO grade 2 meningiomas: RTOG 0539 [13] and EORTC 22042-26042 [14]. RTOG 0539 reported a 3-year PFS of 93.8% and local failure (LF) rate of 4.1% for intermediate risk meningiomas, with no grade 3 or greater radiation-related toxicities. EORTC 22042-26042 reported outcomes of 56 patients with newly diagnosed WHO grade 2 meningioma, who received postoperative EBRT (60 Gy) after GTR. The 3-year PFS was 88.7%, and the primary endpoint of improvement compared to a historic control was met. These two trials represented the base for two phase 3 randomized trials of radiotherapy versus observation following gross total resection of WHO grade 2 meningioma: NRG BN-003 [15] and ROAM/EORTC-1308 [16].

WHO grade 3 meningiomas, also known as anaplastic or malignant meningiomas, represent 2–3% of all meningiomas. These meningiomas are rare, aggressive tumors, prone to multiple recurrences, invasion, and even extra-cranial metastasis: metastases outside the blood–brain barrier are uncommon but can occur to the lungs, liver, bone, and skin [17, 18]. Given the high recurrence rates after surgery alone, combined therapy with maximal safe resection and adjuvant radiotherapy is uniformly recommended by current guidelines [1]. The aim is to achieve optimal local control without risk of normal tissue radiation-induced toxicity. This can be a challenge, considering the propensity of these tumors to invade adjacent dura and bone, which implies the use of larger margins in the radiation treatment plan.

In regards of the correct timing of radiotherapy treatment, to date there is still little information and only prospective randomized trials can assess if the timing of RT influences survival outcomes.

4 Radiation Therapy Treatment Schemes

4.1 Conventional Fractionated Radiation Therapy

Conventional fractionated RT (cFRT) remains an important component in the therapeutic management of meningiomas. For patients with meningiomas not safely amenable to surgery, or after incomplete surgical resection, the current EANO guidelines give class III evidence on the use of fractionated RT [1].

Historically, cFRT was the most used technique for adjuvant treatment of benign meningiomas. With regard to the dose and fractionation, most of published series showed no significant difference in tumor control with a range of dose between 50 and 60 Gy given in 1.8–2.0 Gy fraction, while a dose <50 Gy is associated with higher recurrence rates [19]. In clinical practice, a dose limit of 50–54 Gy is employed for large meningiomas involving the optic pathway.

Local control outcomes for grade 2 and 3 meningioma have been less impressive, as demonstrated by the high-risk arm of RTOG 0539, although dose escalation beyond 60 Gy for WHO grade 3 meningiomas remain under investigation given the predominance of in-field failures [20]. Lee et al. [21] used dose escalation to treat gross residual meningiomas to 66 Gy with a simultaneous integrated boost, resulting in local control in three of four WHO grade 3 meningiomas and no reported serious toxicity.

Zeng et al. [22] compared the outcomes of 118 meningiomas grade 2 and 3. A dose escalation cohort (≥66 Gy) (45.8% of patients) was compared with a standard-dose cohort (<66 Gy) (54.2% of patients). PFS at 3, 4, and 5 years in the dose-escalated versus standard-dose cohort was 78.9%, 72.2%, and 64.6% versus 57.2%, 49.1%, and 40.8%, respectively (P = 0.030). Seven patients (5.9%) developed symptomatic radiation necrosis (RN) with no significant difference between the two cohorts, proving that dose-escalated RT with ≥66 Gy for high-grade meningioma is associated with improved LC and PFS with an acceptable risk. Dose escalation beyond 60 Gy for grade 2 and 3 tumors is being also investigated in a phase I/II study of intensity-modulated proton therapy (IMPT) (NCT02693990).

Other than treatment protocols, tumor site and size have been reported as predictors of tumor control. Connell et al. observed 5-year control rates of 93% for 53 skull base meningiomas less than 5 cm versus 40% for tumors more than 5 cm [23]; similar findings have been reported by others [24].

The reported late toxicities of radiation treatment are hypopituitarism, optic neuropathy, cavernous cranial deficits, brain radiation necrosis and cognitive deficits. With cFRT the rate incidence ranges from 0% to 24%, being less than 5% in most series. Radiation-induced optic neuropathy (RON), appearing as decreased visual acuity or visual field defects, has been reported in 0–5% of patients receiving RT and rarely for doses lower than 54 Gy to the optic pathways with conventional fractionation (1.8–2 Gy/fr) [19, 25]. The development of hypopituitarism is particularly at risk when treating parasellar meningiomas and these patients need a long-life assessment of the pituitary function.

Cranial nerves deficits and brain necrosis have been reported in 0–3% of patients, while neurocognitive dysfunction can be a side effect of large, especially for large irradiated volumes. A recent literature review [26] pointed out how, after RT in meningiomas, cognitive abilities were impaired in 18 studies (81.8%), with 1 study showing also deterioration in all dimensions; memory was the most affected.

High dose radiation may be also associated with the development of a second brain tumor, but it should not preclude, the use of RT as an effective treatment option as the risk at 20 years is about 2.4% [27].

4.2 Advanced Fractionated Radiation Therapy

New techniques have evolved from cFRT to more sophisticated conformal and stereotactic techniques, including fractionated stereotactic radiotherapy (FSRT), intensity-modulated radiotherapy (IMRT), and volumetric modulated arc therapy (VMAT).

Minniti et al. [28] reported an overall survival time of 100% at a median follow-up of 42 months in 52 patients with large skull base meningiomas treated with FSRT (50 Gy in 30 fractions). Similar outcomes have been reported in other studies [7, 29, 30] confirming the efficacy and safety of this technique.

Combs et al. [30] analyzed 506 meningioma patients treated with FSRT (376) or IMRT (131). At a median follow-up of 107 months, no significative differences between the two cohorts were observed in term of LC, proving the efficacy of both techniques.

To date, there is no prospective data comparing outcomes of SRS with FSRT for the treatment of benign intracranial meningiomas. A comprehensive literature review attempted to extrapolate an answer from the available published data [31]. In the review, both SRS (1–5 fractions) and FSRT resulted effective treatments for benign intracranial meningiomas, with similar outcomes in achieving tumor control and 5-year PFS rates. Han et al. [32] retrospectively compared SRS (single and multi-fraction) and FSRT, obtaining no significant difference in terms of clinical and radiological response according to the technique used in the treatment of skull base meningioma. Although without prospective data, the optimal radiation technique is often based on clinical and institutional experience, considering tumor location and size at presentation.

The application of VMAT technique brought concern about the irradiation of temporal lobes. Trembecki et al. [33] evaluated retrospectively 27 meningioma patients, in which the mean values of the V45 Gy for both temporal lobes, the right temporal lobe and the left temporal lobe were lower for VMAT than for 3D: 7.54% and 7.90%, 6.82% and 9.47%, 5.67% and 7.14%, respectively. Thus, meningioma treatments using VMAT do not seem to correlate with higher doses in the temporal lobe area, compared with the conformal technique.

4.3 Single and Multisession Stereotactic Radiosurgery (SRS)

According to the American College of Radiology (ACR) and American Society for Radiation Oncology (ASTRO) Practice Guidelines for the Performance of Stereotactic Radiosurgery (SRS) [34], SRS is defined as "radiation therapy delivered via stereotactic guidance with approximately 1 mm targeting accuracy to intracranial targets in 1 to 5 fractions." Several terms have been used interchangeably for SRS delivered in 2–5 fractions, including "multisession SRS," "fractionated SRS," "hypofractionated SRS," "hypofractionated stereotactic radiotherapy." Instead, SRT refers to the treatment delivered in more than 5 fractions with the same level of accuracy.

Single session SRS (SS-SRS) is administered using a dose of 12–18 Gy in a single fraction, with the treatment usually being associated with low toxicity [35]; doses <12 Gy were associated with inferior local control [36]. SS-SRS provides good local control for small (<3 cm) meningiomas, located away from critical structures; however, with larger volumes (>3 cm), there are higher rates of treatment failure and more volume of normal brain that receives high doses of radiation, resulting in a greater risk of acute and long-term neurotoxicity (e.g., cranial nerves deficiency).

Taking this into consideration, the use of multisession SRS (MS/SRS) may improve the balance between tumor control and normal tissue toxicity over single fraction SRS, particularly for large lesions or for those located in close proximity to critical anatomical structures, such as the optic apparatus and the brainstem.

Pinzi et al. [37] have recently conducted a prospective phase 2 trial of 166 grade 1 intracranial meningioma, larger than 3 cm in size and/or located <3 mm from critical structures, treated with MS-SRS (25 Gy given in 5 fractions). With a median follow-up of 59 months and a median volume of 14 ml, the reported local control of 150 patients was 95% at 5 years. The overall toxicity rate was 12.7% (21 patients), mainly being trigeminal numbness (4%) and visual deterioration (5%).

In a single-institution review of van Buren et al. [38], MS-SRS was associated with excellent tumor control for upfront treated meningiomas, with no grade 3 or higher toxicity, except for a small fraction of patients who experienced radiation necrosis requiring longer-term steroid therapy.

Other therapeutic schedules utilized for meningioma MS-SRS treatment are 18 Gy/3fr [39] and 21 Gy/3fr [40] both carrying low late grade III/IV toxicity rates, as well low RN rate.

In general, radiation treatment of tumors growing in the skull base region has a risk of radiation-induced toxicity ranging from 3% to 10% [41].

5 Proton Therapy (PT)

Currently, the standard radiotherapy treatment is fractionated external beam irradiation with photons or radiosurgery for small tumors not close to critical brain structures. Currently, there is a renewed interest in particle therapy using protons or heavier ions, because of their intrinsic physical and biological properties that consent to maintain highly conformal target coverage while sparing normal surrounding tissues and reducing the integral dose to the patient compared to most modern photon techniques [42]. In the last decade, the expansion of proton beam facilities has led to increased use of this modality. The main difference between photons and protons is the noteworthy dose distribution obtained with the latter, where the dose is delivered at a narrow area at the distal end of the proton trajectory (Bragg Peak). For small target volumes, it may be debatable if PT gives a better dose distribution than radiosurgery [43], while for larger tumors protons usually achieve an improved dose-conformation when compared to photons.

A recent review of Weber et al. highlights excellent tumor control rates and rare radiation-induced adverse events for 500 WHO grade I meningioma patients treated with PT. Currently, no studies have prospectively compared photon-RT and PT in patients with grades 2 and 3 meningiomas. In a review by Eekers et al. [44] of retrospective studies published between 2000 and 2021, the 5-year local control and overall survival rates of 171 patients with atypical and malignant meningiomas treated with PT ranged from 38% to 71% and 53% to 100%, respectively, for grade 2 tumors and from 35.6% to 52% and from 44.4% to 51%, respectively, for grade 3 tumors. Because of the retrospective nature of the studies and the small number of patients, it is complicated to assess the efficacy of PT and photon-RT in grades 2 and 3 meningiomas. Randomized, comparative and prospective studies are necessary to prove the superiority of a technique over another in terms of clinical efficacy and safety.

Florijn et al. [45] conducted a study on 20 patients with meningiomas >3 cm, comparing IMRT, VMAT, and IMPT (intensity-modulated proton therapy) treatment plan. Automated plan generation was used to calculate a VMAT plan with three non-coplanar arcs, an IMRT plan with nine non-coplanar beams, and an IMPT plan. Compared to VMAT and IMRT, IMPT significantly improved dose conformity to the target volume, with a consequent dose reduction in OARs. With respect to VMAT, the mean dose and $D_{40\%}$ in the bilateral hippocampus were reduced by 48% and 74%, respectively ($p \leq 0.005$). With IMPT, the mean dose in the normal brain and volumes receiving 20–30 Gy were about 47% lower ($p \leq 0.01$).

Thus, for skull base meningiomas, when available IMPT should be considered in order to preserve neurocognitive function, especially in younger patients, and to decrease the probability of tumor induction. In fact, radiation-induced meningioma (RIM), diagnosed according to the modified Cahan's criteria [46], is an uncommon late risk of photon irradiation.

6 The Importance of Biomolecular Trim in the Indication for Radiation Therapy of Meningiomas

Radiotherapy has become an essential therapeutic option in the treatment of meningiomas, especially when complete surgical resection is not possible due to the position of the lesion for example, or when the Meningioma is high grade. As we know, meningiomas are classified as follows:

(a) Grade 1, benign meningiomas, are characterized by slow growth and a low likelihood of recurrence;
(b) Grade II, atypical meningiomas, have more aggressive characteristics than the previous ones and tend to recur more frequently;
(c) Grade III, anaplastic or malignant meningiomas, characterized by high aggressiveness and with a high possibility of recurrence and metastasis.

We have known for some time that the biomolecular characteristics of meningiomas can influence the response to treatments.

Recent discoveries have led to a greater understanding of meningiomas at the genomic level, not only from the point of view of treatment but also of the genesis of the tumor [47].

The possibility of performing genetic sequencing has allowed us to identify a series of SSVs (somatic short variants), such as NF2, TRAF7, PIK3CA, AKT1, the TERT promoter or CDKN2A. These somatic short variants can also become the target for systemic therapies. No SSV is currently recognized as a marker that can predict response to radiotherapy or resistance to this treatment [48].

The weight of SSVs in the genesis of the tumor, in the resistance or in the response to treatment, is currently unknown. In Fig. 1, we can see the most commonly detected mutations in meningiomas.

The biological and molecular characteristics of a meningioma can influence radiotherapy. Patients with high-risk mutations or aggressive epigenetic profiles may benefit more from (postoperative radiotherapy (PORT) regarding the doses and the possible use of systemic therapy combined with radiation therapy.

Mutations in some genes can influence the indication for radiotherapy but also the response to treatment.

Let us see some of these:

(a) *NF2 gene*, the most frequently found mutation in meningiomas is NF2, most often associated with grades II and III. This mutation, with or without loss of the long arm of chromosome 22, has an estimated prevalence of around 60% [47]. Only TRAF7/KLF4 and probably POLR2A exclude the possibility of

Gene	Form of Mutation	Clinical Associations
NF2	Loss of function	Sporadic mutations found in 40-60% of meningioma patients 50-75% of patients with germline mutations develop meningiomas Associated with increased tumor size and cell proliferation
SWI/SNF	Frameshift deletion	ARID1A mutation are independently prognostic of significantly increased hazard of death
KLF4	KLF4 K409 Q missense	The mutation results in upregulation of HIF-1a pathway Meningiomas with this mutation may respond to mTOR inhibition
TRAF7	WD40 domain mutation	TRAF7 mutations are found in nearly one-fourth of all meningiomas Meningiomas harboring TRAF7 mutations tend to be benign, chromosome-stable, and originating from medial skull base
TERT	TERT promoter chr5:1,295,228 (C228T) and chr5:1,295,250 (C250T) regional mutations	TERT promoter mutations are more commonly seen in higher grade meningiomas, particularly WHO grade 3 TERT promoter mutations are associated with significantly shorter time to progression, shorter overall survival, and higher chances of recurrence
AKT1	Gain of function	AKT1 mutations occur with higher frequency among skull base meningiomas and are associated with shorter time to recurrence
SMO/SUFU	Gain of function	Associated with higher recurrence rates among olfactory groove meningiomas Associated with larger tumor volume among anterior skull base meningiomas
PIK3CA	Gain of function	Mutations in PIK3CA are estimated to occur in 7% of non-NF2 mutated meningiomas PIK3CA mutations tend to be mutually exclusive with mutations in AKT1 and SMO
CDKN2A/B	Loss of function mutation	Mutations in CDKN2A/B are associated with shorter time recurrence CDKN2A/B alterations are now included as part of the classification criteria for WHO grade 3 meningiomas
POLR2A	Gain of function mutation	POLR2A-mutant tumors harbor distinct characteristics, including meningothelial histology, and a tendency to originate from tuberculum sellae region POLR2A mutations are found almost exclusively in WHO grade 1 meningiomas

Fig. 1 Common mutations in meningioma. (Source: Wang et al. [81])

the NF2 mutation, and this suggests a different process of tumor genesis. Patel et al. in 2021 performed sequencing and histopathological analysis of 121 grade II meningiomas and found that the NF2 gene mutation is associated with necrosis within the tumor mass. This would explain the poor radiosensitivity of NF2-mutated meningiomas [49]. In meningiomas with NF2 mutation, the administration of radiotherapy is associated with a lower PFS [50]. In fact, it seems that the presence of this mutation does not make it possible to exploit the protective effect of PORT, which instead is associated with better results in the case of NF2 wild type. This element can be considered when deciding on the possibility of adjuvant radiotherapy treatment;

(b) *TERT gene*, TERT mutation is rare in meningiomas and is associated with not only more aggressive disease but also with recurrence and progression [51]. TERT mutation is indicative of anaplasia and has a direct effect on adjuvant therapy. TERT mutation can be predicted by radiomics. The identification of this mutation can influence therapeutic decisions regarding not only radiation therapy but also systemic therapy such as target therapy. Deng et al. retrospectively examined 37 cases of patients with newly diagnosed high-grade meningiomas evaluating OS and PFS, finding that TERT mutation is associated with a worse prognosis after PORT [52]. In a future scenario of personalized treatments, is postoperative radiotherapy useful in case of TERT mutation? There is a need for further studies. However, the TERT promoter mutation may open new therapeutic options for progressive meningiomas [53].

    The presence of a TERT mutation or a CDKN2A/B mutation allows for the diagnosis of WHO grade 3 meningioma, and this requires PORT [54, 55];

(c) *TRAF7* is the second most frequently mutated gene in meningiomas. TRAF7 mutation is often associated with KLF4 mutation in secretory meningiomas [54]. The co-presence of TRAF7, AKT1, KLF4, and/or SMO mutations is generally associated with a lower risk of progression [56, 57]. Some TRAF7 mutation-associated meningiomas have heterozygous mutations suggesting partial loss of function of the gene, which may promote meningioma development [58]. Could this data inform a therapeutic decision regarding the use of adjuvant radiotherapy? We need further studies on this issue;

(d) *CDKN2A/B*, The rare CDKN2A/B gene mutation on chromosome 9p21 has been associated with a significant decrease in the time to progression of meningioma. In fact, the analysis of this mutation is recommended in the latest WHO classification [59]. CDKN2A/B plays an important role in the development of malignant meningioma and cell proliferation, making it a negative prognostic factor on the PFS of patients. Current adjuvant therapies may not be sufficient for these patients [60]. CDKN2A/B selection has a strong link with particular tumor aggressiveness [55, 61]. Radiotherapy is considered a first-line treatment for aggressive meningiomas or tumors that cannot be completely resected surgically.

Some epigenetic alterations can influence responses to treatments, just like gene mutations:

(a) **DNA methylation** may predict prognosis and response to radiotherapy, and is often associated with more aggressive behavior.

    Recent discoveries regarding DNA methylation in particular and molecular profiles in general, allow us to move toward personalized treatments [62–65]. DNA methylation combined with WHO histological classification enables a more accurate prediction of neoplasia behavior than histological classification alone [66]. It has been shown that DNA methylation can identify different subgroups within tumors that belong to the same WHO grade [64, 66]. For example, the presence of DNA methylation has allowed the identification of a subtype of WHO grade I meningiomas (2016 criteria) with a high risk of progression and another subgroup of patients affected by grade II meningioma with a low risk of progression [64]. The aggressiveness characteristics that can be extrapolated from the DNA methylation profile and which are therefore associated with a higher risk of recurrence require adjuvant therapies, such as radiotherapy and chemotherapy, since surgery alone is not necessary for tumor control [66–68];

(b) **MicroRNAs**, can modulate tumor radiosensitivity.

Circulating microRNA (miRNA) appears to be a biomarker that predicts or helps differentiate various types of disease [69, 70].

Zhanhìg et al. in 2020 published a study that highlighted the expression of different types of miRNAs in radiosensitive and radioresistant atypical meningiomas [71]. Between 2008 and 2015, 59 patients with atypical meningioma who had undergone total resection were identified. The cases included in the study were 55. The patients also underwent adjuvant radiotherapy with a dose of 50–60 Gy. The last follow-up was in December 2018 (median follow up of 57 months). A total of 43 patients met the characteristics of radiosensitivity and 12 of radio resistance, with no differences in age, location of the meningioma, size or Ki-67. In the Fig. 2, we can see the different types of miRNAs associated with sensitivity or resistance to radiation.

In this study, 14 miRNAs with different expression between radiosensitive and radioresistant atypical meningiomas were found and some of these, already known, seem to have a role in tumor malignancy (miR-7977, miR-4286, miR-1275, miR-30c-1-3p). The miRNA deregulation found in this study may help to define radioresistance. The use of miRNAs as predictors of radioresistance or radiosensitivity has several outlets. Those who are radioresistant and therefore subject to the adverse effects of radiotherapy may choose not to undergo PORT and to continue with observation only after surgery. On the other hand, the path of radiosensitization (such as immunotherapy), to improve the effect of radiation therapy is under study.

Finally, we recall:

(a) 72.2% of meningiomas express the *progesterone receptor*, while 11.3% express the estrogen receptor [72]. Finally, 45.5% of meningiomas show the androgen receptor. 78.5% of meningiomas develop in women and express the progesterone receptor. Paradoxically, the presence of progesterone receptors is associated with more favorable prognostic factors. Grade I meningiomas are

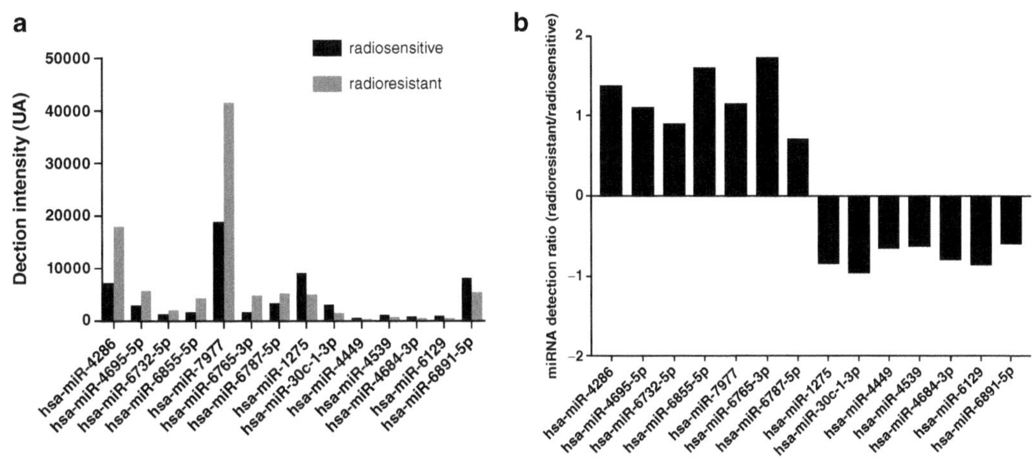

Fig. 2 Differential expression microRNAs (miRNAs) in the radiosensitive and radioresistant groups of atypical meningiomas with gross total rsection plus radiotherapy. (Source: Zhang et al. [71])

expressed in more than 75% of cases, while grade III are expressed in approximately 20%. Similarly, meningiomas that recur show a lower expression of progestin receptors. Ren et al. in 2023 published a study that included 231 patients (120 males and 11 females) with a new diagnosis of meningioma and who underwent neurosurgery [73]. All patients with atypical meningioma received radiotherapy at a dose of 54–60 Gy, (1.8–2 Gy per fraction). The last follow-up was October 30, 2021 with a median time of 70 months. 138 patients received adjuvant radiotherapy approximately 2–3 months after surgery. In the subgroup of patients with positive progestin receptor, PFS was the same for those who underwent ART (adjuvant radiotherapy) and those who did not. The hazard ratio heterogeneity analysis showed that patients who did not express estrogen receptors tended to have a greater sensitivity to PORT. This study demonstrated that adjuvant radiotherapy prolonged PFS in the PR-negative cohort but not in the PR-positive cohort. Determining progestin receptor status can guide the choice of radiotherapy;

(b) the *Ki-67* proliferation index, associated with more aggressive behavior and can guide the choice of radiotherapy dose. Ki-67 is a marker of cell growth and can be used to estimate the aggressiveness of many tumors [74]. Ki-67 is a marker of cell growth and can be used to estimate the aggressiveness of many tumors. The possibility of applying Ki-67 to meningiomas is controversial. However, there have been several studies that have suggested the idea that an elevated Ki-67 is synonymous with a higher risk of recurrence [75–77]. Ki-67 value can be used as a choice parameter for adjuvant therapy, including PORT.

As already mentioned, it is possible to use radiosensitizers to improve the outcome of radiotherapy. Among these we recall:

- Panobinostat in combination with stereotactic radiation, a protocol used for brain tumors, including high-grade meningiomas [78]. High-grade meningiomas will be treated with 54 Gy and 60 Gy in fractions of either 1.8 Gy or 2 Gy. Panobinostat will be administered orally three times a week for 2 weeks, starting on the day of first fraction of radiation. We are awaiting the results of the study.
- Nivolumab with or without ipilimumab in combination with radiosurgery for the treatment of intracranial meningiomas [79];
- Pembrolizumab in combination with radiotherapy, stereotactic surgery, may improve survival for patients affected by recurrent meningioma [80].

7 Conclusion

The biomolecular characteristics of meningiomas have now taken on an important role in the radiotherapy treatment of this pathology. Knowing the molecular profile of a meningioma makes it possible to profile an increasingly individualized therapy with better clinical and prognostic results.

References

1. Goldbrunner R, Stavrinou P, Jenkinson MD, Sahm F, Mawrin C, Weber DC, et al. EANO guideline on the diagnosis and management of meningiomas. Neurooncology [Internet]. 2021;23(11):1821–34. https://doi.org/10.1093/neuonc/noab150.
2. Simpson D. The recurrence of intracranial meningiomas after surgical treatment. J Neurol Neurosurg Psychiatry [Internet]. 1957;1;20(1):22–39. https://doi.org/10.1136/jnnp.20.1.22.
3. Behling F, Fodi C, Hoffmann E, Renovanz M, Skardelly M, Tabatabai G, et al. The role of Simpson grading in meningiomas after integration of the updated WHO classification and adjuvant radiotherapy. Neurosurg Rev [Internet]. 2020;44(4):2329–36. https://doi.org/10.1007/s10143-020-01428-7.
4. Scaringi C, Agolli L, Minniti G. Technical advances in radiation therapy for brain tumors. Anticancer Res [Internet]. 2018;38(11):6041–5. https://doi.org/10.21873/anticanres.12954.
5. Tseng CL, Chen H, Stewart J, Lau AZ, Chan RW, Lawrence LSP, et al. High grade glioma radiation therapy on a high field 1.5 Tesla MR-Linac—workflow and initial experience with daily adapt-to-

position (ATP) MR guidance: a first report. Front Oncol [Internet]. 2022;12 https://doi.org/10.3389/fonc.2022.1060098.
6. Hardcastle N, Tomé WA, Foo K, Miller A, Carolan M, Metcalfe P. Comparison of prostate IMRT and VMAT Biologically Optimised Treatment plans. Medical Dosimetry [Internet]. 2011;36(3):292–8. https://doi.org/10.1016/j.meddos.2010.06.001.
7. Combs SE, Baumert BG, Bendszus M, Bozzao A, Brada M, Fariselli L, et al. ESTRO ACROP guideline for target volume delineation of skull base tumors. Radiot Oncol [Internet]. 2021;156:80–94. https://doi.org/10.1016/j.radonc.2020.11.014.
8. Gehler B, Paulsen F, Öksüz MÖ, Hauser TK, Eschmann SM, Bares R, et al. [68Ga]-DOTATOC-PET/CT for meningioma IMRT treatment planning. Rad Oncol [Internet]. 2009;4(1) https://doi.org/10.1186/1748-717x-4-56.
9. Combs SE, Welzel T, Habermehl D, Rieken S, Dittmar JO, Kessel K, et al. Prospective evaluation of early treatment outcome in patients with meningiomas treated with particle therapy based on target volume definition with MRI and 68Ga-DOTATOC-PET. Acta Oncol [Internet]. 2013;52(3):514–20. https://doi.org/10.3109/0284186x.2013.762996.
10. Rogers CL, Pugh SL, Vogelbaum MA, Perry A, Ashby LS, Modi JM, et al. Low-risk meningioma: initial outcomes from NRG Oncology/RTOG 0539. Neuro-oncology [Internet]. 2022;25(1):137–45. https://doi.org/10.1093/neuonc/noac137.
11. Chun SW, Kim KM, Kim MS, Kang H, Dho YS, Seo Y, et al. Adjuvant radiotherapy versus observation following gross total resection for atypical meningioma: a systematic review and meta-analysis. Radiat Oncol [Internet]. 2021;16(1) https://doi.org/10.1186/s13014-021-01759-9.
12. Palmieri M, Armocida D, De Pietro R, Chiarello G, Rizzo F, Garbossa D, et al. Clinical efficacy of adjuvant radiotherapy for World Health Organization Grade II intracranial meningioma. World Neurosurg [Internet]. 2023;175:e1117–23. https://doi.org/10.1016/j.wneu.2023.04.075.
13. Rogers CL, Won M, Vogelbaum MA, Perry A, Ashby LS, Modi JM, et al. High-risk meningioma: initial outcomes from NRG Oncology/RTOG 0539. Int J Radiat Oncol Biol Phys [Internet]. 2020;106(4):790–9. https://doi.org/10.1016/j.ijrobp.2019.11.028.
14. Weber DC, Ares C, Villa S, Peerdeman SM, Renard L, Baumert BG, et al. Adjuvant postoperative high-dose radiotherapy for atypical and malignant meningioma: a phase-II parallel non-randomized and observation study (EORTC 22042-26042). Radiot Oncol [Internet]. 2018;128(2):260–5. https://doi.org/10.1016/j.radonc.2018.06.018.
15. https://www.nrgoncology.org/BN003
16. Jenkinson MD, Javadpour M, Haylock BJ, Young B, Bulbeck H, Farrell M, et al. P12.10 The ROAM/EORTC 1308 trial: radiation versus observation following surgical resection of atypical meningioma—study update. Neuro-oncology [Internet]. 2017;19(suppl_3):iii97. https://doi.org/10.1093/neuonc/nox036.368.
17. Dalle Ore CL, Magill ST, Yen AJ, Shahin MN, Lee DS, Lucas CHG, et al. Meningioma metastases: incidence and proposed screening paradigm. J Neurosurg [Internet]. 2020;132(5):1447–55. https://doi.org/10.3171/2019.1.jns181771.
18. Modha A, Gutin PH. Diagnosis and treatment of atypical and anaplastic meningiomas: a review. Neurosurgery/Neurosurgery Online [Internet]. 2005;57(3):538–50. https://doi.org/10.1227/01.neu.0000170980.47582.a5.
19. Miralbell R, Linggood RM, De La Monte S, Convery K, Munzenrider JE, Mirimanoff RO. The role of radiotherapy in the treatment of subtotally resected benign meningiomas. J Neuro-oncol [Internet]. 1992;13(2):157–64. https://doi.org/10.1007/bf00172765.
20. Press RH, Prabhu RS, Appin CL, Brat DJ, Shu HKG, Hadjipanayis C, et al. Outcomes and patterns of failure for Grade 2 meningioma treated with reduced-margin intensity modulated radiation therapy. Int J Radiat Oncol Biol Phys [Internet]. 2014;88(5):1004–10. https://doi.org/10.1016/j.ijrobp.2013.12.037.
21. Lee JJB, Lee J, Yoon HI, Kim SH, Cho J, Lee KS, et al. Analysis of patterns of failure and appraisal of postoperative radiation field for grade II–III meningioma. J Neuro-Oncol [Internet]. 2019;144(2):333–41. https://doi.org/10.1007/s11060-019-03232-w.
22. Zeng KL, Soliman H, Myrehaug S, Tseng CL, Detsky J, Chen H, et al. Dose-escalated radiation therapy is associated with improved outcomes for High-Grade meningioma. Int J Radiat Oncol Biol Phys [Internet]. 2024;118(3):662–71. https://doi.org/10.1016/j.ijrobp.2023.09.026.
23. Tumor size predicts control of benign meningiomas treated with radiotherapy [Internet]. PubMed. 1999. Available from: https://pubmed.ncbi.nlm.nih.gov/10371618/
24. Goldsmith BJ, Wara WM, Wilson CB, Larson DA. Postoperative irradiation for subtotally resected meningiomas. J Neurosurg [Internet]. 1994;80(2):195–201. https://doi.org/10.3171/jns.1994.80.2.0195.
25. Pirzkall A, Debus J, Haering P, Rhein B, Grosser KH, Höss A, et al. Intensity modulated radiotherapy (IMRT) for recurrent, residual, or untreated skull-base meningiomas: preliminary clinical experience. Int J Radiat Oncol Biol Phys [Internet]. 2003;55(2):362–72. https://doi.org/10.1016/s0360-3016(02)03809-9.
26. Gondar R, Patet G, Schaller K, Meling TR. Meningiomas and cognitive impairment after treatment: a systematic and narrative review. Cancers [Internet]. 2021;13(8):1846. https://doi.org/10.3390/cancers13081846.
27. Minniti G, Traish D, Ashley S, Gonsalves A, Brada M. Risk of second brain tumor after conservative surgery and radiotherapy for pituitary adenoma: update

after an additional 10 years. J Clin Endocrinol Metab [Internet]. 2005;90(2):800–4. https://doi.org/10.1210/jc.2004-1152.
28. Minniti G, Clarke E, Cavallo L, Osti MF, Esposito V, Cantore G, et al. Fractionated stereotactic conformal radiotherapy for large benign skull base meningiomas. Rad Oncol [Internet]. 2011;6(1) https://doi.org/10.1186/1748-717x-6-36.
29. Debus J, Wuendrich M, Pirzkall A, Hoess A, Schlegel W, Zuna I, et al. High efficacy of fractionated stereotactic radiotherapy of large base-of-skull meningiomas: long-term results. J Clin Oncol [Internet]. 2001;19(15):3547–53. https://doi.org/10.1200/jco.2001.19.15.3547.
30. Combs SE, Adeberg S, Dittmar JO, Welzel T, Rieken S, Habermehl D, et al. Skull base meningiomas: long-term results and patient self-reported outcome in 507 patients treated with fractionated stereotactic radiotherapy (FSRT) or intensity modulated radiotherapy (IMRT). Radiot Oncol [Internet]. 2013;106(2):186–91. https://doi.org/10.1016/j.radonc.2012.07.008.
31. Chung LK, Mathur I, Lagman C, Bui TT, Lee SJ, Voth BL, et al. Stereotactic radiosurgery versus fractionated stereotactic radiotherapy in benign meningioma. J Clin Neurosci [Internet]. 2017;(36):1–5. https://doi.org/10.1016/j.jocn.2016.10.009.
32. Han J, Girvigian MR, Chen JCT, Miller MJ, Lodin K, Rahimian J, et al. A comparative study of stereotactic radiosurgery, hypofractionated, and fractionated stereotactic radiotherapy in the treatment of skull base meningioma. Am J Clin Oncol [Internet]. 2014;37(3):255–60. https://doi.org/10.1097/coc.0b013e318271b36a.
33. Trembecki Ł, Siudziński T, Sztuder A, Maciejczyk A. Dosimetric comparison of conformal technique (3D) with volumetric modulated arc therapy with respect to doses obtained in the temporal lobe area in patients irradiated for brain meningioma. Reports of Practical Oncology & Radiotherapy [Internet]. 2019;24(4):325–30. https://doi.org/10.1016/j.rpor.2019.05.007.
34. Seung SK, Larson DA, Galvin JM, Mehta MP, Potters L, Schultz CJ, et al. American College of Radiology (ACR) and American Society for Radiation Oncology (ASTRO) Practice Guideline for the Performance of Stereotactic Radiosurgery (SRS). Am J Clin Oncol [Internet]. 2013;36(3):310–5. https://doi.org/10.1097/coc.0b013e31826e053d.
35. Amichetti M, Amelio D, Minniti G. Radiosurgery with photons or protons for benign and malignant tumours of the skull base: a review. Rad Oncol [Internet]. 2012;7(1) https://doi.org/10.1186/1748-717x-7-210.
36. Kuhn EN, Taksler GB, Dayton O, Loganathan AG, Vern-Gross TZ, Bourland JD, et al. Patterns of recurrence after stereotactic radiosurgery for treatment of meningiomas. Neurosurg FOCUS [Internet]. 2013;35(6):E14. https://doi.org/10.3171/2013.8.focus13283.
37. Pinzi V, Marchetti M, Viola A, Tramacere I, Cane I, Iezzoni C, et al. Hypofractionated radiosurgery for large or in critical-site intracranial meningioma: results of a phase 2 prospective study. Int J Radiat Oncol Biol Phys [Internet]. 2023;115(1):153–63. https://doi.org/10.1016/j.ijrobp.2022.08.064.
38. Van Buren A, Quinsey C, Sasaki-Adams D, Jaikumar S, Ewend M, Tepper JE, et al. Local control and toxicity following hypofractionated stereotactic radiotherapy for intracranial meningioma. Int J Radiat Oncol Biol Phys [Internet]. 2020;108(3):e672–3. https://doi.org/10.1016/j.ijrobp.2020.07.2007.
39. Nguyen EK, Pond GR, Greenspoon JN, Whitton AC, Hann C. Hypofractionated stereotactic radiotherapy for the treatment of benign intracranial meningiomas: long-term safety and efficacy. Curr Oncol [Internet]. 2021;28(5):3683–91. https://doi.org/10.3390/curroncol28050314.
40. Huang SH, Wang CC, Wei KC, Chang CN, Chuang CC, Chen HC, et al. Treatment of intracranial meningioma with single-session and fractionated radiosurgery: a propensity score matching study. Sci Rep [Internet]. 2020;10(1) https://doi.org/10.1038/s41598-020-75559-8.
41. Albano L, Losa M, Flickinger J, Mortini P, Minniti G. Radiotherapy of parasellar tumours. Neuroendocrinology [Internet]. 2020;110(9–10):848–58. https://doi.org/10.1159/000506902.
42. Combs SE, Kessel K, Habermehl D, Haberer T, Jäkel O, Debus J. Proton and carbon ion radiotherapy for primary brain tumors and tumors of the skull base. Acta Oncol [Internet]. 2013;52(7):1504–9. https://doi.org/10.3109/0284186x.2013.818255.
43. Weber DC, Bogner J, Verwey J, Georg D, Dieckmann K, Escudé L, et al. Proton beam radiotherapy versus fractionated stereotactic radiotherapy for uveal melanomas: a comparative study. Int J Radiat Oncol Biol Phys [Internet]. 2005;63(2):373–84. https://doi.org/10.1016/j.ijrobp.2005.01.057.
44. Eekers DBP, Zegers CML, Ahmed KA, Amelio D, Gupta T, Harrabi SB, et al. Controversies in neuro-oncology: focal proton versus photon radiation therapy for adult brain tumors. Neuro-Oncol Pract [Internet]. 2024; https://doi.org/10.1093/nop/npae040.
45. Florijn MA, Sharfo AWM, Wiggenraad RGJ, Van Santvoort JPC, Petoukhova AL, Hoogeman MS, et al. Lower doses to hippocampi and other brain structures for skull-base meningiomas with intensity modulated proton therapy compared to photon therapy. Radiot Oncol [Internet]. 2020;142:147–53. https://doi.org/10.1016/j.radonc.2019.08.019.
46. Singh GK, Yadav V, Singh P, Bhowmik KT. Radiation-induced malignancies making radiotherapy a "two-edged sword": a review of literature. World J Oncol [Internet]. 2017;8(1):1–6. https://doi.org/10.14740/wjon996w.
47. Chen WC, Lucas CHG, Magill ST, Rogers CL, Raleigh DR. Radiotherapy and radiosurgery for meningiomas. Neuro-oncol Adv [Internet].

2023;5(Supplement_1):i67–83. https://doi.org/10.1093/noajnl/vdac088.
48. Chen WC, Perlow HK, Choudhury A, Nguyen MP, Mirchia K, Youngblood MW, et al. Radiotherapy for meningiomas. J Neuro-oncol [Internet]. 2022;160(2):505–15. https://doi.org/10.1007/s11060-022-04171-9.
49. Patel B, Mahlokozera T, Lu HC, Yang A, Mao D, Sinha N, et al. CBIO-24. NF2 loss-of-function cooperates with hypoxia to drive radiation resistance in grade 2 meningiomas. Neuro-oncology [Internet]. 2021;23(Supplement_6):vi32. https://doi.org/10.1093/neuonc/noab196.123.
50. Vasan V, Dullea JT, Devarajan A, Ali M, Rutland JW, Gill CM, et al. NF2 mutations are associated with resistance to radiation therapy for grade 2 and grade 3 recurrent meningiomas. J Neuro-oncol [Internet]. 2022;161(2):309–16. https://doi.org/10.1007/s11060-022-04197-z.
51. Akkurt BH, Spille DC, Peetz-Dienhart S, Kiolbassa NM, Mawrin C, Musigmann M, et al. Radiomics-Based prediction of TERT promotor mutations in intracranial high-grade meningiomas. Cancers [Internet]. 2023;15(17):4415. https://doi.org/10.3390/cancers15174415.
52. Deng J, Sun S, Chen J, Wang D, Cheng H, Chen H, et al. TERT alterations predict tumor progression in de novo high-grade meningiomas following adjuvant radiotherapy. Front Oncol [Internet]. 2021;11 https://doi.org/10.3389/fonc.2021.747592.
53. Spiegl-Kreinecker S, Lötsch D, Neumayer K, Kastler L, Gojo J, Pirker C, et al. TERTpromoter mutations are associated with poor prognosis and cell immortalization in meningioma. Neuro-oncology [Internet]. 2018;20(12):1584–93. https://doi.org/10.1093/neuonc/noy104.
54. Sahm F, Schrimpf D, Olar A, Koelsche C, Reuss D, Bissel J, et al. TERT promoter mutations and risk of recurrence in meningioma. J Natl Cancer Inst [Internet]. 2015;108(5):djv377. https://doi.org/10.1093/jnci/djv377.
55. Sievers P, Hielscher T, Schrimpf D, Stichel D, Reuss DE, Berghoff AS, et al. CDKN2A/B homozygous deletion is associated with early recurrence in meningiomas. Acta Neuropathol [Internet]. 2020;140(3):409–13. https://doi.org/10.1007/s00401-020-02188-w.
56. Reuss DE, Piro RM, Jones DTW, Simon M, Ketter R, Kool M, et al. Secretory meningiomas are defined by combined KLF4 K409Q and TRAF7 mutations. Acta Neuropathol [Internet]. 2013;125(3):351–8. https://doi.org/10.1007/s00401-013-1093-x.
57. Berghoff AS, Hielscher T, Ricken G, Furtner J, Schrimpf D, Widhalm G, et al. Prognostic impact of genetic alterations and methylation classes in meningioma. Brain Pathol [Internet]. 2022;32(2) https://doi.org/10.1111/bpa.12970.
58. Najm P, Zhao P, Steklov M, Sewduth RN, Baietti MF, Pandolfi S, et al. Loss-of-function mutations in TRAF7 and KLF4 cooperatively activate RAS-like GTPASE signaling and promote meningioma development. Cancer Res [Internet]. 2021;81(16):4218–29. https://doi.org/10.1158/0008-5472.can-20-3669.
59. Louis D, Perry A, Wesseling P, Brat D, Cree I, Figarella-Branger D, et al. The 2021 WHO classification of tumors of the central nervous system: a summary. Neuro-oncology [Internet]. 2021;23(8):1231–51. https://doi.org/10.1093/neuonc/noab106.
60. Wach J, Basaran AE, Arlt F, Vychopen M, Seidel C, Barrantes-Freer A, et al. CDKN2A/B deletions are strongly associated with meningioma progression: a meta-analysis of individual patient data. Acta Neuropatholog Commun [Internet]. 2023;11(1) https://doi.org/10.1186/s40478-023-01690-y.
61. Perry AMD, Banerjee R, Lohse CM, Kleinschmidt-DeMasters BK, Scheithauer BW. A role for chromosome 9P21 deletions in the malignant progression of meningiomas and the prognosis of anaplastic meningiomas. Brain Pathol [Internet]. 2002;12(2):183–90. https://doi.org/10.1111/j.1750-3639.2002.tb00433.x.
62. Wang JZ, Nassiri F, Landry AP, Patil V, Liu J, Aldape K, et al. The multiomic landscape of meningiomas: a review and update. J Neuro-oncol [Internet]. 2023;161(2):405–14. https://doi.org/10.1007/s11060-023-04253-2.
63. Ehret F, Perez E, Teichmann D, Meier S, Geiler C, Zeus C, et al. Clinical implications of DNA methylation-based integrated classification of histologically defined grade 2 meningiomas. Acta Neuropatholog Commun [Internet]. 2024;12(1) https://doi.org/10.1186/s40478-024-01739-6.
64. Sahm F, Schrimpf D, Stichel D, Jones DTW, Hielscher T, Schefzyk S, et al. DNA methylation-based classification and grading system for meningioma: a multicentre, retrospective analysis. Lancet Oncol [Internet]. 2017;18(5):682–94. https://doi.org/10.1016/s1470-2045(17)30155-9.
65. Choudhury A, Magill ST, Eaton CD, Prager BC, Chen WC, Cady MA, et al. Meningioma DNA methylation groups identify biological drivers and therapeutic vulnerabilities. Nat Genet [Internet]. 2022;54(5):649–59. Available from: https://www.nature.com/articles/s41588-022-01061-8
66. Driver J, Hoffman SE, Tavakol S, Woodward E, Maury EA, Bhave V, et al. A molecularly integrated grade for meningioma. Neuro-oncology [Internet]. 2021;24(5):796–808. https://doi.org/10.1093/neuonc/noab213.
67. Brastianos PK, Galanis E, Butowski N, Chan JW, Dunn IF, Goldbrunner R, et al. Advances in multidisciplinary therapy for meningiomas. Neuro-oncology [Internet]. 2019;21(Supplement_1):i18–31. https://doi.org/10.1093/neuonc/noy136.
68. Maggio I, Franceschi E, Tosoni A, Di Nunno V, Gatto L, Lodi R, et al. Meningioma: not always a benign tumor. A review of advances in the treatment of meningiomas. CNS Oncol [Internet]. 2021;10(2) https://doi.org/10.2217/cns-2021-0003.
69. Summerer I, Unger K, Braselmann H, Schuettrumpf L, Maihoefer C, Baumeister P, et al. Circulating

microRNAs as prognostic therapy biomarkers in head and neck cancer patients. Br J Cancer [Internet]. 2015;113(1):76–82. Available from: https://www.nature.com/articles/bjc2015111
70. Lu J, Getz G, Miska EA, Alvarez-Saavedra E, Lamb J, Peck D, et al. MicroRNA expression profiles classify human cancers. Nature [Internet]. 2005;435(7043):834–8. Available from: https://www.nature.com/articles/nature03702
71. Zhang X, Zhang G, Huang H, Li H, Lin S, Wang Y. Differentially expressed MicroRNAs in radioresistant and radiosensitive atypical meningioma: a clinical study in Chinese patients. Front Oncol [Internet]. 2020;10 https://doi.org/10.3389/fonc.2020.00501.
72. Agopiantz M, Carnot M, Denis C, Martin E, Gauchotte G. Hormone receptor expression in meningiomas: a systematic review. Cancers [Internet]. 2023;15(3):980. https://doi.org/10.3390/cancers15030980.
73. Ren L, Cheng H, Chen J, Deng J, Wang D, Xie Q, et al. Progesterone receptor expression and prediction of benefits of adjuvant radiotherapy in de novo atypical meningiomas after gross-total resection. J Neurosurg [Internet]. 2022;1–10 https://doi.org/10.3171/2022.9.jns221530.
74. Li LT, Jiang G, Chen Q, Zheng JN. Ki67 is a promising molecular target in the diagnosis of cancer (Review). Mol Med Rep [Internet]. 2014;11(3):1566–72. https://doi.org/10.3892/mmr.2014.2914.
75. Bender L, Somme F, Lhermitte B, Ahle G, Boone M, Blonski M, et al. High risk of recurrence for grade II meningioma: a 10-year multicenter analysis of prognosis factors. Chin Clin Oncol [Internet]. 2021;10(3):26. https://doi.org/10.21037/cco-20-226.
76. Haddad AF, Young JS, Kanungo I, Sudhir S, Chen JS, Raleigh DR, et al. Recurrence: identifying high risk patients using histopathological features and the MIB-1 Index. Front Oncol [Internet]. 2020;10 https://doi.org/10.3389/fonc.2020.01522.
77. Nowak-Choi K, Palmer JD, Casey J, Chitale A, Kalchman I, Buss E, et al. Resected WHO grade I meningioma and predictors of local control. J Neurooncol [Internet]. 2021;152(1):145–51. https://doi.org/10.1007/s11060-020-03688-1.
78. https://ichgcp.net/clinical-trials-registry/NCT01324635
79. Huang J, Gao F, Shimony J, Johanns TM, Mantica M, Gershon TR, et al. The interim result of a phase I/II study of nivolumab with or without ipilimumab in combination with multi-fraction stereotactic radiosurgery for recurrent, high-grade, radiation-relapsed meningioma. J Clin Oncol [Internet]. 2022;40(16_suppl):2068. https://doi.org/10.1200/jco.2022.40.16_suppl.2068.
80. https://braintumorcenter.ucsf.edu/clinical-trial/stereotactic-radiosurgery-and-immunotherapy-pembrolizumab-treatment-recurrent
81. Wang EJ, Haddad AF, Young JS, Morshed RA, Wu JPH, Salha DM, Butowski N, Aghi MK. Recent advances in the molecular prognostication of meningiomas. Front Oncol. 2023;12:910199. https://doi.org/10.3389/fonc.2022.910199.

Biomolecular-Related Medical Therapy for Meningiomas

Alessia Pellerino, Francesco Bruno, Edoardo Pronello, Roberta Rudà, and Riccardo Soffietti

1 Introduction

Patients with progressive or recurrent meningiomas failing all surgical and radiotherapeutic treatment options or patients with meningiomas in surgically challenging areas, such as the skull base, represent an unmet need group as the efficacy of systemic treatments is limited.

A variety of agents, alone or in combination, have been employed and responses have been very few. Unfortunately, the interpretation of meningioma trials can be challenging as most studies are small or underpowered, and lack control arms. Moreover, the traditional criteria for evaluation of response are unable to fully catch the potential activity of new compounds on tumors with such an indolent growth over many years.

This chapter will be focused on a critical review of molecular basis of new treatments and ongoing trials.

A. Pellerino (✉) · F. Bruno · E. Pronello · R. Rudà
Division of Neuro-Oncology, Department of Neuroscience "Rita Levi Montalcini", University and City of Health and Science Hospital, Turin, Italy
e-mail: alessia.pellerino@unito.it

R. Soffietti
Candiolo Cancer Institute, FPO-IRCCS, Turin, Italy

2 Molecular Features and Correlation with Prognosis

TERT promoter mutations have been correlated with short progression-free survival in a retrospective series of 252 meningiomas, with 10.1 months in patients with TERT promoter mutations compared with 179 months in patients without a TERT promoter mutations regardless of the histological grading [1]. Other studies have demonstrated the negative prognostic role of TERT mutations regardless of the WHO grading, suggesting that TERT-mutated meningiomas should be followed carefully or treated aggressively, and include TERT analysis in the routine diagnostic assessment [2, 3]. Furthermore, the loss of function of dystrophin-encoding and muscular dystrophy-associated gene (DMD) has been considered an additional negative prognostic factor in TERT-mutated meningiomas [4]. SMO- and AKT-1-mutated meningiomas have been shown to recur more frequently, compared with meningiomas lacking SMO and AKT1 mutations [5]. Furthermore, AKT1 p.Glu17Lys mutation confers a reduced time to tumor recurrence [6]. Conversely, a larger study on 3031 meningioma samples from 514 individual cases has shown that TRAF7, AKT1, and/or KLF4 mutations were significantly associated with a lower risk of progression [7].

DNA methylation analyses distinguish six clinically different meningioma groups, i.e., three

F. Maiuri, M. Del Basso De Caro (eds.), *Meningiomas*,
https://doi.org/10.1007/978-3-031-76680-0_17

with favorable outcome, two with intermediate outcome, and one with a poor outcome representing a new approach for decisions regarding postoperative therapeutic interventions, in particular whether to treat patients with adjuvant radiotherapy versus observation alone [8–10]. Recently, Berghoff et al. [11] have investigated meningioma-relevant mutations and their correlation with DNA methylation clusters and patient survival. TRAKL pattern (any of the following mutations: TRAF7, AKT1, and KLF4) was predominantly found in methylation classes with favorable outcome, while NF2 was associated with methylation classes with poor outcome. TRAF7, KLF4, and TRAKL mutation genotypes were associated with improved progression-free survival (PFS) and overall survival (OS), whereas TERT promoter methylation, and intermediate- and poor outcome methylation classes, were associated with impaired PFS and OS. Methylation clustering showed better prognostic discrimination for PFS and OS than each of the individual mutations, where TERT mutation remained the unique independent significant prognostic factor for PFS in multivariable analysis [11]. Loss of H3K27me3 has been reported as prognostically unfavorable alteration in meningiomas: 13.9% (21/151) of meningiomas displayed the H3K27me3 loss by immunohistochemistry (IHC) in a multicenter study and identified a subset of WHO grades 1 and 2 meningiomas with increased risk of recurrence [12].

3 Driver Signaling Mutations in Meningiomas: Potential New Targets of Therapy

Different growth factor receptors and kinases may promote the meningioma growth, including epidermal growth factor receptor (EGFR), platelet-derived growth factor receptor (PDGFR), vascular endothelial growth factor receptor (VEGFR), and insulin-like growth factor receptor (IGFR) [13]. PDGFR, EGFR, and VEGFR may have a dual activity on the RAS–RAF–MEK–MAPK or FAK–PI3K–AKT pathway, resulting in growth-favoring signals in meningiomas. Moreover, after the activation of AKT, an intracellular signaling pathway acts on mTORC1 and 2 and regulates DNA replication. Other signaling pathways shown to be activated in meningiomas, are the phospholipase A2–arachidonic acid–cyclooxygenase pathway [14], the phospholipase C 1 (PLC1)–protein kinase C pathway (PKC) [15], and the transforming growth factor-beta (TGF-β) SMAD signaling pathway [16, 17], which act as inhibitory mechanisms of meningioma growth, thus representing potential targets of treatment.

4 Role of Tumor Microenvironment in Meningiomas: Is It Druggable?

The meningioma microenvironment seems to play a role in tumor growth, and some data suggest that WHO grades 2 and 3 meningiomas represent a relatively immunosuppressed status. In particular, grading in meningioma negatively correlates with the amount of CD4+, CD8+, and PD-1+ lymphocytes, along with increased numbers of Treg (FOXP3+) cells in the tumor. Moreover, programmed death-ligand 1 (PD-L1) expression has been correlated with the grading of meningiomas, with PD-L1 protein detection in 40% of grade 1, 60% of grade 2, and 77–88% of grade 3 meningiomas [18]. Karimi et al. [19] reported that PD-L1 protein expression has a patchy pattern, along with peri-vascular and perinecrotic, membranous, and cytoplasmic immunoreactivity in both tumor and immune cells of the microenvironment. Furthermore, PD-L1 expression was confined to a small subpopulation of cells (median < 1% of cells, range 0–20% of cells), and correlated with a higher risk of early recurrence regardless of grading, the extent of resection, and tumor diameter. Tumor-infiltrating T lymphocytes (TILs) around meningiomas may influence the prognosis. Rapp et al. [20] analyzed the presence of TILs in 97 newly diagnosed and 62 recurrent high-grade meningiomas, reporting that a higher number of cytotoxic TILs (CD3+ CD8+ FOXP3−) were associated with an improved PFS, while recurrent meningiomas were characterized by lower numbers of TILs and proportions of PD-1+CD8+ T cells. Some findings support the hypothesis that somatic genetic alterations in meningiomas can potentially affect PD-L1 or other checkpoint protein expression. For instance, an increased PD-L1 expression was found in TRAF7-mutated, compared with wild-type skull base meningiomas [21], and an increased number of CTLA4+/CD3+ lymphocytes was found in grades 2 and 3 meningiomas harboring the PI3K–AKT–mTOR pathway or SMO mutations [22]. Lastly, 40% of NF2-mutated meningiomas express PD-L1 in the surrounding microenvironment [23], suggesting that the therapeutic role of checkpoint inhibitors is worth-investigating in progressive/refractory meningiomas after the failure of surgery and/or radiation therapy [24].

As M2 macrophages are the most prevalent immune cell type in meningiomas, Yeung et al. [25] have targeted with a specific monoclonal antibody the colony-stimulating factor 1 (CSF1) and CFS1 receptor (CSF1R) expressed in myeloid cells, reporting a significant reduction in tumor growth in a murine meningioma model. This provides a strong rationale for future human clinical trials targeting the CSF1–CSF1R pathway in malignant meningiomas.

5 Systemic Therapy for Progressive/Recurrent Meningiomas: Present and Future

Patients with progressive/recurrent meningiomas, in whom surgery and/or radiation are not feasible anymore, have a limited PFS, ranging from 6-month PFS of 29% for grade 1 to 26% for grades 2 and 3 tumors [26]. To date, there is no evidence regarding the standard of care, and enrollment in clinical trials is recommended in case of disease progression [27]. A matter of debate is the choice of the best endpoint in clinical trials in surgery- and radiation-refractory meningioma: in 2019, the Response Assessment in Neuro-Oncology (RANO) group stated that an appropriate endpoint for medical therapy trials is either a 6-month PFS rate alone or in combination with radiological response [28].

Different cytotoxic chemotherapies and targeted agents have been investigated with poor results in terms of disease control and survival: they include hydroxyurea [29–31], temozolomide [32], irinotecan [33], trabectedin [34, 35], IFN-α [36–38], somatostatin analogs (pasireotide) [39] and octreotide [40, 41], VEGF/VEGFR inhibitors [42–45], EGFR inhibitors (erlotinib and gefitinib) [46], imatinib [47], and mifepristone [48].

As meningiomas are highly vascularized, anti-VEGF drugs have been largely investigated.

The most employed compound was bevacizumab, which has displayed some benefit in terms of PFS (median PFS 16.8 months, range 6.5–22 months; 6-month PFS: 73%, range 44–93%), with particular advantage in patients with high-grade and/or multiple and/or radiation-induced meningiomas [49]. The clinical and radiological benefit of bevacizumab may derive from a pronounced inhibitory effect on tumor growth, as well as some anti-edema activity, in comparison with other targeted therapies and cytotoxic agents [50]. Given these favorable properties, bevacizumab was investigated in combination with other compounds. Shih et al. [43] evaluated the activity of bevacizumab and the mTORC1 inhibitor everolimus and reported the best radiological response for stable disease (SD) in 15 patients (88%), and 6 of these patients had SD for >12 months. Median PFS was 22 months (95% CI 4.5–26.8) and was longer for patients with grade 2 and 3 than for those with grade 1 meningiomas (22.0 months vs. 17.5 months, respectively).

The Combination of Everolimus and Octreotide LAR in Aggressive Recurrent Meningiomas (CEVOREM) phase 2 trial reported a 6-month PFS of 55% (95% CI 31.3–73.5%), and 6-month- and 12-month OS of 90% (95% CI 65.6–97.4%) and 75% (95% CI, 50.0–88.7%), respectively, in 20 patients with progressive meningiomas. A radiological response (decrease >50%) was achieved in 78% of patients, with a median tumor growth rate decreasing from 16.6% 3 months before inclusion to 0.02% after 3 months and 0.48% at 6 months after treatment [51].

In the era of precision medicine, we may select appropriate therapy based on specific genetic mutations. In this regard, the Alliance/NCI A071401 study initiated a genomically-driven meningioma phase 2 trial in which the targeted therapy is delivered according to the mutation found in the tissue. Thus, different compounds are under investigation, including the SMO inhibitor vismodegib in SMO-mutant tumors, the AKT inhibitor capivasertib (AZD5363) for AKT/PIK3CA-mutant tumors, the CDK inhibitor abemaciclib for NF2 or CDK-mutant tumors, and the FAK inhibitor GSK2256098 for SMO/PTCH1-mutant and NF2-mutant meningiomas, respectively (NCT02523014). For latter group the FAK inhibitor arm has completed the accrual with 37 patients enrolled (12 grade 1 and 25 grades 2 and 3 meningiomas) in the trial [52]. Most patients received prior radiotherapy (75.7%) and chemotherapy (40.5%) before the start of the FAK inhibitor. One patient had a partial response, and 24 had SD as the best response to treatment. In grade 1 meningiomas, the 6-month PFS was 83% (10/12 patients; 95% CI: 52–98%) while in grade 2 and 3 meningiomas, the 6–month PFS was 33% (8/24 patients; 95% CI: 16–55%). The study met the 6-month PFS endpoint both for the grade 1 and the grade 2 and 3 cohorts with excellent tolerability.

However, a major concern for all targeted therapies is the resistance owing to tumor heterogeneity. Indeed, malignant meningiomas have been shown significant molecular heterogeneity between the original tumor and recurrence [53].

High-grade meningiomas may harbor an immunosuppressive microenvironment. In this regard, some studies reported that a subset of high-grade meningiomas have a high somatic mutation burden [54], that could represent a predisposing factor for response to immune-checkpoint inhibitors [54, 55]. Brastianos et al. [56] designed a phase 2 trial evaluating pembrolizumab in 26 patients with recurrent high-grade meningiomas (23 grade 2, and 3 grade 3 tumors). The study met the primary endpoint and achieved a 6-month PFS of 48% (90% CI 31–66), a median PFS of 7.6 months (90% CI 3.4–12.9 months), and a median OS of 20.2 months (90% CI 14.8–25.8 months). Eighteen patients had SD as the best radiological response, while no patient had complete or partial response according to RANO criteria. PD-L1 expression in pretreatment tissue was not correlated with outcome. Notably, the trial enrolled seven patients with metastatic extracranial meningiomas, and one patient had PFS lasting for nearly 20 months. While the trial met the primary endpoint, these results will require additional validation, and further studies are needed to identify which meningioma subtypes or tumor microenvironment patterns are correlated with the efficacy of immune-checkpoint inhibitors [56].

Conversely, a phase 2 trial on 20 patients using anti-PD1 nivolumab failed to improve 6-month PFS (42.4%, 95% CI 22.8, 60.7), although a subset of patients appeared to derive benefit (one patient obtained a partial response) [57].

Novel targeted agents are under investigation in clinical trials, including MEK inhibitor alone (selumetinib, NCT03095248) or in combination with Pi3K_ inhibitor (alpelisib, NCT03631953), VEGF inhibitor apatinib (NCT04501705), CDK-p16-Rb inhibitor ribociclib (NCT02933736), immune-checkpoint inhibitors, such as nivolumab alone or in combination with ipilimumab (NCT02648997) or with stereotactic radiosurgery (NCT03604978), or sintilimab (NCT04728568) (Table 1).

Table 1 Ongoing clinical trials in meningiomas

Trial ID	Type of study	Arm of treatment	*N*	Endpoints
NCT02648997	Phase 2	Nivolumab alone (Cohort 1) or in combination with ipilimumab (Cohort 2)	50	Primary: 6-month PFS Secondary: median PFS, median OS, ORR, safety
NCT04728568	Prospective	Sintilimab	15	Primary: PFS Secondary: OS
NCT04501705	Prospective	Apatinib	29	Primary: 6-month PFS Secondary: ORR, OS
NCT03604978	Phase 1–2	Nivolumab alone or plus ipilimumab in combination with fractionated SRS	15	Primary: DLT, safety, ORR Secondary: median PFS, median OS, changes in peripheral T cells
NCT02933736	Early phase 1	Ribociclib	48	Primary: plasma exposure, CSF penetration, brain accumulation of ribociclib
NCT02523014	Phase 2	Vismodegib or FAK inhibitor, or GSK2256098 or capivasertib or abemaciclib based on molecular screening	124	Primary: 6-month PFS, ORR Secondary: median PFS, median OS, safety
NCT04659811	Phase 2	Pembrolizumab plus SRS	90	Primary: 12-month PFS Secondary: median PFS, median OS
NCT04374305	Phase 2	Brigatinib	80	Primary: radiological response rate Secondary: safety
NCT03095248	Phase 2	Selumetinib	34	Primary: change in hearing response, response rate of other NF2 related tumors (including meningiomas)
NCT04541082	Phase 1	ONC206	102	Primary: MTD
NCT05425004	Phase 2	Cabozantinib	24	Primary: PFS Secondary: ORR, OS, safety, quality of life
NCT04997317	Early phase 1	– 177Lu-DOTA-JR11 (Phase 0); Cycle 1 and Cycle 2 (cross-over); – 177Lu-DOTATOC (Phase 0); Cycle 1 and Cycle 2 (cross-over), Cycle 3 and 4; – Drug: 177Lu-DOTA-JR11 (Phase I/II)	18	Primary: Change in tumor-to-dose limiting organ dose ratio T-to-bone marrow: therapeutic index (phase 0); Change in tumor-to-dose limiting organ dose ratio T-to-kidney: therapeutic index (phase 0); Safety Secondary: tumor absorbed dose (Gy), organ absorbed dose (Gy), organ and tumor residence time, tumor-to-remaining organ dose ratio, evaluation radiation doses, early onset toxicity, change in plasma concentration of 177Lu-DOTA-JR11, whole body and organ radiation doses
NCT05023018	Phase 2	NEO100 (perillyl alcohol)	30	Primary: 6-month PFS Secondary: ORR, OS
NCT03971461	Phase 2	Lutathera (177Lu-DOTATATE radionuclide)	32	Primary: 6-month PFS Secondary: ORR, 12-month OS, median OS, PFS

(continued)

Table 1 (continued)

Trial ID	Type of study	Arm of treatment	*N*	Endpoints
NCT02847559	Phase 2	Bevacizumab plus Electric Field Therapy (NovoTTF-200A device)	27	Primary: 6-month PFS Secondary: ORR, OS, quality of life
NCT03180268	Phase 3	Radiation therapy versus observation after surgery	148	Primary: PFS Secondary: median OS, 5-year OS, 3-year PFS, neurocognitive functions, patient reported outcomes, safety
NCT02693990	Phase 1/2	IMPT	60	Primary: safety and utility of increased dose IMPT (DLT) Secondary: PFS, OS, calculate linear energy transfer using computer simulations based on the treatment plan

Gy gray, *IMPT* intensity modulated proton therapy, *PFS* progression-free survival, *OS* overall survival, *ORR* objective response rate, *DLT* dose limiting toxicity, *SRS* stereotactic radiosurgery, *MTD* maximum tolerated dose

6 Conclusions

Despite multiple studies on different cytotoxic agents performed on recurrent meningiomas.

in the past several years, limited evidence of efficacy and clinical benefit exist thus far. Since 2013, a genomic revolution in the biology and genomic landscape of meningiomas is underway, where the identification of molecular alterations driving the aggressiveness is translated into more reliable preclinical models that allow for rapid translation of discoveries into clinical trials. These key molecular alterations are refining the histological and molecular classification of meningiomas and allow for the stratification of patients with different outcomes and tailored treatments.

References

1. Sahm F, Schrimpf D, Olar A, Koelsche C, Reuss D, Bissel J, Kratz A, Capper D, Schefzyk S, Hielscher T, Wang Q, Sulman EP, Adeberg S, Koch A, Okuducu AF, Brehmer S, Schittenhelm J, Becker A, Brokinkel B, Schmidt M, Ull T, Gousias K, Kessler AF, Lamszus K, Debus J, Mawrin C, Kim YJ, Simon M, Ketter R, Paulus W, Aldape KD, Herold-Mende C, von Deimling A. TERT promoter mutations and risk of recurrence in meningioma. J Natl Cancer Inst. 2015;108(5):djv377.
2. Mirian C, Duun-Henriksen AK, Juratli T, Sahm F, Spiegl-Kreinecker S, Peyre M, Biczok A, Tonn JC, Goutagny S, Bertero L, Maier AD, Møller Pedersen M, Law I, Broholm H, Cahill DP, Brastianos P, Poulsgaard L, Fugleholm K, Ziebell M, Munch T, Mathiesen T. Poor prognosis associated with TERT gene alterations in meningioma is independent of the WHO classification: an individual patient data meta-analysis. J Neurol Neurosurg Psychiatry. 2020;91(4):378–87.
3. Mirian C, Grell K, Juratli TA, Sahm F, Spiegl-Kreinecker S, Peyre M, Biczok A, Tonn JC, Goutagny S, Bertero L, Maier AD, Jensen LR, Schackert G, Broholm H, Scheie D, Cahill DP, Brastianos PK, Skjøth-Rasmussen J, Fugleholm K, Ziebell M, Munch TN, Kristensen BW, Mathiesen T. Implementation of TERT promoter mutations improve prognostication of the WHO classification in meningioma. Neuropathol Appl Neurobiol. 2022;48(3):e12773.
4. Juratli TA, McCabe D, Nayyar N, Williams EA, Silverman IM, Tummala SS, Fink AL, Baig A, Martinez-Lage M, Selig MK, Bihun IV, Shankar GM, Penson T, Lastrapes M, Daubner D, Meinhardt M, Hennig S, Kaplan AB, Fujio S, Kuter BM, Bertalan MS, Miller JJ, Batten JM, Ely HA, Christiansen J, Baretton GB, Stemmer-Rachamimov AO, Santagata S, Rivera MN, Barker FG 2nd, Schackert G, Wakimoto H, Iafrate AJ, Carter SL, Cahill DP, Brastianos PK. DMD genomic deletions characterize a subset of progressive/higher-grade meningiomas with poor outcome. Acta Neuropathol. 2018;136(5):779–92.
5. Boetto J, Bielle F, Sanson M, Peyre M, Kalamarides M. SMO mutation status defines a distinct and frequent molecular subgroup in olfactory groove meningiomas. Neuro-Oncology. 2017;19(3):345–51.
6. Yesilöz Ü, Kirches E, Hartmann C, Scholz J, Kropf S, Sahm F, Nakamura M, Mawrin C. Frequent AKT1E17K mutations in skull base meningiomas are associated with mTOR and ERK1/2 activation and reduced time to tumor recurrence. Neuro-Oncology. 2017;19(8):1088–96.
7. Maas SLN, Stichel D, Hielscher T, Sievers P, Berghoff AS, Schrimpf D, Sill M, Euskirchen P, Blume C, Patel A, Dogan H, Reuss D, Dohmen H, Stein M, Reinhardt A, Suwala AK, Wefers AK, Baumgarten P, Ricklefs F, Rushing EJ, Bewerunge-Hudler M, Ketter

R, Schittenhelm J, Jaunmuktane Z, Leu S, Greenway FEA, Bridges LR, Jones T, Grady C, Serrano J, Golfinos J, Sen C, Mawrin C, Jungk C, Hänggi D, Westphal M, Lamszus K, Etminan N, Jungwirth G, Herold-Mende C, Unterberg A, Harter PN, Wirsching HG, Neidert MC, Ratliff M, Platten M, Snuderl M, Aldape KD, Brandner S, Hench J, Frank S, Pfister SM, Jones DTW, Reifenberger G, Acker T, Wick W, Weller M, Preusser M, von Deimling A, Sahm F, German Consortium on Aggressive Meningiomas (KAM). Integrated molecular-morphologic meningioma classification: a multicenter retrospective analysis, retrospectively and prospectively validated. J Clin Oncol. 2021;39(34):3839–52.
8. Sahm F, Schrimpf D, Stichel D, Jones DTW, Hielscher T, Schefzyk S, Okonechnikov K, Koelsche C, Reuss DE, Capper D, Sturm D, Wirsching HG, Berghoff AS, Baumgarten P, Kratz A, Huang K, Wefers AK, Hovestadt V, Sill M, Ellis HP, Kurian KM, Okuducu AF, Jungk C, Drueschler K, Schick M, Bewerunge-Hudler M, Mawrin C, Seiz-Rosenhagen M, Ketter R, Simon M, Westphal M, Lamszus K, Becker A, Koch A, Schittenhelm J, Rushing EJ, Collins VP, Brehmer S, Chavez L, Platten M, Hänggi D, Unterberg A, Paulus W, Wick W, Pfister SM, Mittelbronn M, Preusser M, Herold-Mende C, Weller M, von Deimling A. DNA methylation-based classification and grading system for meningioma: a multicentre, retrospective analysis. Lancet Oncol. 2017;18(5):682–94.
9. Olar A, Wani KM, Wilson CD, Zadeh G, DeMonte F, Jones DT, Pfister SM, Sulman EP, Aldape KD. Global epigenetic profiling identifies methylation subgroups associated with recurrence-free survival in meningioma. Acta Neuropathol. 2017;133(3):431–44.
10. Nassiri F, Mamatjan Y, Suppiah S, Badhiwala JH, Mansouri S, Karimi S, Saarela O, Poisson L, Gepfner-Tuma I, Schittenhelm J, Ng HK, Noushmehr H, Harter P, Baumgarten P, Weller M, Preusser M, Herold-Mende C, Tatagiba M, Tabatabai G, Sahm F, von Deimling A, International Consortium on Meningiomas, Zadeh G, Aldape KD. DNA methylation profiling to predict recurrence risk in meningioma: development and validation of a nomogram to optimize clinical management. Neuro-Oncology. 2019;21(7):901–10.
11. Berghoff AS, Hielscher T, Ricken G, Furtner J, Schrimpf D, Widhalm G, Rajky U, Marosi C, Hainfellner JA, von Deimling A, Sahm F, Preusser M. Prognostic impact of genetic alterations and methylation classes in meningioma. Brain Pathol. 2022;32(2):e12970.
12. Nassiri F, Wang JZ, Singh O, Karimi S, Dalcourt T, Ijad N, Pirouzmand N, Ng HK, Saladino A, Pollo B, Dimeco F, Yip S, Gao A, Aldape KD, Zadeh G, International Consortium on Meningiomas. Loss of H3K27me3 in meningiomas. Neuro-Oncology. 2021;23(8):1282–91.
13. Preusser M, Brastianos PK, Mawrin C. Advances in meningioma genetics: novel therapeutic opportunities. Nat Rev Neurol. 2018;14(2):106–15.
14. Kang HC, Kim IH, Park CI, Park SH. Immunohistochemical analysis of cyclooxygenase-2 and brain fatty acid binding protein expression in grades I-II meningiomas: correlation with tumor grade and clinical outcome after radiotherapy. Neuropathology. 2014;34(5):446–54.
15. Johnson MD, Horiba M, Winnier AR, Arteaga CL. The epidermal growth factor receptor is associated with phospholipase C-gamma 1 in meningiomas. Hum Pathol. 1994;25(2):146–53.
16. Johnson MD, Shaw AK, O'Connell MJ, Sim FJ, Moses HL. Analysis of transforming growth factor β receptor expression and signaling in higher grade meningiomas. J Neuro-Oncol. 2011;103(2):277–85.
17. Johnson MD, O'Connell MJ, Vito F, Pilcher W. Bone morphogenetic protein 4 and its receptors are expressed in the leptomeninges and meningiomas and signal via the Smad pathway. J Neuropathol Exp Neurol. 2009;68(11):1177–83.
18. Du Z, Abedalthagafi M, Aizer AA, McHenry AR, Sun HH, Bray MA, Viramontes O, Machaidze R, Brastianos PK, Reardon DA, Dunn IF, Freeman GJ, Ligon KL, Carpenter AE, Alexander BM, Agar NY, Rodig SJ, Bradshaw EM, Santagata S. Increased expression of the immune modulatory molecule PD-L1 (CD274) in anaplastic meningioma. Oncotarget. 2015;6(7):4704–16.
19. Karimi S, Mansouri S, Mamatjan Y, Liu J, Nassiri F, Suppiah S, Singh O, Aldape K, Zadeh G. Programmed death ligand-1 (PD-L1) expression in meningioma; prognostic significance and its association with hypoxia and NFKB2 expression. Sci Rep. 2020;10(1):14115.
20. Rapp C, Dettling S, Liu F, Ull AT, Warta R, Jungk C, Roesch S, Mock A, Sahm F, Schmidt M, Jungwirth G, Zweckberger K, Lamszus K, Gousias K, Kessler AF, Grabe N, Loehr M, Ketter R, Urbschat S, Senft C, Westphal M, Abdollahi A, Debus J, von Deimling A, Unterberg A, Simon M, Herold-Mende CC. Cytotoxic T cells and their activation status are independent prognostic markers in meningiomas. Clin Cancer Res. 2019;25(17):5260–70.
21. Hao S, Huang G, Feng J, Li D, Wang K, Wang L, Wu Z, Wan H, Zhang L, Zhang J. Non-NF2 mutations have a key effect on inhibitory immune checkpoints and tumor pathogenesis in skull base meningiomas. J Neuro-Oncol. 2019;144(1):11–20.
22. Proctor DT, Patel Z, Lama S, Resch L, van Marle G, Sutherland GR. Identification of PD-L2, B7-H3 and CTLA-4 immune checkpoint proteins in genetic subtypes of meningioma. Onco Targets Ther. 2018;8(1):e1512943.
23. Wang S, Liechty B, Patel S, Weber JS, Hollmann TJ, Snuderl M, Karajannis MA. Programmed death ligand 1 expression and tumor infiltrating lymphocytes in neurofibromatosis type 1 and 2 associated tumors. J Neuro-Oncol. 2018;138(1):183–90.
24. Karimi S, Mansouri S, Nassiri F, Bunda S, Singh O, Brastianos PK, Dunn IF, Zadeh G. Clinical significance of checkpoint regulator "Programmed

death ligand-1 (PD-L1)" expression in meningioma: review of the current status. J Neuro-Oncol. 2021;151(3):443–9.
25. Yeung J, Yaghoobi V, Miyagishima D, Vesely MD, Zhang T, Badri T, Nassar A, Han X, Sanmamed MF, Youngblood M, Peyre M, Kalamarides M, Rimm DL, Gunel M, Chen L. Targeting the CSF1/CSF1R axis is a potential treatment strategy for malignant meningiomas. Neuro-Oncology. 2021;23(11):1922–35.
26. Kaley T, Barani I, Chamberlain M, McDermott M, Panageas K, Raizer J, Rogers L, Schiff D, Vogelbaum M, Weber D, Wen P. Historical benchmarks for medical therapy trials in surgery- and radiation-refractory meningioma: a RANO review. Neuro-Oncology. 2014;16(6):829–40.
27. Goldbrunner R, Stavrinou P, Jenkinson MD, Sahm F, Mawrin C, Weber DC, Preusser M, Minniti G, Lund-Johansen M, Lefranc F, Houdart E, Sallabanda K, Le Rhun E, Nieuwenhuizen D, Tabatabai G, Soffietti R, Weller M. EANO guideline on the diagnosis and management of meningiomas. Neuro-Oncology. 2021;23(11):1821–34.
28. Huang RY, Bi WL, Weller M, Kaley T, Blakeley J, Dunn I, Galanis E, Preusser M, McDermott M, Rogers L, Raizer J, Schiff D, Soffietti R, Tonn JC, Vogelbaum M, Weber D, Reardon DA, Wen PY. Proposed response assessment and endpoints for meningioma clinical trials: report from the Response Assessment in Neuro-Oncology Working Group. Neuro-Oncology. 2019;21(1):26–36.
29. Chamberlain MC, Johnston SK. Hydroxyurea for recurrent surgery and radiation refractory meningioma: a retrospective case series. J Neuro-Oncol. 2011;104(3):765–71.
30. Chamberlain MC. Hydroxyurea for recurrent surgery and radiation refractory high-grade meningioma. J Neuro-Oncol. 2012;107(2):315–21.
31. Mazza E, Brandes A, Zanon S, Eoli M, Lombardi G, Faedi M, Franceschi E, Reni M. Hydroxyurea with or without imatinib in the treatment of recurrent or progressive meningiomas: a randomized phase II trial by Gruppo Italiano Cooperativo di Neuro-Oncologia (GICNO). Cancer Chemother Pharmacol. 2016;77(1):115–20.
32. Chamberlain MC, Tsao-Wei DD, Groshen S. Temozolomide for treatment-resistant recurrent meningioma. Neurology. 2004;62(7):1210–2.
33. Chamberlain MC, Tsao-Wei DD, Groshen S. Salvage chemotherapy with CPT-11 for recurrent meningioma. J Neuro-Oncol. 2006;78(3):271–6.
34. Preusser M, Spiegl-Kreinecker S, Lötsch D, Wöhrer A, Schmook M, Dieckmann K, Saringer W, Marosi C, Berger W. Trabectedin has promising antineoplastic activity in high-grade meningioma. Cancer. 2012;118(20):5038–49.
35. Preusser M, Silvani A, Le Rhun E, Soffietti R, Lombardi G, Sepulveda JM, Brandal P, Brazil L, Bonneville-Levard A, Lorgis V, Vauleon E, Bromberg J, Erridge S, Cameron A, Lefranc F, Clement PM, Dumont S, Sanson M, Bronnimann C, Balaná C, Thon N, Lewis J, Mair MJ, Sievers P, Furtner J, Pichler J, Bruna J, Ducray F, Reijneveld JC, Mawrin C, Bendszus M, Marosi C, Golfinopoulos V, Coens C, Gorlia T, Weller M, Sahm F, Wick W. Trabectedin for recurrent WHO grade 2 or 3 meningioma: a randomized phase II study of the EORTC Brain Tumor Group (EORTC-1320-BTG). Neuro-Oncology. 2022;24(5):755–67.
36. Koper JW, Zwarthoff EC, Hagemeijer A, Braakman R, Avezaat CJ, Bergström M, Lamberts SW. Inhibition of the growth of cultured human meningioma cells by recombinant interferon-alpha. Eur J Cancer. 1991;27(4):416–9.
37. Chamberlain MC, Glantz MJ. Interferon-alpha for recurrent World Health Organization grade 1 intracranial meningiomas. Cancer. 2008;113(8):2146–51.
38. Chamberlain MC. IFN-α for recurrent surgery- and radiation-refractory high-grade meningioma: a retrospective case series. CNS Oncol. 2013;2(3):227–35.
39. Norden AD, Ligon KL, Hammond SN, Muzikansky A, Reardon DA, Kaley TJ, Batchelor TT, Plotkin SR, Raizer JJ, Wong ET, Drappatz J, Lesser GJ, Haidar S, Beroukhim R, Lee EQ, Doherty L, Lafrankie D, Gaffey SC, Gerard M, Smith KH, McCluskey C, Phuphanich S, Wen PY. Phase II study of monthly pasireotide LAR (SOM230C) for recurrent or progressive meningioma. Neurology. 2015;84(3):280–6.
40. Chamberlain MC, Glantz MJ, Fadul CE. Recurrent meningioma: salvage therapy with long-acting somatostatin analogue. Neurology. 2007;69(10):969–73.
41. Simó M, Argyriou AA, Macià M, Plans G, Majós C, Vidal N, Gil M, Bruna J. Recurrent high-grade meningioma: a phase II trial with somatostatin analogue therapy. Cancer Chemother Pharmacol. 2014;73(5):919–23.
42. Kaley TJ, Wen P, Schiff D, Ligon K, Haidar S, Karimi S, Lassman AB, Nolan CP, DeAngelis LM, Gavrilovic I, Norden A, Drappatz J, Lee EQ, Purow B, Plotkin SR, Batchelor T, Abrey LE, Omuro A. Phase II trial of sunitinib for recurrent and progressive atypical and anaplastic meningioma. Neuro-Oncology. 2015;17(1):116–21.
43. Shih KC, Chowdhary S, Rosenblatt P, Weir AB 3rd, Shepard GC, Williams JT, Shastry M, Burris HA 3rd, Hainsworth JD. A phase II trial of bevacizumab and everolimus as treatment for patients with refractory, progressive intracranial meningioma. J Neuro-Oncol. 2016;129(2):281–8.
44. Lou E, Sumrall AL, Turner S, Peters KB, Desjardins A, Vredenburgh JJ, McLendon RE, Herndon JE 2nd, McSherry F, Norfleet J, Friedman HS, Reardon DA. Bevacizumab therapy for adults with recurrent/progressive meningioma: a retrospective series. J Neuro-Oncol. 2012;109(1):63–70.
45. Nayak L, Iwamoto FM, Rudnick JD, Norden AD, Lee EQ, Drappatz J, Omuro A, Kaley TJ. Atypical and anaplastic meningiomas treated with bevacizumab. J Neuro-Oncol. 2012;109(1):187–93.
46. Norden AD, Raizer JJ, Abrey LE, Lamborn KR, Lassman AB, Chang SM, Yung WK, Gilbert MR, Fine

HA, Mehta M, Deangelis LM, Cloughesy TF, Robins HI, Aldape K, Dancey J, Prados MD, Lieberman F, Wen PY. Phase II trials of erlotinib or gefitinib in patients with recurrent meningioma. J Neuro-Oncol. 2010;96(2):211–7.
47. Wen PY, Yung WK, Lamborn KR, Norden AD, Cloughesy TF, Abrey LE, Fine HA, Chang SM, Robins HI, Fink K, Deangelis LM, Mehta M, Di Tomaso E, Drappatz J, Kesari S, Ligon KL, Aldape K, Jain RK, Stiles CD, Egorin MJ, Prados MD. Phase II study of imatinib mesylate for recurrent meningiomas (North American Brain Tumor Consortium study 01-08). Neuro-Oncology. 2009;11(6):853–60.
48. Ji Y, Rankin C, Grunberg S, Sherrod AE, Ahmadi J, Townsend JJ, Feun LG, Fredericks RK, Russell CA, Kabbinavar FF, Stelzer KJ, Schott A, Verschraegen C. Double-blind phase III randomized trial of the antiprogestin agent mifepristone in the treatment of unresectable meningioma: SWOG S9005. J Clin Oncol. 2015;33(34):4093–8.
49. Franke AJ, Skelton WP IV, Woody LE, Bregy A, Shah AH, Vakharia K, Komotar RJ. Role of bevacizumab for treatment-refractory meningiomas: a systematic analysis and literature review. Surg Neurol Int. 2018;9:133.
50. Furtner J, Schöpf V, Seystahl K, Le Rhun E, Rudà R, Roelcke U, Koeppen S, Berghoff AS, Marosi C, Clement P, Faedi M, Watts C, Wick W, Soffietti R, Weller M, Preusser M. Kinetics of tumor size and peritumoral brain edema before, during, and after systemic therapy in recurrent WHO grade II or III meningioma. Neuro-Oncology. 2016;18(3):401–7.
51. Graillon T, Sanson M, Campello C, Idbaih A, Peyre M, Peyrière H, Basset N, Autran D, Roche C, Kalamarides M, Roche PH, Fuentes S, Tabouret E, Barrie M, Cohen A, Honoré S, Boucekine M, Baumstarck K, Figarella-Branger D, Barlier A, Dufour H, Chinot OL. Everolimus and octreotide for patients with recurrent meningioma: results from the phase II CEVOREM trial. Clin Cancer Res. 2020;26(3):552–7.
52. Brastianos PK, Twohy EL, Gerstner ER, Kaufmann TJ, Iafrate AJ, Lennerz J, Jeyapalan S, Piccioni DE, Monga V, Fadul CE, Schiff D, Taylor JW, Chowdhary SA, Bettegowda C, Ansstas G, De La Fuente M, Anderson MD, Shonka N, Damek D, Carrillo J, Kunschner-Ronan LJ, Chaudhary R, Jaeckle KA, Senecal FM, Kaley T, Morrison T, Thomas AA, Welch MR, Iwamoto F, Cachia D, Cohen AL, Vora S, Knopp M, Dunn IF, Kumthekar P, Sarkaria J, Geyer S, Carrero XW, Martinez-Lage M, Cahill DP, Brown PD, Giannini C, Santagata S, Barker FG 2nd, Galanis E. Alliance A071401: phase II trial of focal adhesion kinase inhibition in meningiomas with somatic *NF2* mutations. J Clin Oncol. 2023;41(3):618–28.
53. Bi WL, Greenwald NF, Abedalthagafi M, Wala J, Gibson WJ, Agarwalla PK, Horowitz P, Schumacher SE, Esaulova E, Mei Y, Chevalier A, Ducar M, Thorner AR, van Hummelen P, Stemmer-Rachamimov A, Artyomov M, Al-Mefty O, Dunn GP, Santagata S, Dunn IF, Beroukhim R. Genomic landscape of high-grade meningiomas. NPJ Genom Med. 2017;2:15.
54. Dunn IF, Du Z, Touat M, Sisti MB, Wen PY, Umeton R, Dubuc AM, Ducar M, Canoll PD, Severson E, Elvin JA, Ramkissoon SH, Lin JR, Cabrera L, Acevedo B, Sorger PK, Ligon KL, Santagata S, Reardon DA. Mismatch repair deficiency in high-grade meningioma: a rare but recurrent event associated with dramatic immune activation and clinical response to PD-1 blockade. JCO Precis Oncologia. 2018;2018:PO.18.00190.
55. Nebot-Bral L, Brandao D, Verlingue L, Rouleau E, Caron O, Despras E, El-Dakdouki Y, Champiat S, Aoufouchi S, Leary A, Marabelle A, Malka D, Chaput N, Kannouche PL. Hypermutated tumours in the era of immunotherapy: the paradigm of personalised medicine. Eur J Cancer. 2017;84:290–303.
56. Brastianos PK, Kim AE, Giobbie-Hurder A, Lee EQ, Wang N, Eichler AF, Chukwueke U, Forst DA, Arrillaga-Romany IC, Dietrich J, Corbin Z, Moliterno J, Baehring J, White M, Lou KW, Larson J, de Sauvage MA, Evancic K, Mora J, Nayyar N, Loeffler J, Oh K, Shih HA, Curry WT, Cahill DP, Barker FG, Gerstner ER, Santagata S. Phase 2 study of pembrolizumab in patients with recurrent and residual high-grade meningiomas. Nat Commun. 2022;13(1):1325.
57. Bi WL, Nayak L, Meredith DM, Driver J, Du Z, Hoffman S, Li Y, Lee EQ, Beroukhim R, Rinne M, McFaline-Figueroa R, Chukwueke U, McCluskey C, Gaffey S, Cherniack AD, Stefanik J, Doherty L, Taubert C, Cifrino M, LaFrankie D, Graillon T, Wen PY, Ligon KL, Al-Mefty O, Huang RY, Muzikansky A, Chiocca EA, Santagata S, Dunn IF, Reardon DA. Activity of PD-1 blockade with nivolumab among patients with recurrent atypical/anaplastic meningioma: phase II trial results. Neuro-Oncology. 2022;24(1):101–13.

Part V

Meningioma Pathology and Recurrences

Time to Recurrence of Meningiomas

Sergio Corvino

1 Introduction

Intracranial meningiomas account for more than one third of primary intracranial tumors [1]. Despite their predominantly benign WHO histology (~80%), a relevant percentage of patients (10–25%) harbor one or more recurrences during lifetime, even after many years, requiring even multiple reoperations and adjuvant treatments [2–4].

Since the pioneering study provided by Simpson in 1957 [5], in which the risk of tumor recurrence was estimated according to the extent of resection, many other risk factors of recurrence have been identified; they include sex and age of patients, WHO grade, proliferation index Ki67-MIB1, progesterone receptor (PR) expression, tumor size, shape, location and brain invasion, and molecular features [4–14].

The time to recurrence has scarcely been investigated [15, 16]. Despite the well-defined histopathological criteria in establishing the grading, there is large variation of the time to recurrence within the same WHO grade among patients with intracranial meningiomas; moreover, patients undergoing similar extent of surgical resection often recur after very different intervals.

S. Corvino (✉)
Department of Neurosciences, Reproductive and Odontostomatological Sciences, Neurosurgical Division, University of Naples "Federico II", Naples, Italy

This chapter discusses the main factors affecting the time to recurrence of meningiomas and the clinical implications.

2 Factors Affecting the Time to Recurrence

The extent of surgical resection and the WHO histological grade are the main factors for predicting the meningioma recurrence; nevertheless, many cases of incongruence between the WHO grade and clinical course, in which low-grade meningiomas have a worse recurrence-free survival than high-grade ones, are observed in the clinical practice [15–18]. Thus, other risk factors have been investigated, such as demographics, neuroradiological, pathological, surgical, and biomolecular, some of which have been correlated to the time to recurrence.

2.1 Patients and Tumor Features

Patient age and sex, tumor size and morphology, loss of the brain-tumor interface, peritumoral edema, tumor vascularity, and bone invasion [15, 16, 18–22] are investigated in patient as tumor-related risk factors of meningioma recurrences. However, no significant association between time to recurrence and patient age (<65 vs ≥65 years) [23], preoperative tumor size,

F. Maiuri, M. Del Basso De Caro (eds.), *Meningiomas*,
https://doi.org/10.1007/978-3-031-76680-0_18

tumor shape (flat vs. round), and peritumoral edema, has been found.

2.2 Meningioma Location

Several studies report a longer median time to recurrence in skull base meningiomas than in non-skull base ones [24], and in midline than lateral skull base locations [21]. These data can be explained by several reasons. First, the embryology of the meninges and biomolecular expression of the meningiomas differ according to the location. In detail, meninges of the cerebral convexity and falx develop from the neural crest while the dura mater of skull base develops from the mesoderm; furthermore, most of non-skull base meningiomas, as well as those of lateral skull base, exhibit NF2 mutation and/or loss of chromosome 22 which are considered molecular biomarkers of greater aggressivity; conversely, most meningiomas of the medial skull base are non-NF2. Finally, paravenous sinus locations, like parasagittal/falx regions [7], show higher incidence of atypical meningiomas WHO grade II characterized by a more aggressive biological behavior. Besides, these locations show higher rates of subtotal resection due to sinus invasion and spread of tumor cells along the dura layer of the falx.

In conclusion, the locations of the meningiomas, also because of the different dural origin, not only show different recurrence rates but also different time to recurrence. Medial skull base meningiomas show longer time to recurrence also because of the lesser rate of atypical forms.

2.3 Extent of Resection and Adjuvant Therapy

Although over the last years, the assessment through the Simpson grading has been debated in several studies [25–27], the extent of resection still represents one of the main risk factors of meningioma recurrence. Over the years, many preoperative and intraoperative tools have been developed to achieve complete resection while preserving the neurological functions [28–30].

Several recent studies have shown that the EOR also affects the time to recurrence [21, 23, 24, 31]: Maiuri et al. [24] observed longer median time to recurrence for patients undergoing Simpson grades I and II compared to those who underwent grade III and grade IV resections. Traylor et al. [21] report a median recurrence-free survival of 160 months in patients who underwent GTR, compared to 106 months for those who underwent STR.

The role of preoperative embolization of intracranial meningiomas is still debated in terms of benefits and limits [32]. The contribution of this procedure to the time to recurrence is also controversial, with some authors asserting that it does not affect the recurrence-free survival [33], and others who report its prolongation up to twice without an increase of complications [34].

Adjuvant radiotherapy seems to prolong the recurrence-free survival in all patients with subtotally resected meningiomas regardless of WHO grade [21], thus providing a better local control of the disease.

2.4 Pathological Findings

Pathological factors investigated for correlation with the time to recurrence of meningiomas include WHO grade, proliferation index and PR expression.

The proliferation index of the meningioma at initial diagnosis is the most relevant factor. Olar et al. [15] found that the mitotic index, defined using the anti-pHH3 antibody, was significantly associated to the recurrence-free survival. Most studies use the Ki67-MIB1 as pathological prognostic factor [35–38]; it represents an immunohistochemical marker for cell proliferation. A Ki67 cut-off of 4% is suggested to differentiate between high and low risk of recurrence. In a recent meta-analysis including 43 studies, Liu et al. [38] found significant negative prognostic value of high Ki67 levels for both overall survival and recurrence-free survival of meningiomas, especially for Ki67 index higher than 4%. Mirian et al. [16] found a significant correlation between higher Ki67 values and shorter time to recurrence, whereas they did not find the Ki67 an appropriate predictor of recurrence. Maiuri et al. [24] have shown strong correlation between Ki67 values >4% and shorter time to recurrence, whereas the difference between higher values (5–9% vs. ≥10%) was not significant. Besides, the Ki67-MIB1 was more significantly correlated than the WHO grade. This reflects the presence of meningiomas with different proliferation potential within the same WHO grade.

Many studies have focused on PR expression and recurrence of meningiomas [35, 36, 39–43]; two [35, 42] have found significant inverse correlation, with higher recurrence rates in meningiomas with low PR expression at initial surgery. PR is expressed by healthy meninges and between 33% and 89% of meningiomas. Its value is determined with monoclonal antibody against the progesterone receptor and expressed as percentage of positive nuclei among 100 cells. This must be included in the routinary histological diagnosis of meningiomas. Maiuri et al. [24] first correlated PR expression and time to recurrence; the data show statistical significance only for low PR values. This confirms that the PR expression is more important for predicting the risk of recurrence than the time to recurrence.

2.5 Recurrence-Related Findings

Several recent studies [8, 24, 44–47] have discussed the correlation of the time to recurrence with some recurrence-related findings, including their number (single versus multiple) and growth pattern (localized versus diffuse) and the WHO grade as compared to the initial grade.

Albeit most recurrent meningiomas are completely cured after one reoperation without or with adjuvant radiotherapy, a not negligible percentage requires two or more reoperations even over many years [24, 44, 45, 48]. Nevertheless, there is not unanimous consent about the impact of multiple reoperations on the time to recurrence: Maiuri et al. [24, 48] found that the median time to recurrence after initial surgery does not significantly differ in patients who underwent one reoperation versus those who underwent two or more surgical procedures; Lemée et al. [44] found that the time to retreatment significantly decreases between surgeries in patients requiring repeated resections; in addition, they noted that it was significantly longer in recurrent meningiomas treated with surgery than in those treated by radiation therapy.

Few studies focused on the topographic pattern of recurrences of intracranial meningiomas (localized-peripheral versus multicentric-diffuse) [8, 24, 46, 47]. Two investigated its correlation with the time to recurrence [24, 46] and both confirm that the time to recurrence after initial surgery is not correlated to the topography of the first recurrence.

One study [24] has also shown that meningiomas which show progression of the WHO grade and increase of Ki67-MIB1 at the first recurrence do not recur earlier than those showing stable pathological findings. This suggests that the time to recurrence mainly depends on the WHO grade and Ki67-MIB1 at the initial diagnosis, whereas it is not correlated to the pathological progression of the recurrent tumor.

Several studies [49, 50] have delineated molecular classifications of meningiomas based on DNA methylation that predict outcomes more accurately than the WHO histopathological classification. This is mainly important for some WHO grade I meningiomas which recur earlier than expected from the initial pathological profile.

Further studies will confirm whether the biomolecular profile of meningiomas may reflect different recurrence time in cases with similar pathological findings.

3 Clinical Implication and Conclusion

The extent of resection and the Ki67-MIB1 represent the most important factors predicting shorter recurrence time of intracranial meningiomas. Patients with incomplete (grades III and IV) resection and high Ki67-MIB1 values, mainly with non-skull base meningiomas and low PR values, need for a closer short-term clinical and radiological follow-up in the first years after the operation, because they are at risk of a short-term recurrence.

References

1. Ostrom QT, Patil N, Cioffi G, Waite K, Kruchko C, Barnholtz-Sloan JS. CBTRUS statistical report: primary brain and other central nervous system tumors diagnosed in the United States in 2013–2017. Neuro-Oncology. 2020;22(12 Suppl 2):iv1–iv96. https://doi.org/10.1093/neuonc/noaa200.
2. Lüthge S, Spille DC, Steinbicker AU, Schipmann S, Streckert EMS, Hess K, et al. The applicability of established clinical and histopathological risk factors for tumor recurrence during long-term postoperative care in meningioma patients. Neurosurg Rev. 2022;45(2):1635–43. https://doi.org/10.1007/s10143-021-01697-w.
3. Goldbrunner R, Stavrinou P, Jenkinson MD, Sahm F, Mawrin C, Weber DC, et al. EANO guideline on the diagnosis and management of meningiomas. Neuro-Oncology. 2021;23(11):1821–34. https://doi.org/10.1093/neuonc/noab150.
4. Lam Shin Cheung V, Kim A, Sahgal A, Das S. Meningioma recurrence rates following treatment: a systematic analysis. J Neuro-Oncol. 2018;136(2):351–61. https://doi.org/10.1007/s11060-017-2659-6.
5. Simpson D. The recurrence of intracranial meningiomas after surgical treatment. J Neurol Neurosurg Psychiatry. 1957;20(1):22–39.
6. Maiuri F, Mariniello G, Somma T, Guadagno E, Corvino S, Pagano S, et al. Meningiomas in premenopausal women: role of the hormone related conditions. Front Oncol. 2020;10:556701. https://doi.org/10.3389/fonc.2020.556701.
7. Maiuri F, Mariniello G, Guadagno E, Barbato M, Corvino S, Del Basso De Caro M. WHO grade, proliferation index, and progesterone receptor expression are different according to the location of meningioma. Acta Neurochir. 2019;161(12):2553–61. https://doi.org/10.1007/s00701-019-04084-z.
8. Maiuri F, Mariniello G, Peca C, Guadagno E, Corvino S, d'Avanzo S, et al. Multicentric and diffuse recurrences of meningiomas. Br J Neurosurg. 2020;34:1–8. https://doi.org/10.1080/02688697.2020.1754335.
9. Mariniello G, de Divitiis O, Corvino S, Strianese D, Iuliano A, Bonavolontà G, et al. Recurrences of spheno-orbital meningiomas: risk factors and management. World Neurosurg. 2022;161:e514. https://doi.org/10.1016/j.wneu.2022.02.048.
10. Haddad AF, Young JS, Kanungo I, Sudhir S, Chen JS, Raleigh DR, et al. WHO grade I meningioma recurrence: identifying high risk patients using histopathological features and the MIB-1 index. Front Oncol. 2020;10:1522. https://doi.org/10.3389/fonc.2020.01522.
11. Vaubel RA, Kumar R, Weiskittel TM, Jenkins S, Dasari S, Uhm JH, et al. Genomic markers of recurrence risk in atypical meningioma following gross total resection. Neurooncol Adv. 2023;5(1):vdad004. https://doi.org/10.1093/noajnl/vdad004.
12. Spille DC, Adeli A, Sporns PB, Heß K, Streckert EMS, Brokinkel C, et al. Predicting the risk of postoperative recurrence and high-grade histology in patients with intracranial meningiomas using routine preoperative MRI. Neurosurg Rev. 2021;44(2):1109–17. https://doi.org/10.1007/s10143-020-01301-7.
13. Maiuri F, Corvino S, Lorenzetti M, Franca RA, Esposito F, Del Basso De Caro M. Intracranial meningiomas in patients aged ≥80 years: pathological features and surgical problems. World Neurosurg. 2023;173:e498–508. https://doi.org/10.1016/j.wneu.2023.02.086.
14. Franca RA, Della Monica R, Corvino S, Chiariotti L, Del Basso De Caro M. WHO grade and pathological markers of meningiomas: clinical and prognostic role. Pathol Res Pract. 2023;243:154340. https://doi.org/10.1016/j.prp.2023.154340.
15. Olar A, Wani KM, Sulman EP, Mansouri A, Zadeh G, Wilson CD, et al. Mitotic index is an independent predictor of recurrence-free survival in meningioma. Brain Pathol. 2015;25(3):266–75. https://doi.org/10.1111/bpa.12174.
16. Mirian C, Skyrman S, Bartek J, Jensen LR, Kihlström L, Förander P, et al. The Ki-67 proliferation index as a marker of time to recurrence in intracranial meningioma. Neurosurgery. 2020;87(6):1289–98. https://doi.org/10.1093/neuros/nyaa226.
17. Sahm F, Schrimpf D, Olar A, Koelsche C, Reuss D, Bissel J, et al. TERT promoter mutations and risk of recurrence in meningioma. J Natl Cancer Inst. 2016;108(5):djv377. https://doi.org/10.1093/jnci/djv377.
18. Sievers P, Hielscher T, Schrimpf D, Stichel D, Reuss DE, Berghoff AS, et al. CDKN2A/B homozygous deletion is associated with early recurrence in meningiomas. Acta Neuropathol. 2020;140(3):409–13. https://doi.org/10.1007/s00401-020-02188-w.
19. Spille DC, Hess K, Bormann E, Sauerland C, Brokinkel C, Warneke N, et al. Risk of tumor recurrence in intracranial meningiomas: comparative analyses of the predictive value of the postoperative tumor volume and the Simpson classification. J Neurosurg. 2020;134(6):1764–71. https://doi.org/10.3171/2020.4.JNS20412.
20. Hwang WL, Marciscano AE, Niemierko A, Kim DW, Stemmer-Rachamimov AO, Curry WT, et al. Imaging and extent of surgical resection predict risk of meningioma recurrence better than WHO histopathological grade. Neuro-Oncology. 2016;18(6):863–72. https://doi.org/10.1093/neuonc/nov285.
21. Traylor JI, Plitt AR, Hicks WH, Mian TM, Mickey BE, Barnett SL. Evaluating risk of recurrence in patients

with meningioma. J Neurosurg. 2023;138(3):621–8. https://doi.org/10.3171/2022.6.JNS221162.
22. Meling TR, Da Broi M, Scheie D, Helseth E. Meningiomas: skull base versus non-skull base. Neurosurg Rev. 2019;42(1):163–73. https://doi.org/10.1007/s10143-018-0976-7.
23. Tang AR, Chotai S, Guidry BS, Sun L, Ye F, Kelly PD, et al. Conditional recurrence-free survival after surgical resection of meningioma. Neurosurgery. 2023;93(2):339–47. https://doi.org/10.1227/neu.0000000000002416.
24. Maiuri F, Corvino S, Corazzelli G, Berardinelli J, Di Crescenzo RM, Del Basso De Caro M. Time to recurrence of intracranial meningiomas from a monoinstitutional surgical series. World Neurosurg. 2024; https://doi.org/10.1016/j.wneu.2024.02.087.
25. Gousias K, Schramm J, Simon M. The Simpson grading revisited: aggressive surgery and its place in modern meningioma management. J Neurosurg. 2016;125(3):551–60. https://doi.org/10.3171/2015.9.JNS15754.
26. Schwartz TH, McDermott MW. The Simpson grade: abandon the scale but preserve the message. J Neurosurg. 2020;135:1–8. https://doi.org/10.3171/2020.6.JNS201904.
27. Chotai S, Schwartz TH. The Simpson grading: is it still valid? Cancers (Basel). 2022;14(8) https://doi.org/10.3390/cancers14082007.
28. Dijkstra BM, Jeltema HJR, Kruijff S, Groen RJM. The application of fluorescence techniques in meningioma surgery-a review. Neurosurg Rev. 2019;42(4):799–809. https://doi.org/10.1007/s10143-018-01062-4.
29. Bin-Alamer O, Abou-Al-Shaar H, Gersey ZC, Huq S, Kallos JA, McCarthy DJ, et al. Intraoperative imaging and optical visualization techniques for brain tumor resection: a narrative review. Cancers (Basel). 2023;15(19) https://doi.org/10.3390/cancers15194890.
30. Maiuri F, Corvino S. Meningiomas of the rolandic region: risk factors for motor deficit and role of intra-operative monitoring. Acta Neurochir. 2023;165(7):1707–16. https://doi.org/10.1007/s00701-023-05630-6.
31. Przybylowski CJ, Suki D, Raza SM, DeMonte F. Volumetric extent of resection and survival for recurrent atypical meningioma. J Neurosurg. 2023;139(3):769–79. https://doi.org/10.3171/2022.12.JNS221815.
32. Singla A, Deshaies EM, Melnyk V, Toshkezi G, Swarnkar A, Choi H, et al. Controversies in the role of preoperative embolization in meningioma management. Neurosurg Focus. 2013;35(6):E17. https://doi.org/10.3171/2013.9.FOCUS13351.
33. Wirsching HG, Richter JK, Sahm F, Morel C, Krayenbuehl N, Rushing EJ, et al. Post-operative cardiovascular complications and time to recurrence in meningioma patients treated with versus without preoperative embolization: a retrospective cohort study of 741 patients. J Neuro-Oncol. 2018;140(3):659–67. https://doi.org/10.1007/s11060-018-2996-0.
34. Akimoto T, Ohtake M, Miyake S, Suzuki R, Iida Y, Shimohigoshi W, et al. Preoperative tumor embolization prolongs time to recurrence of meningiomas: a retrospective propensity-matched analysis. J Neurointerv Surg. 2023;15(8):814–20. https://doi.org/10.1136/neurintsurg-2022-019080.
35. Maiuri F, De Caro MB, Esposito F, Cappabianca P, Strazzullo V, Pettinato G, et al. Recurrences of meningiomas: predictive value of pathological features and hormonal and growth factors. J Neuro-Oncol. 2007;82(1):63–8. https://doi.org/10.1007/s11060-005-9078-9.
36. Roser F, Nakamura M, Bellinzona M, Rosahl SK, Ostertag H, Samii M. The prognostic value of progesterone receptor status in meningiomas. J Clin Pathol. 2004;57(10):1033–7. https://doi.org/10.1136/jcp.2004.018333.
37. Abry E, Thomassen I. The significance of Ki-67/MIB-1 labeling index in human meningiomas: a literature study. Pathol Res Pract. 2010;206:810.
38. Liu N, Song SY, Jiang JB, Wang TJ, Yan CX. The prognostic role of Ki-67/MIB-1 in meningioma: a systematic review with meta-analysis. Medicine (Baltimore). 2020;99(9):e18644. https://doi.org/10.1097/MD.0000000000018644.
39. Fewings PE, Battersby RD, Timperley WR. Long-term follow up of progesterone receptor status in benign meningioma: a prognostic indicator of recurrence? J Neurosurg. 2000;92(3):401–5. https://doi.org/10.3171/jns.2000.92.3.0401.
40. Strik HM, Strobelt I, Pietsch-Breitfeld B, Iglesias-Rozas JR, Will B, Meyermann R. The impact of progesterone receptor expression on relapse in the long-term clinical course of 93 benign meningiomas. In Vivo. 2002;16(4):265–70.
41. Pravdenkova S, Al-Mefty O, Sawyer J, Husain M. Progesterone and estrogen receptors: opposing prognostic indicators in meningiomas. J Neurosurg. 2006;105(2):163–73. https://doi.org/10.3171/jns.2006.105.2.163.
42. Maiuri F, Mariniello G, de Divitiis O, Esposito F, Guadagno E, Teodonno G, et al. Progesterone receptor expression in meningiomas: pathological and prognostic implications. Front Oncol. 2021;11:611218. https://doi.org/10.3389/fonc.2021.611218.
43. Agopiantz M, Carnot M, Denis C, Martin E, Gauchotte G. Hormone receptor expression in meningiomas: a systematic review. Cancers (Basel). 2023;15(3) https://doi.org/10.3390/cancers15030980.
44. Lemée JM, Corniola MV, Meling TR. Benefits of re-do surgery for recurrent intracranial meningiomas. Sci Rep. 2020;10(1):303. https://doi.org/10.1038/s41598-019-57254-5.
45. Richardson GE, Gillespie CS, Mustafa MA, Taweel BA, Bakhsh A, Kumar S, et al. Clinical outcomes following re-operations for intracranial meningioma. Cancers (Basel). 2021;13(19) https://doi.org/10.3390/cancers13194792.

46. Ong K, Rizzuto M, Makarenko S. Location pattern of recurrence of fully resected grade 1 meningiomas. Acta Neurochir. 2023;165(10):2865–71. https://doi.org/10.1007/s00701-023-05758-5.
47. Obiri-Yeboah D, Soni P, Oyem PC, Almeida JP, Murayi R, Recinos PF, et al. Beyond the surgical margin: patterns of recurrence in World Health Organization Grade 2 intracranial meningiomas. World Neurosurg. 2023; https://doi.org/10.1016/j.wneu.2023.06.100.
48. Maiuri F, Corvino S, Corazzelli G, Del Basso De Caro M. Single versus multiple reoperations for recurrent intracranial meningiomas. J Neuro-Oncol. 2024;
49. Sahm F, Schrimpf D, Stichel D, Jones DTW, Hielscher T, Schefzyk S, et al. DNA methylation-based classification and grading system for meningioma: a multicentre, retrospective analysis. Lancet Oncol. 2017;18(5):682–94. https://doi.org/10.1016/S1470-2045(17)30155-9.
50. Nassiri F, Liu J, Patil V, Mamatjan Y, Wang JZ, Hugh-White R, et al. A clinically applicable integrative molecular classification of meningiomas. Nature. 2021;597(7874):119–25. https://doi.org/10.1038/s41586-021-03850-3.

Topography of Meningioma Recurrences: Localized Versus Diffuse

Sergio Corvino and Francesco Maiuri

1 Introduction

Meningiomas show variable tendency to recur even after gross-total resection with or without adjuvant radiotherapy; the risk of recurrence or symptomatic progression must be considered during the planning of the treatment strategy as well as during the follow-up [1].

Many risk factors of recurrence have been identified and include patient (sex and age), pathological findings (WHO grade, proliferation index Ki67-MIB1, progesterone receptor (PR) expression), tumor-related factors (size, shape, location, peritumoral edema, vascularity, brain invasion), extent of resection, biomolecular expression [1–17]. On the other hand, the topographic patterns of recurrences have poorly been investigated.

This chapter focuses on the spatial clustering patterns of intracranial meningioma recurrences, the related risk factors, and the treatment options.

S. Corvino (✉) · F. Maiuri
Department of Neurosciences, Reproductive and Odontostomatological Sciences, Neurosurgical Division, University of Naples "Federico II", Naples, Italy
e-mail: frmaiuri@unina.it

2 Etiopathogenetic Hypotheses of Recurrences

Several etiopathogenetic hypotheses have been proposed to explain the recurrences after apparently macroscopic gross-total resection of meningiomas.

Some studies focused on the histological pattern, seeking clues of malignancy, but providing contradictory conclusions [18, 19]. Others suggested the extent of resection at first surgery as the more relevant factor of the recurrence [18–21]; in this way, the recurrence can be justified by unintentionally incomplete resection of infiltrated dural sinus wall, brain tissue, or bone. In this setting, the sometimes precocious recurrences of WHO grade I meningiomas after apparently complete resection (Simpson grade I) at first surgery is the most interesting and debated problem. Two main hypotheses have been proposed by Borovich et al. [22] (localized dural multifocality of the tumor) and by von Deimling et al. [23] (the clonal origin of recurrent meningiomas).

Borovich et al. [22] found in 9 out of 14 patients operated on for intracranial meningiomas, macroscopic nodules of neoplastic cells at the inner layer of the dura at 1–3 cm surrounding the site of the tumor attachment; they were considered the cause of unexpected recurrences. Therefore, Borovich et al. [22] and Kinjo et al. [24] suggested to introduce to Simpson's scale

F. Maiuri, M. Del Basso De Caro (eds.), *Meningiomas*,
https://doi.org/10.1007/978-3-031-76680-0_19

the grade 0, which indicates a wide dural excision for at least 4 cm around the initial dural attachment, when possible. This thesis was supported by Qi et al. [25] who identified tumor tissue in the dura surrounding the primary tumor attachment for 2.5 cm in 90% of the examined cases.

von Deimling et al. [23], in a small series of intracranial meningiomas, found that all the recurrences were clonal with respect to the primary tumors; thus, they conclude that most of recurrent meningiomas are direct descendants from the primary tumors, arising locally from residual tumor or distantly from dissemination of tumor fragments either by surgical manipulation or by subarachnoid spread via cerebrospinal fluid.

3 Topographic Classifications of Meningioma Recurrences

Different classifications of growth patterns and topographic distribution of the meningioma recurrences have been proposed in surgical studies.

Nakasu et al. [26] classified the patterns of the meningioma recurrences into three types: "local," at the site of the initial tumor; "peripheral," at the dura adjacent to the site of primary tumor attachment, which was removed or coagulated; "distant," ≥4 cm away from the original tumor.

Maiuri et al. [2, 27] classified recurrent meningiomas in four types: type 1, "local," confined to the previous dural site (Fig. 1); type 2, "peripheral," at the surrounding dura, contiguous to the previous site (Fig. 2); type 3, "multicentric," with multiple nodules both at the dural site and distant, with seemly normal interposed dura mater (Fig. 3); type 4, "diffuse," with multiple nodules with interposed dural infiltration or diffuse extradural infiltration (Fig. 4).

Obiri-Yeboah et al. [28] identified the following three zones of recurrence on T1-weighted volumetric contrast-enhanced MRI: (1) central growth, inside the area of the previously resected tumor greater than 1 cm inside the original tumor margin; (2) marginal growth, observed within 1 cm (inside or outside) of the original tumor margin; and (3) remote growth, observed >1 cm outside the original tumor margin.

Ong et al. [29] classified recurrences on follow-up MRI in three types: type A, within the

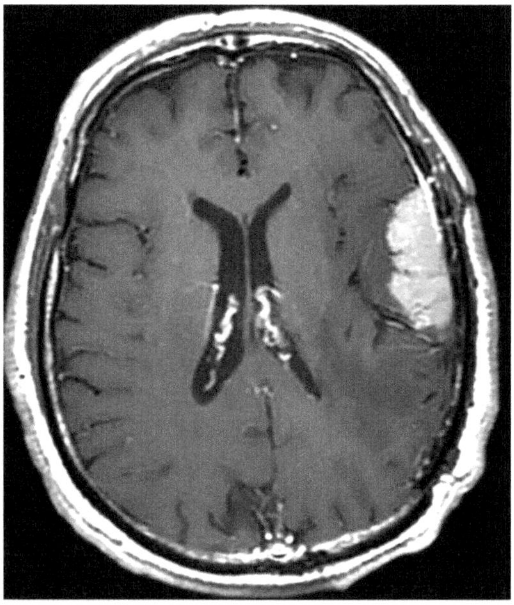

Fig. 1 MRI of the brain, axial sequence. *Local* recurrence of a left parieto-temporal convexity meningioma 5 years after initial surgery

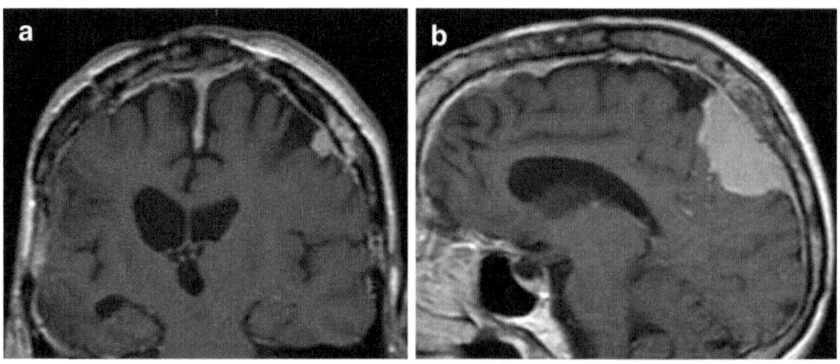

Fig. 2 MRI of the brain, coronal (**a**) and sagittal (**b**) sequences. *Peripheral* recurrence of the middle third parasagittal meningioma, posterior to the initial site

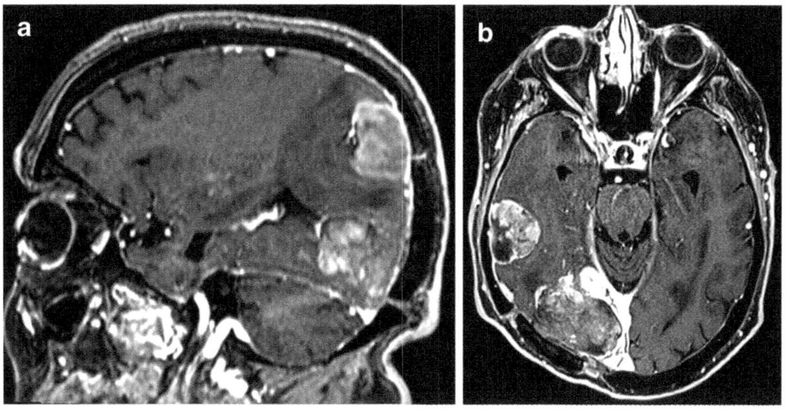

Fig. 3 Contrast-enhanced MRI of the brain. Sagittal (**a**) and axial (**b**) sequences: *Multicentric* recurrence of a meningioma of the right parietal-occipital convexity (WHO II) 5 years after first surgery (WHO III at recurrence)

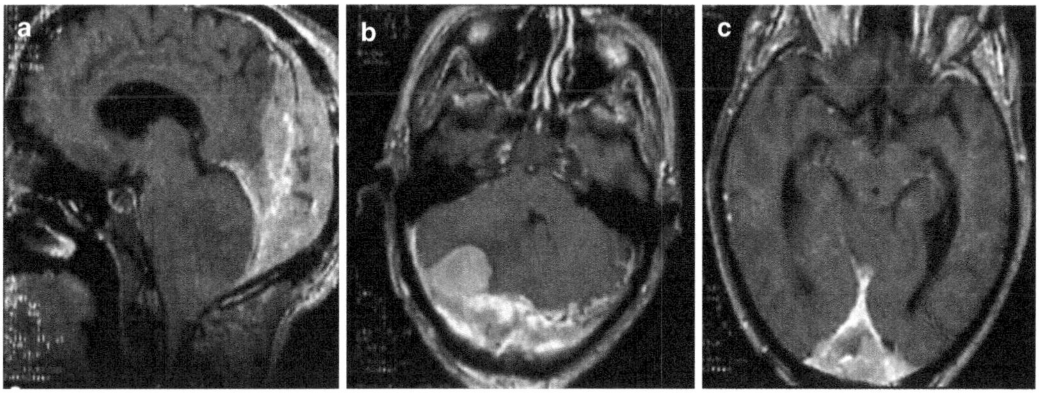

Fig. 4 Contrast-enhanced MRI of the brain. Sagittal (**a**) and Axial (**b**, **c**) sequences: 6 years after resection of a WHO grade II meningioma of the right occipital convexity: extensive *diffuse* recurrence extending in supra- and infratentorial compartments and to the contralateral side. No further reoperation

surgical bed, either on the dura side or on the brain side; type B, outside of the surgical bed but within 1 cm from the initial site; type C, distal (≥1 cm) from the resection site.

Some topographic patterns of meningioma recurrences have also been reported after different types of radiation treatments [30–32] in terms of treatment failure, with the aim to better define the risk factors of recurrence and to define the optimal target volume, boundaries of the field and radiation dose to increase the local and distant control of disease. Kuhn et al. [30] classified recurrences after stereotactic radiosurgery as "local" or "distant failure." "Local failure" indicates the recurrence either within the radiosurgical prescription volume ("central failure") or just outside the radiosurgical prescription margin within 2 cm of the tumor margin on MRI ("marginal failure"); "distant failure" defined recurrences occurring more than 2 cm outside the tumor margin. In this study, multifocal disease and high-grade histology (WHO grades 2 and 3) were associated to higher risk of local recurrence and male sex was associated to higher risk of distant failure; radiation doses greater than or equal to 12 Gy were associated to decreased local failure risk. This study also suggests 12 Gy as minimal sufficient margin dose for management of meningiomas. Rajkrishna et al. [32] classify recurrences after conformal radiation therapy as "in-field," when tumor recurred within the 90% isodose line, and "out-of-field," when tumor recurrence occurred outside the 90% isodose line: the authors suggest that increased clinical target volume margin, escalated dose up to 59.4 Gy and 3D conformal RT may be helpful in preventing local recurrences in grade 2 and grade 3 meningiomas.

4 Topographic Patterns of Recurrences: Risk Factors

4.1 Clinical-Radiological Findings

All four reviewed surgical studies [2, 26, 28, 29] focusing on the different topographic patterns of recurrence have failed to show statistically significative correlation with patients' sex and age.

The initial meningioma location affects the extent of resection and the pathological findings (with medial skull base meningiomas more frequently exhibiting lower WHO grades and proliferation index than those of lateral and no-skull base [33]). Nakasu et al. [26] found that skull base meningiomas only recurred locally or peripherally, but never distant, while non-skull base meningiomas, in particular those of the brain convexity, also recurred distant in a small percentage of cases. Obiri et al. [28] did not find significant differences of recurrences topography according to initial location; Maiuri et al. [2] provided the same conclusion, although spheno-orbital and parasagittal meningiomas showed slightly higher tendency to recur in diffuse pattern. Spheno-orbital meningiomas variably extend to the sphenoid bone, skull base and orbital structures, ethmoid-sphenoid sinuses and infratemporal fossa [15]; this affects the extent of resection and the pattern of recurrence. Parasagittal meningiomas show tendency to invade the superior sagittal sinus and to recur in local-peripheral [26] or diffuse pattern [2].

According to tumor shape, meningiomas are classified as round or flat.

In the study by Maiuri et al. [2] flat-shaped morphology at initial diagnosis was associated to significantly higher risk of multicentric-diffuse recurrences. This aspect could be explained by the prevalent dural involvement in flat-shaped meningiomas which would account for various degrees of even distant microscopic dural infiltration.

4.2 Pathological Findings

WHO grade [34, 35], Ki67-MIB1 [36, 37] and PR expression [38] represent well-demonstrated risk factors of meningioma recurrence; in particular, high-grades (WHO II-III) meningiomas, values of Ki67-MIB1 $\geq$ 4% and lower values of PR expression are related to higher recurrence risk.

The correlation between WHO grade, PR expression and topographic pattern of recurrence has been investigated only in one study [2], which failed to evidence significant association.

The role of proliferation index has been evaluated in two studies [2, 26]: Nakasu et al. [26] found that mean values of Ki67 of 3.3% were associated to the "peripheral" recurrences. Maiuri et al. [2] found higher rate of "multicentric-diffuse" recurrences for values of Ki67-MIB1 $\geq$4%, suggesting that small, even distant tumor foci in the surrounding dura, may diffusely regrow, if their growth potential is higher.

4.3 Surgical Findings

The extent of resection at initial diagnosis, according to Simpson grade, is not significantly associated to the pattern of growth at recurrence [2, 26, 28, 29].

The loss of CSF-vascular cleft or brain-tumor interface, which represents a risk factor of recurrence [39], was investigated in one study and was not correlated to the topographic pattern of regrowth [2]. No significant difference of time to recurrence was evidenced in patients who developed localized-peripheral versus multicentric-diffuse recurrences.

5 Management of Multicentric-Diffuse Recurrences

No studies defining the guidelines for reoperation of multicentric and diffuse recurrences of intracranial meningiomas have been published. Thus, the decision on surgical versus conservative management is based on several factors, including KPS, tumor location (critical versus non-critical), significant intradural mass versus prevalent dural infiltration, entity of extradural extension, WHO grade of the initial tumor, time to recurrence, surgeon's opinion. Patients with significant intradural tumor (>3 cm) and no extensive en-plaque dural infiltration are the best candidate to re-surgery. However, a gross-total resection (Simpson grades I and II) is obtained only in 20% of the cases versus 76% of local-peripheral recurrence.

Radiation therapy of diffuse-multicentric recurrences of meningiomas is mandatory, if not already administered after the initial operation, both because of the more often partial resection at reoperation (80% of our cases) and the WHO II and III histology in most cases (21/25 or 84% in our series) [2].

The need for radiation therapy in patients with WHO II grade recurrent meningiomas is well recognized, particularly for incompletely resected tumors [40–43]. Stereotactic radiosurgery is reported for tumors ≤3 cm, with a local control at 2 years from 50% to 80% and not infrequent recurrence outside the radiosurgical target [41, 44]. Besides, even re-radiosurgery for recurrent meningiomas has been recommended if the first radiosurgical treatment was unsatisfactory [45]. However, all studies include the overall group of recurrences; on the other hand, there are not studies concerning the diffuse recurrences and the best radiation modality. We suggest external brain radiation therapy. It is necessary for diffuse-multicentric recurrences, due to their extensive growth. We reserve stereotactic radiosurgery to selected cases in association with the external radiotherapy to further control smaller nodules.

6 Conclusion

Meningiomas may sometimes present multicentric or diffuse patterns of regrowth at the first recurrence, although this finding is more frequent at further recurrences. The flat-shaped morphology and the high Ki67-MIB1 values at initial diagnosis are risk factors for multicentric-diffuse pattern of regrowth; however, further studies will investigate the role of different biomolecular profiles.

The reoperation is advisable in cases with intradural tumor and not extensive "en-plaque" dural infiltration. In selected cases, even multiple reoperations over a period of several years may obtain long survivals, mainly in not anaplastic and less aggressive forms.

References

1. Lam Shin Cheung V, Kim A, Sahgal A, Das S. Meningioma recurrence rates following treatment: a systematic analysis. J Neuro-Oncol. 2018;136(2):351–61. https://doi.org/10.1007/s11060-017-2659-6.
2. Maiuri F, Mariniello G, Peca C, Guadagno E, Corvino S, d'Avanzo S, et al. Multicentric and diffuse recurrences of meningiomas. Br J Neurosurg. 2020;34:1–8. https://doi.org/10.1080/02688697.2020.1754335.
3. Mariniello G, de Divitiis O, Corvino S, Strianese D, Iuliano A, Bonavolontà G, et al. Recurrences of spheno-orbital meningiomas: risk factors and management. World Neurosurg. 2022;161:e514. https://doi.org/10.1016/j.wneu.2022.02.048.
4. Haddad AF, Young JS, Kanungo I, Sudhir S, Chen JS, Raleigh DR, et al. WHO Grade I meningioma recurrence: identifying high risk patients using histopathological features and the MIB-1 index. Front Oncol. 2020;10:1522. https://doi.org/10.3389/fonc.2020.01522.
5. Spille DC, Adeli A, Sporns PB, Heß K, Streckert EMS, Brokinkel C, et al. Predicting the risk of postoperative recurrence and high-grade histology in patients with intracranial meningiomas using routine preoperative MRI. Neurosurg Rev. 2021;44(2):1109–17. https://doi.org/10.1007/s10143-020-01301-7.
6. Vaubel RA, Kumar R, Weiskittel TM, Jenkins S, Dasari S, Uhm JH, et al. Genomic markers of recurrence risk in atypical meningioma following gross total resection. Neurooncol Adv. 2023;5(1):vdad004. https://doi.org/10.1093/noajnl/vdad004.
7. Olar A, Wani KM, Sulman EP, Mansouri A, Zadeh G, Wilson CD, et al. Mitotic index is an independent predictor of recurrence-free survival in meningioma. Brain Pathol. 2015;25(3):266–75. https://doi.org/10.1111/bpa.12174.
8. Maiuri F, Mariniello G, de Divitiis O, Esposito F, Guadagno E, Teodonno G, et al. Progesterone receptor expression in meningiomas: pathological and prognostic implications. Front Oncol. 2021;11:611218. https://doi.org/10.3389/fonc.2021.611218.
9. Sahm F, Schrimpf D, Stichel D, Jones DTW, Hielscher T, Schefzyk S, et al. DNA methylation-

based classification and grading system for meningioma: a multicentre, retrospective analysis. Lancet Oncol. 2017;18(5):682–94. https://doi.org/10.1016/S1470-2045(17)30155-9.
10. Youngblood MW, Miyagishima DF, Jin L, Gupte T, Li C, Duran D, et al. Associations of meningioma molecular subgroup and tumor recurrence. Neuro-Oncology. 2021;23(5):783–94. https://doi.org/10.1093/neuonc/noaa226.
11. Patel AJ, Wan YW, Al-Ouran R, Revelli JP, Cardenas MF, Oneissi M, et al. Molecular profiling predicts meningioma recurrence and reveals loss of DREAM complex repression in aggressive tumors. Proc Natl Acad Sci USA. 2019;116(43):21715–26. https://doi.org/10.1073/pnas.1912858116.
12. Maiuri F, Corvino S, Lorenzetti M, Franca RA, Esposito F, Del Basso De Caro M. Intracranial meningiomas in patients aged ≥80 years: pathological features and surgical problems. World Neurosurg. 2023;173:e498–508. https://doi.org/10.1016/j.wneu.2023.02.086.
13. Maiuri F, Corvino S. Meningiomas of the rolandic region: risk factors for motor deficit and role of intra-operative monitoring. Acta Neurochir. 2023;165(7):1707–16. https://doi.org/10.1007/s00701-023-05630-6.
14. Maiuri F, Corvino S, Corazzelli G, Berardinelli J, Di Crescenzo RM, Del Basso De Caro M. Time to recurrence of intracranial meningiomas from a monoinstitutional surgical series. World Neurosurg. 2024;185:e612. https://doi.org/10.1016/j.wneu.2024.02.087.
15. Mariniello G, Bonavolontà G, Tranfa F, Iuliano A, Corvino S, Teodonno G, et al. Management of the skull base invasion in spheno-orbital meningiomas. Br J Neurosurg. 2023:1–8. https://doi.org/10.1080/02688697.2022.2161472.
16. Franca RA, Della Monica R, Corvino S, Chiariotti L, Del Basso De Caro M. WHO grade and pathological markers of meningiomas: clinical and prognostic role. Pathol Res Pract. 2023;243:154340. https://doi.org/10.1016/j.prp.2023.154340.
17. Maiuri F, Corvino S, Corazzelli G, Del Basso De Caro M. Single versus multiple reoperations for recurrent intracranial meningiomas. J Neuro-Oncol. 2024;168:527.
18. Adegbite AB, Khan MI, Paine KW, Tan LK. The recurrence of intracranial meningiomas after surgical treatment. J Neurosurg. 1983;58(1):51–6. https://doi.org/10.3171/jns.1983.58.1.0051.
19. Crompton MR, Gautier-Smith PC. The prediction of recurrence in meningiomas. J Neurol Neurosurg Psychiatry. 1970;33(1):80–7. https://doi.org/10.1136/jnnp.33.1.80.
20. Chan RC, Thompson GB. Morbidity, mortality, and quality of life following surgery for intracranial meningiomas. A retrospective study in 257 cases. J Neurosurg. 1984;60(1):52–60. https://doi.org/10.3171/jns.1984.60.1.0052.
21. Melamed S, Sahar A, Beller AJ. The recurrence of intracranial meningiomas. Neurochirurgia (Stuttg). 1979;22(2):47–51. https://doi.org/10.1055/s-0028-1090287.
22. Borovich B, Doron Y. Recurrence of intracranial meningiomas: the role played by regional multicentricity. J Neurosurg. 1986;64(1):58–63. https://doi.org/10.3171/jns.1986.64.1.0058.
23. von Deimling A, Larson J, Wellenreuther R, Stangl AP, van Velthoven V, Warnick R, et al. Clonal origin of recurrent meningiomas. Brain Pathol. 1999;9(4):645–50. https://doi.org/10.1111/j.1750-3639.1999.tb00546.x.
24. Kinjo T, al-Mefty O Kanaan I Grade zero removal of supratentorial convexity meningiomas. Neurosurgery 1993;33(3):394–9; discussion 9. doi: https://doi.org/10.1227/00006123-199309000-00007.
25. Qi ST, Liu Y, Pan J, Chotai S, Fang LX. A radiopathological classification of dural tail sign of meningiomas. J Neurosurg. 2012;117(4):645–53. https://doi.org/10.3171/2012.6.JNS111987.
26. Nakasu S, Nakasu Y, Nakajima M, Matsuda M, Handa J. Preoperative identification of meningiomas that are highly likely to recur. J Neurosurg. 1999;90(3):455–62. https://doi.org/10.3171/jns.1999.90.3.0455.
27. Maiuri F, Corvino S, Guadagno E, Del Basso De Caro M. Topographic distribution of intracranial meningioma's recurrences: localized versus diffuse-multicentric [Internet]. Front Clin Neurosurg IntechOpen. 2021; https://doi.org/10.5772/intechopen.97120.
28. Obiri-Yeboah D, Soni P, Oyem PC, Almeida JP, Murayi R, Recinos PF, et al. Beyond the surgical margin: patterns of recurrence in World Health Organization Grade 2 intracranial meningiomas. World Neurosurg. 2023;177:e593. https://doi.org/10.1016/j.wneu.2023.06.100.
29. Ong K, Rizzuto M, Makarenko S. Location pattern of recurrence of fully resected grade 1 meningiomas. Acta Neurochir. 2023;165(10):2865–71. https://doi.org/10.1007/s00701-023-05758-5.
30. Kuhn EN, Taksler GB, Dayton O, Loganathan AG, Vern-Gross TZ, Bourland JD, et al. Patterns of recurrence after stereotactic radiosurgery for treatment of meningiomas. Neurosurg Focus. 2013;35(6):E14. https://doi.org/10.3171/2013.8.FOCUS13283.
31. Hoffmann E, Clasen K, Frey B, Ehlers J, Behling F, Skardelly M, et al. Retrospective analysis of recurrence patterns and clinical outcome of grade II meningiomas following postoperative radiotherapy. Radiat Oncol. 2021;16(1):116. https://doi.org/10.1186/s13014-021-01825-2.
32. Rajkrishna B, Rajesh B, Patricia S, Selvamani B. Patterns of recurrence in patients receiving conformal radiation for intracranial meningioma: a single-institution experience. J Radiother Pract. 2021;20(4):406–12.
33. Maiuri F, Mariniello G, Guadagno E, Barbato M, Corvino S, Del Basso De Caro M. WHO grade, proliferation index, and progesterone receptor expression are different according to the location of meningioma.

Acta Neurochir. 2019;161(12):2553–61. https://doi.org/10.1007/s00701-019-04084-z.

34. Maiuri F, De Caro MB, Esposito F, Cappabianca P, Strazzullo V, Pettinato G, et al. Recurrences of meningiomas: predictive value of pathological features and hormonal and growth factors. J Neuro-Oncol. 2007;82(1):63–8. https://doi.org/10.1007/s11060-005-9078-9.
35. Matsuno A, Fujimaki T, Sasaki T, Nagashima T, Ide T, Asai A, et al. Clinical and histopathological analysis of proliferative potentials of recurrent and non-recurrent meningiomas. Acta Neuropathol. 1996;91(5):504–10.
36. Oya S, Kawai K, Nakatomi H, Saito N. Significance of Simpson grading system in modern meningioma surgery: integration of the grade with MIB-1 labeling index as a key to predict the recurrence of WHO Grade I meningiomas. J Neurosurg. 2012;117(1):121–8. https://doi.org/10.3171/2012.3.JNS111945.
37. Mirian C, Skyrman S, Bartek J, Jensen LR, Kihlström L, Förander P, et al. The Ki-67 proliferation index as a marker of time to recurrence in intracranial meningioma. Neurosurgery. 2020;87(6):1289–98. https://doi.org/10.1093/neuros/nyaa226.
38. Maiuri F, Mariniello G, Somma T, Guadagno E, Corvino S, Pagano S, et al. Meningiomas in premenopausal women: role of the hormone related conditions. Front Oncol. 2020;10:556701. https://doi.org/10.3389/fonc.2020.556701.
39. Champeaux C, Dunn L. World Health Organization grade II meningioma: a 10-year retrospective study for recurrence and prognostic factor assessment. World Neurosurg. 2016;89:180–6. https://doi.org/10.1016/j.wneu.2016.01.055.
40. Mair R, Morris K, Scott I, Carroll TA. Radiotherapy for atypical meningiomas. J Neurosurg. 2011;115(4):811–9. https://doi.org/10.3171/2011.5.JNS11112.
41. Hardesty DA, Wolf AB, Brachman DG, McBride HL, Youssef E, Nakaji P, et al. The impact of adjuvant stereotactic radiosurgery on atypical meningioma recurrence following aggressive microsurgical resection. J Neurosurg. 2013;119(2):475–81. https://doi.org/10.3171/2012.12.JNS12414.
42. Rogers L, Gilbert M, Vogelbaum MA. Intracranial meningiomas of atypical (WHO grade II) histology. J Neuro-Oncol. 2010;99(3):393–405. https://doi.org/10.1007/s11060-010-0343-1.
43. Klinger DR, Flores BC, Lewis JJ, Hatanpaa K, Choe K, Mickey B, et al. Atypical meningiomas: recurrence, reoperation, and radiotherapy. World Neurosurg. 2015;84(3):839–45. https://doi.org/10.1016/j.wneu.2015.04.033.
44. Kano H, Takahashi JA, Katsuki T, Araki N, Oya N, Hiraoka M, et al. Stereotactic radiosurgery for atypical and anaplastic meningiomas. J Neuro-Oncol. 2007;84(1):41–7. https://doi.org/10.1007/s11060-007-9338-y.
45. Kim M, Lee DH, Kim Rn HJ, Cho YH, Kim JH, Kwon DH. Analysis of the results of recurrent intracranial meningiomas treated with re-radiosurgery. Clin Neurol Neurosurg. 2017;153:93–101. https://doi.org/10.1016/j.clineuro.2016.12.014.

Management of Recurrences of Non-Skull Base Meningiomas

Francesco DiMeco, Edoardo Porto, and Lapo Bonosi

Abbreviations

CK-SRS	CyberKnife Stereotactic Radiosurgery
EANO	European Association of Neuro-oncology
EBRT	External beam radiotherapy
EGF	Epidermal growth factor
EOR	Extent of resection
GTR	Gross total resection
OS	Overall survival
PDGF	Platelet-derived growth factor
PFS	Progression-free survival
RT	Radiotherapy
SBM	Skull base meningioma
TTR	Time to recurrence
VEGF	Vascular endothelial growth factor

1 Background

Meningiomas are among the most common intracranial tumors accounting for 55% of nonmalignant primary central nervous system lesions [1]. Their first-line management does not usually present intrinsic decisional complexity since surgical treatment is recommended when symptoms are patent or radiological progression is confirmed [2]. In this context, other factors, such as patient age and general conditions, should always be taken into account in the decision-making process. In the scenario of surgical indication, safe maximal resection and, if possible, removal of the dural tail and drilling of bone involvement, should be always pursued to reduce the risk of recurrence [3]. Nevertheless, tumor size, specific locations, and neurovascular relationships may

F. DiMeco (✉)
Department of Neurosurgery, IRCCS Fondazione Istituto Neurologico Carlo Besta, Milan, Italy

Department of Oncology and Hemato-Oncology, University of Milan, Milan, Italy

Department of Neurological Surgery, Hunterian BrainTumor Research Laboratory, Johns Hopkins Medical School, Baltimore, MD, USA
e-mail: francesco.dimeco@istituto-besta.it

E. Porto
Department of Neurosurgery, IRCCS Fondazione Istituto Neurologico Carlo Besta, Milan, Italy

Department of Neurosurgery, School of Medicine, Emory University, Atlanta, GA, USA

L. Bonosi
Department of Neurosurgery, IRCCS Fondazione Istituto Neurologico Carlo Besta, Milan, Italy

Department of Oncology and Hemato-Oncology, University of Milan, Milan, Italy

Neurosurgical Clinic, AOUP "Paolo Giaccone", Post Graduate Residency Program in Neurologic Surgery, Department of Experimental Biomedicine and Clinical Neurosciences, School of Medicine, University of Palermo, Palermo, Italy

F. Maiuri, M. Del Basso De Caro (eds.), *Meningiomas*,
https://doi.org/10.1007/978-3-031-76680-0_20

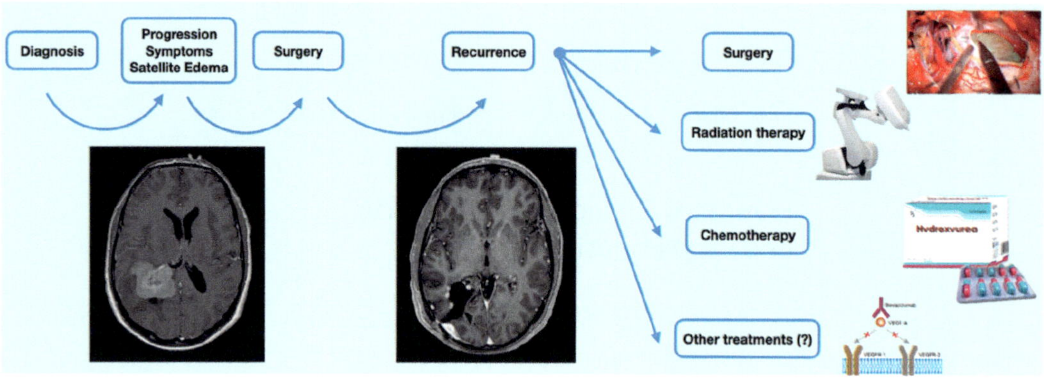

Fig. 1 Flowchart showing the standard of care for meningiomas

limit the possibilities of obtaining a gross total resection (GTR) [3–5] or determining a non-tolerable risk of surgical complication and functional impairment [6]. For instance, skull base meningiomas (SBM) may represent a surgical challenge for which radical resection may be often unlikely or even expose patients to the risk of neurovascular injury [7–9]. Furthermore, Corniola et al. have recently noted how meningiomas may be regarded as a 'chronic condition' to be managed rather than a tumor to be radically resected at any cost [10].

In the neurosurgical tradition, the definition of extent of resection (EOR) for meningiomas resides in the 1957 Simpson classification which is based on intraoperative surgical observation and provides a classification into biopsy, subtotal resection, complete resection, and complete resection with removal of dural and bone infiltration [11]. As the Simpson score is based on subjective assessments, it may be characterized by low sensitivity levels [12–14]. This is confirmed by the fact that the risk of residual tumor remains even in cases of Simpson 1 removals [15]. Notwithstanding, the EOR represents the favored positive prognostic factor for an increased progression-free survival (PFS) and it is taken for granted that a more extensive resection leads to improved disease control and lower chances for tumor recurrence [16–18]. Additionally, it has been proven that it may offer longer overall survivals (OS) in higher-grade meningiomas [19].

In conclusion, meningioma recurrence is a frequent circumstance, despite an optimal initial EOR, for which second surgery or the use of adjuvant treatments is often required [18, 20]. Although the published literature regarding initial meningioma management is abundant, there is scarce evidence on the optimal management strategy for recurrent meningiomas [21, 22]. This chapter aims to outline the principles of non-skull base meningioma recurrence management and define some future perspectives (Fig. 1).

2 Risk Factors of Recurrence

Intracranial meningiomas are estimated to recur in 10–32% of cases at 10 years [23, 24]. Several variables have been investigated as risk factors for meningioma recurrence. Some studies demonstrated how the WHO grade and tumor location may be strongly related to the risk of recurrence [25, 26]. With this regard, grade II and III meningiomas present a risk of recurrence which amounts to up to 29% after the first recurrence and skull base locations seem to increase by more than twofold the risk of tumor relapses compared to non-skull base locations [26], probably due to the surgical difficulty in obtaining a GTR.

In recent years, the role of specific genomic signatures has been increasingly addressed. For instance, TERT promoter mutations in the hotspot regions C228T and C250T have been associated with a shorter time to progression [27]. Moreover, Sahm et al. recently described how DNA methylation profiling may distinguish six different clinically relevant methylation classes associated with typical mutational, cytogenetic, and gene expression patterns that, compared with WHO grading, may more accurately identify patients at high risk of disease progression in tumors with WHO grade I histology, and patients at lower risk of recurrence among WHO grade II tumors [28]. Similarly, another study by Youngblood et al. described how some molecular meningioma subgroups (NF2, PI3K, HH, tumor necrosis factor receptor-associated factor 7 [TRAF7]) can recur at an average rate of 22 times higher than others [29].

Disease-free survival is also related to the EOR in any WHO grade and, in this context, Simpson grades from 1 to 3 are regarded as GTR by the EANO [2]. Nevertheless, although grade 3 is considered complete resection, the risk of tumor residual along surgical margins and meninges still represents a high risk of recurrence [12]. Borovich and Doron showed that the primary mass is surrounded in roughly 60% of cases by micro- and macroscopic meningotheliomatous dural nodules as far as 4 cm from the tumor margin that may account for the recurrence [30]. Regarding this concern, some new tools may help detect microscopic pathological foci that can trigger the recurrence. For instance, Haslund-Vinding and colleagues have devised a new grading system that can detect microscopic tumor infiltration in the surgical margins using DOTATOC PET scanning [31]. Moreover, the presence of symptoms at presentation, location in the skull base, and bone infiltration may predict a subtotal resection. Some studies have further underlined how partial resections may increase the rate of retreatment at 10 years of follow-up [21].

In addition, a shorter PFS was encountered in patients with multiple localizations, larger tumors, and higher-grade tumors although, in this context, adjuvant radiotherapy seems to prolong PFS. This suggests how the role of upfront radiation therapy for grade 2 meningiomas in skull base locations is a matter of controversy [32]. On the other side of the spectrum, some evidence shows that radiation-induced meningiomas may have a more aggressive behavior and a very high rate of recurrence along with a tendency to evolve to higher WHO grade in consideration of specific molecular signatures [33]. Similarly, skull base meningiomas can fare worse and behave more aggressively after radiosurgery in the long run [34]. This consideration raises the question of whether observation may be an option for this circumstance.

3 The Role of Surgery

Surgery has been demonstrated to be an effective tool to manage meningioma recurrence. Some studies have highlighted how PFS may be significantly longer after surgery as compared to other treatment modalities, [26, 35] and this is especially true for the cohort of SBM [36]. A recent study was the first to describe the role of repeated surgeries in meningioma recurrence and the time to recurrence (TTR) between the different surgeries for the recurrence of intracranial meningiomas. Interestingly, TTR decreases significantly between surgeries in patients requiring repeated resections, while the risk of complications remains stable. This continuous and consistent decrease in TTR between surgeries means that the impact of surgical treatments, albeit beneficial, decreases with successive surgeries. Nevertheless, this trend may also be correlated to the poorer physical condition of patients undergoing several surgeries [26].

A paper by Maiuri et al. classified meningiomatous recurrences according to their topography on post-contrast MRI, dividing them into four types (local, peripheral, multicentric, and diffuse). Intriguingly, the authors pointed out that patients reoperated for multicenter or diffuse recurrences showed significantly lower rates of tumor control and higher rates of tumor progression and death than those reoperated for local or peripheral recurrences, emphasizing that radiotherapy treatment rather than surgical reintervention may be recommended in these cases. However, further studies are needed to confirm this finding [37].

A special condition to bear in mind is the case of superior sagittal sinus meningiomas which may be subjected to an increased risk of recurrence [38]. The risks of complications and recurrence are indeed significantly higher in patients with higher Sindou grades [39]. Although the tumor portion of parasagittal meningiomas that involves the sinus wall should be actively resected to the largest extent possible to reduce the recurrence rate, in this circumstance, the benefit-to-risk ratio should be carefully evaluated before posing surgical indication or deciding the adequate surgical strategy. For higher-grade parasagittal meningiomas, it might be necessary to use stereotactic radiosurgery (SRS) for residual tumor portions in addition to surgical resection [40].

When dealing with higher-grade meningiomas, recurrence management becomes more difficult since GTR is more unlikely to be achieved. In this scenario, tailoring the approach to the specific patient becomes of paramount importance keeping the aim of preserving the patient functional status and quality of life. Multidisciplinary discussions play an important role according to the EANO guidelines [2].

Surgery for meningioma recurrence should always be tailored based on the patient's intrinsic risk factors. For instance, in the elderly, the surgical option may be less pertinent in case of a shorter residual life expectancy or in case of increased complication risks. Thus, the benefit-to-risk ratio should always be considered. In this scenario, surgery may be an option only in selected patients [5, 26, 41]. Concerning postoperative complications, there is evidence that postoperative neurological worsening may be more frequent for recurrent meningioma surgery [26]. Similarly, the rate of postoperative hematomas and infection seems to be increased after recurrence surgery [42–45].

4 The Role of Radiation Therapy

4.1 Adjuvant Radiotherapy for Higher-Grade Meningiomas

Determining the adequate management strategy in case of meningioma relapse is a multidisciplinary decision [11]. Some studies have underlined the role of adjuvant radiotherapy (RT) for WHO grade-2 meningiomas demonstrating an increased tumor control at 5 years after initial surgery followed by adjuvant RT [46]. Similarly, Byun et al. in a multi-institutional study showed how adjuvant RT significantly improved PFS and the rate of recurrence in patients with grade-2 meningioma, regardless of the surgical extent, and can be recommended after GTR. There is evidence that also favors the use of adjuvant RT for grade III meningiomas [47], conversely, it has not been demonstrated that in grade I meningioma recurrences depend on whether patients are treated or not with adjuvant RT [48]. In this context, RTOG 0539 is an ongoing clinical trial that studies recurrent meningiomas establishing that adjuvant RT should be a standard of care for resected higher-grade meningiomas [47]. Nevertheless, whether adjuvant RT decreases tumor recurrence after GTR grade II meningioma remains a matter of debate. As a matter of fact, RT may reduce the risk of recurrence but may increase the risk of chronic radiation side effects [2]. Finally, Kim et al. analyzed the results of re-radiosurgery for recurrent meningiomas and observed that WHO grade II and grade III meningiomas were significantly associated with a lower PFS rate than grade I meningiomas, underscoring how re-radiosurgery could be a viable option for patients with grade I meningiomas [49].

4.2 Radiotherapeutic Modalities for Recurrent Meningioma

Stereotactic radiosurgery (SRS) may be considered as the standard of care when RT is employed for recurrent meningiomas as an alternative to surgery. This is particularly true for small recurrent tumors in elderly or critically ill patients. Also, this may be a primary therapeutic choice for recurrent SBMs where cranial nerve function is a major concern. In contrast, some studies have demonstrated a significantly higher rate of symptomatic worsening of peri-tumoral brain edema and neurological worsening after SRS for parasagittal or falcine meningiomas, probably because of their association with unique biological features, such as a higher frequency of immune infiltration and more aggressive behavior [50, 51].

4.3 Alternative Radiotherapeutic Modalities for Recurrent Meningioma

Hypofractionated stereotactic radiotherapy can be used as the first treatment at recurrence. There is evidence that shows how control was obtained in 81% of cases in primary treatments and 39% in the re-irradiation group. Furthermore, the rate of radionecrosis was significantly lower in 2% of cases for primary treatments after 24 months [52].

The use of brachytherapy with I-125 has been explored in the past to treat recurrent previously irradiated high-grade meningiomas. Despite a moderate median time to progression, the rate of complications was around 40%. Thus, the use of brachytherapy should be confined to salvage treatments [53].

Particle-therapy RT (either protons or carbon ions) can be a viable option for side effects and recent irradiation. Nevertheless, its effect, if compared to standard RT, is to be investigated with future trials [54, 55]. Moreover, proton re-irradiation has been proven to be safe since the detected rate of radionecrosis amounted to 13% obtaining a good local control. In conclusion, despite the lack of strong data supporting it, particle-therapy RT seems to be a viable therapeutic option that allows high-volume treatments and low toxicity rates [54–59].

5 Systemic Treatments

5.1 Chemotherapy

Systemic therapies for recurrent meningioma are usually favored for situations in which surgery or RT is not a viable option anymore. Although a consensus or a gold standard has not been reached, the use of several drugs has been investigated with poor results in terms of OS and disease control. Moreover, due to the paucity of controlled trials on the topic, the evidence reported in the literature is scarce.

Several chemotherapeutic drugs have been used and described to treat recurrent meningiomas, among the classes more frequently employed protein kinase inhibitors, sunitinib malate, hydroxyurea, and somatostatin analogs among others are the more widely described. Hydroxyurea is the medication on which most of the data available focus; nonetheless its efficacy has not been proven by any clinical trial. In a study in which 60 recurrent meningiomas were included, no radiological response to hydroxyurea was noted with a PFS of just 2 months [60]. Another prospective study showed a hydroxyurea response rate of just 5% [61]. Temozolomide was also investigated and showed modest results with an average progression after the second cycle with a median OS of 7 months [62]. Another molecule that was tested for recurrent meningiomas was irinotecan (campothecin family) which was employed in a phase II study. In this study, no partial or complete response to treatment was highlighted with a median of 7 months of OS and a time to progression of 5 months [63].

There is evidence that the majority of meningiomas express somatostatin receptors. Thus, the possibility of using somatostatin analogs to treat recurrent meningiomas has been explored. A small prospective trial has confirmed that octreotide may be used after the detection of somatostatin receptors using SPECT scanning. The response rate was 31% with a PFS of 7.5 months after treatment [63]. Another study tested other somatostatin analogs such as (pasireotide-LAR) in patients with higher-grade meningiomas showing worse results [64].

Due to the presence of progesterone receptors in meningioma tissue, mifepristone has also been employed in several studies as a rescue therapy for tumor recurrence. Six different trials have been conducted, showing a low rate of response [65].

Trabectedin is another antineoplastic drug that has been explored due to its mechanism of binding with the DNA's minor groove and subsequent formation of DNA adducts, resulting in the interruption of the cellular cycle [66]. Despite its mechanism, several studies have proven the inefficacy of this drug in improving PFS and OS on recurrent meningiomas with worse toxicity [67].

5.2 Target Therapy

There is novel evidence that shows that many meningiomas express growth factors, such as the epidermal growth factor (EGF), the platelet-derived growth factor (PDGF), and the vascular endothelial growth factor (VEGF). Nonetheless, little is known about their role in tumor pathogenesis [68]. With this regard, the role of protein kinase inhibitors has been explored as a treatment for recurrent meningiomas in several studies including clinical trials [69–71]. A phase II study has recently evaluated the role of EGF receptor inhibitors (gefitinib, erlotinib) as a third-line treatment. The treatment was well-tolerated but did not have an impact on tumor progression, not improving PFS and OS, although tumor stability was proven in one-third of patients [72].

Since the majority of meningiomas express specific ligands of PDGF and the PDGF-beta receptor supporting meningioma proliferation, a phase II study has recently explored the impact of imatinib, a PDGF receptor inhibitor, as a treatment for heavily pre-treated recurrent meningiomas of all grades. In this case, the treatment was also well-tolerated but did not show efficacy as a radiological response was not noted [73].

A multi-kinase inhibitor, sunitib, active in targeting VEGF and PDGF receptors has been studied as a treatment for recurrent meningioma and a prospective phase II study proved its capability to increase the 6-month PFS [74]. Similarly, the efficacy of vatalinib, another VEGF and PDGF receptor inhibitor, was evaluated in a clinical trial for meningiomas of all grades. This study demonstrated radiological stability in two-thirds of cases as the best response to treatment [69]. The role of VEGF pathway in meningiomas has been widely studied in the published literature [75]. Nevertheless, just a few phases II investigated its use and, among these, Kumthekar et al. demonstrated the obtainment of pathology stability with bevacizumab in the majority of patients who showed good tolerance to this drug [76].

As a consequence of the Alliance/NCI A071401 study, FAK inhibitors have also been evaluated for recurrent meningiomas showing an improved PFS for grade I meningiomas and stable disease for grade II and III meningiomas as the best response.

Alterations of the mTOR pathway and the decreased meningioma proliferation in relation to its inhibition are well-established concepts [77]. Thus, everolimus, an mTOR inhibitor, in association with octreotide has been explored by the CEVOREM study in recurrent meningioma patients who were not candidates for surgery or radiation therapy, demonstrating a prolonged PFS and OS [71]. Furthermore, everolimus has been studied in combination with bevacizumab showing encouraging results [70].

5.3 Immunotherapy

Little has been studied about the role of immunotherapy and checkpoint inhibitors, on meningiomas, although several trials are ongoing at present. Interferon-α has been demonstrated to interfere with meningioma cells in vitro [78] while a prospective study has not confirmed its impact on tumor progression [79]. Similarly, pembrolizumab, an anti-PD-1, has been tested in meningioma patients showing a prolonged median OS, although a partial or complete response has not been obtained. Another antiPD-1, nivolumab, has been studied for grade II and III meningiomas in a prospective study obtaining partial response.

6 Conclusions and Treatment Algorithm

The published literature on meningioma recurrence is scarce. Nevertheless, nowadays the standard of care for non-skull base meningioma recurrence is based on a surgical strategy. Limits to the surgical indication are represented by poor general conditions, unfavorable anatomical features or proximity and invasion of venous sinuses. Radiosurgery may be proposed in some cases of recurrent meningioma where radiological progression is confirmed in the absence of clinical symptoms. Of interest, the role of systematic

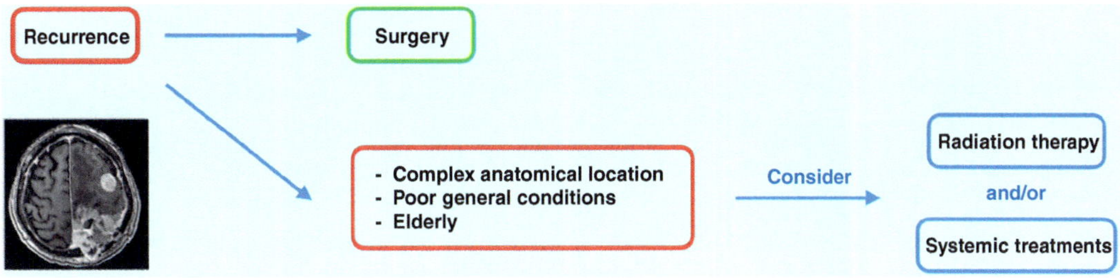

Fig. 2 Flowchart showing the most accepted management strategy for meningioma recurrence

postoperative adjuvant RT in intracranial meningiomas has yet to be evaluated in larger clinical trials (Fig. 2). Finally, recent advances in the understanding of meningioma biology and genomic profiling may improve clinical risk stratification of patients, refining patient selection for postoperative RT, radiosurgery, chemotherapy, or imaging surveillance.

6.1 Illustrative Cases

Case 1—Management of Peri-site and Extra-site Recurrence

This is a 44-year-old man who presented to our attention for the onset of left homonymous lateral hemianopsia. Brain contrast MRI showed the presence of a voluminous endoventricular meningioma of the right atrium. The patient underwent surgical resection with subsequent histological diagnosis of atypical grade II meningioma according to WHO classification (2007). Approximately 3 years after surgery, multiple right atrial endoventricular nodular lesions, consistent with recurrence, and relapse of disease also at the level of the dura mater in the left posterior parietal location (extraventricular seeding) were revealed on follow-up MRI. Given the patient's young age and good general condition, a second surgery and external beam RT (EBRT) on the operating cavity (total dose of 56 Gy) were performed. Seven years after the second surgery, the patient presented again to our attention with extensive recurrence in the right parieto-occipital site with falco-tentorial involvement and the presence of a second localization at the level of the fourth ventricle. After a multidisciplinary discussion, it was decided to perform debulking of the parieto-occipital lesion (grade II atypical meningioma with MIB-1 of 9–10%). The intraventricular lesion was treated using CyberKnife Stereotactic Radiosurgery (CK-SRS), given its deep-seated location.

Nine years after surgery, the fourth ventricle lesion was smaller than previous scans and a new endoventricular lesion appeared at the level of the right temporal horn of the lateral ventricle. Considering the diagnosis, the previously performed treatments, the patient's poor general conditions, and the rapid and multicentric growth of the disease, the patient was started on systemic chemotherapy treatment, and the new lesion was treated using CK-SRS. At the last follow-up, the patient presented with radiologically confirmed disease stability (Fig. 3).

Case 2—Management of Intra- and Extra-site Recurrence

This is a 61-year-old woman who came to our attention for the finding of a duro-elastic mass in the left parietal bone of the calvarium. A brain CT scan documented an erosive lesion of the bone developing intra- and extracranially, with invasion of the underlying dura mater. She underwent surgery with a subsequent diagnosis of atypical transitional meningioma with partial chordoid aspects (grade II WHO 2021) and MIB-1 of 5–6%.

The patient was disease-free for about 7 years. Afterward, during radiological follow-up, recurrences of disease were found at both the surgical margins (intra-site) and left temporal convexity (extra-site). Given the progressive dimensional

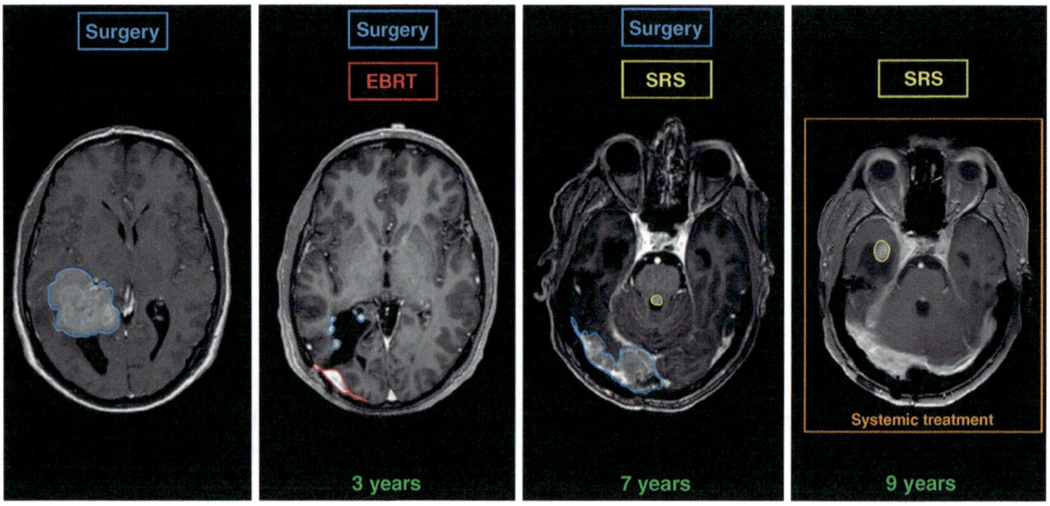

Fig. 3 Post-gadolinium T1 magnetic resonance axial images of different time points in the clinical history of case 1. *EBRT* external beam radiotherapy, *SRS* stereotactic radiosurgery

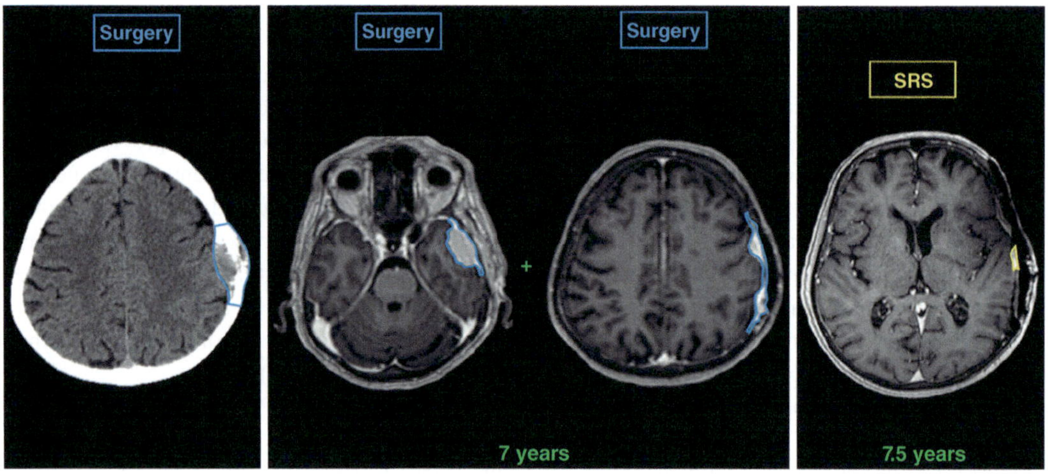

Fig. 4 CT axial image (first image on the left) and post-gadolinium T1 magnetic resonance axial images (three images on the right) of different time points in the clinical history of case 2. *SRS* stereotactic radiosurgery

growth of the recurrences, a re-operation was proposed to the patient and performed achieving a Simpson grade I resection. Nevertheless, 6 months after the second surgery, the patient presented with a new intra-site recurrence which was approached using CK-SRS after multidisciplinary discussion in consideration of the poor general conditions. At the last follow-up, the patient presented free of disease (Fig. 4).

References

1. Ostrom QT, Patil N, Cioffi G, Waite K, Kruchko C, Barnholtz-Sloan JS. CBTRUS statistical report: primary brain and other central nervous system tumors diagnosed in the United States in 2013–2017. Neuro-Oncology. 2020;22(12 Suppl 2):iv1–iv96. https://doi.org/10.1093/neuonc/noaa200.
2. Goldbrunner R, Stavrinou P, Jenkinson MD, Sahm F, Mawrin C, Weber DC, et al. EANO guideline on

the diagnosis and management of meningiomas. Neuro-Oncology. 2021;23(11):1821–34. https://doi.org/10.1093/neuonc/noab150.

3. Hasseleid BF, Meling TR, Rønning P, Scheie D, Helseth E. Surgery for convexity meningioma: Simpson Grade I resection as the goal: clinical article. J Neurosurg. 2012;117(6):999–1006. https://doi.org/10.3171/2012.9.Jns12294.
4. Giammattei L, di Russo P, Starnoni D, Passeri T, Bruneau M, Meling TR, et al. Petroclival meningiomas: update of current treatment and consensus by the EANS skull base section. Acta Neurochir. 2021;163(6):1639–63. https://doi.org/10.1007/s00701-021-04798-z.
5. Meling TR, Da Broi M, Scheie D, Helseth E. Skull base versus non-skull base meningioma surgery in the elderly. Neurosurg Rev. 2019;42(4):961–72. https://doi.org/10.1007/s10143-018-1005-6.
6. Lemée JM, Corniola MV, Da Broi M, Joswig H, Scheie D, Schaller K, et al. Extent of resection in meningioma: predictive factors and clinical implications. Sci Rep. 2019;9(1):5944. https://doi.org/10.1038/s41598-019-42451-z.
7. Lemée JM, Joswig H, Da Broi M, Corniola MV, Scheie D, Schaller K, et al. WHO grade I meningiomas: classification-tree for prognostic factors of survival. Neurosurg Rev. 2020;43(2):749–58. https://doi.org/10.1007/s10143-019-01117-0.
8. Mathiesen T, Lindquist C, Kihlström L, Karlsson B. Recurrence of cranial base meningiomas. Neurosurgery. 1996;39(1):2–7.; discussion 8–9. https://doi.org/10.1097/00006123-199607000-00002.
9. Nakao N, Ohkawa T, Miki J, Nishibayahsi H, Ogura M, Uematsu Y, et al. Analysis of factors affecting the long-term functional outcome of patients with skull base meningioma. J Clin Neurosci. 2011;18(7):895–8. https://doi.org/10.1016/j.jocn.2010.10.020.
10. Corniola MV, Meling TR. Management of recurrent meningiomas: state of the art and perspectives. Cancers (Basel). 2022;14(16) https://doi.org/10.3390/cancers14163995.
11. Simpson D. The recurrence of intracranial meningiomas after surgical treatment. J Neurol Neurosurg Psychiatry. 1957;20(1):22–39. https://doi.org/10.1136/jnnp.20.1.22.
12. Schwartz TH, McDermott MW. The Simpson grade: abandon the scale but preserve the message. J Neurosurg JNS. 2020;135:1–8. https://doi.org/10.3171/2020.6.JNS201904.
13. Heros RC. Editorial: Simpson grades. J Neurosurg JNS. 2012;117(6):997–8. https://doi.org/10.3171/2012.5.JNS12865.
14. Voß KM, Spille DC, Sauerland C, Suero Molina E, Brokinkel C, Paulus W, et al. The Simpson grading in meningioma surgery: does the tumor location influence the prognostic value? J Neuro-Oncol. 2017;133(3):641–51. https://doi.org/10.1007/s11060-017-2481-1.
15. Jääskeläinen J. Seemingly complete removal of histologically benign intracranial meningioma: late recurrence rate and factors predicting recurrence in 657 patients. A multivariate analysis. Surg Neurol. 1986;26(5):461–9. https://doi.org/10.1016/0090-3019(86)90259-4.
16. Rydzewski NR, Lesniak MS, Chandler JP, Kalapurakal JA, Pollom E, Tate MC, et al. Gross total resection and adjuvant radiotherapy most significant predictors of improved survival in patients with atypical meningioma. Cancer. 2018;124(4):734–42. https://doi.org/10.1002/cncr.31088.
17. Gousias K, Schramm J, Simon M. The Simpson grading revisited: aggressive surgery and its place in modern meningioma management. J Neurosurg. 2016;125(3):551–60. https://doi.org/10.3171/2015.9.Jns15754.
18. Adegbite AB, Khan MI, Paine KWE, Tan LK. The recurrence of intracranial meningiomas after surgical treatment. J Neurosurg. 1983;58(1):51–6. https://doi.org/10.3171/jns.1983.58.1.0051.
19. Aizer AA, Bi WL, Kandola MS, Lee EQ, Nayak L, Rinne ML, et al. Extent of resection and overall survival for patients with atypical and malignant meningioma. Cancer. 2015;121(24):4376–81. https://doi.org/10.1002/cncr.29639.
20. Mirimanoff RO, Dosoretz DE, Linggood RM, Ojemann RG, Martuza RL. Meningioma: analysis of recurrence and progression following neurosurgical resection. J Neurosurg. 1985;62(1):18–24. https://doi.org/10.3171/jns.1985.62.1.0018.
21. Da Broi M, Borrelli P, Meling TR. Predictors of survival in subtotally resected WHO Grade I skull base meningiomas. Cancers. 2021;13(6):1451.
22. Kent CL, Mowery YM, Babatunde O, Wright AO, Barak I, McSherry F, et al. Long-term outcomes for patients with atypical or malignant meningiomas treated with or without radiation therapy: a 25-year retrospective analysis of a single-institution experience. Adv Radiat Oncol. 2022;7(3):100878. https://doi.org/10.1016/j.adro.2021.100878.
23. Nakasu S, Fukami T, Jito J, Nozaki K. Recurrence and regrowth of benign meningiomas. Brain Tumor Pathol. 2009;26(2):69–72. https://doi.org/10.1007/s10014-009-0251-2.
24. Aghi MK, Carter BS, Cosgrove GR, Ojemann RG, Amin-Hanjani S, Martuza RL, et al. Long-term recurrence rates of atypical meningiomas after gross total resection with or without postoperative adjuvant radiation. Neurosurgery. 2009;64(1):56–60; discussion. https://doi.org/10.1227/01.Neu.0000330399.55586.63.
25. Meling TR, Da Broi M, Scheie D, Helseth E. Meningiomas: skull base versus non-skull base. Neurosurg Rev. 2019;42(1):163–73. https://doi.org/10.1007/s10143-018-0976-7.
26. Lemée J-M, Corniola MV, Meling TR. Benefits of re-do surgery for recurrent intracranial meningiomas. Sci Rep. 2020;10(1):303. https://doi.org/10.1038/s41598-019-57254-5.

27. Sahm F, Schrimpf D, Olar A, Koelsche C, Reuss D, Bissel J, et al. TERT promoter mutations and risk of recurrence in meningioma. JNCI J Natl Cancer Inst. 2015;108(5) https://doi.org/10.1093/jnci/djv377.
28. Sahm F, Schrimpf D, Stichel D, Jones DTW, Hielscher T, Schefzyk S, et al. DNA methylation-based classification and grading system for meningioma: a multicentre, retrospective analysis. Lancet Oncol. 2017;18(5):682–94. https://doi.org/10.1016/s1470-2045(17)30155-9.
29. Youngblood MW, Miyagishima DF, Jin L, Gupte T, Li C, Duran D, et al. Associations of meningioma molecular subgroup and tumor recurrence. Neuro-Oncology. 2021;23(5):783–94. https://doi.org/10.1093/neuonc/noaa226.
30. Borovich B, Doron Y. Recurrence of intracranial meningiomas: the role played by regional multicentricity. J Neurosurg. 1986;64(1):58–63. https://doi.org/10.3171/jns.1986.64.1.0058.
31. Haslund-Vinding J, Skjoth-Rasmussen J, Poulsgaard L, Fugleholm K, Mirian C, Maier AD, et al. Proposal of a new grading system for meningioma resection: the Copenhagen Protocol. Acta Neurochir. 2022;164(1):229–38. https://doi.org/10.1007/s00701-021-05025-5.
32. Kuranari Y, Tamura R, Tsuda N, Kosugi K, Morimoto Y, Yoshida K, et al. Long-term clinical outcome of first recurrence skull base meningiomas. J Clin Med. 2020;9(1):106.
33. Al-Mefty O, Topsakal C, Pravdenkova S, Sawyer JR, Harrison MJ. Radiation-induced meningiomas: clinical, pathological, cytokinetic, and cytogenetic characteristics. J Neurosurg. 2004;100(6):1002–13. https://doi.org/10.3171/jns.2004.100.6.1002.
34. Couldwell WT, Cole CD, Al-Mefty O. Patterns of skull base meningioma progression after failed radiosurgery. J Neurosurg JNS. 2007;106(1):30–5. https://doi.org/10.3171/jns.2007.106.1.30.
35. Richardson GE, Gillespie CS, Mustafa MA, Taweel BA, Bakhsh A, Kumar S, et al. Clinical outcomes following re-operations for intracranial meningioma. Cancers (Basel). 2021;13(19) https://doi.org/10.3390/cancers13194792.
36. Magill ST, Lee DS, Yen AJ, Lucas CG, Raleigh DR, Aghi MK, et al. Surgical outcomes after reoperation for recurrent skull base meningiomas. J Neurosurg. 2018;130(3):876–83. https://doi.org/10.3171/2017.11.Jns172278.
37. Maiuri F, Mariniello G, Peca C, Guadagno E, Corvino S, d'Avanzo S, et al. Multicentric and diffuse recurrences of meningiomas. Br J Neurosurg. 2020;34(4):439–46. https://doi.org/10.1080/02688697.2020.1754335.
38. DiMeco F, Li KW, Casali C, Ciceri E, Giombini S, Filippini G, et al. Meningiomas invading the superior sagittal sinus: surgical experience in 108 cases. Neurosurgery. 2004;55(6):1263–72.; discussion 72–4. https://doi.org/10.1227/01.neu.0000143373.74160.f2.
39. Sindou MP, Alvernia JE. Results of attempted radical tumor removal and venous repair in 100 consecutive meningiomas involving the major dural sinuses. J Neurosurg JNS. 2006;105(4):514–25. https://doi.org/10.3171/jns.2006.105.4.514.
40. Chen WW, Wang Y, Hu YC, Zhao YL. Analysis of the common complications and recurrence-related factors of superior parasagittal sinus meningioma. Front Surg. 2022;9:1023021. https://doi.org/10.3389/fsurg.2022.1023021.
41. Konglund A, Rogne SG, Lund-Johansen M, Scheie D, Helseth E, Meling TR. Outcome following surgery for intracranial meningiomas in the aging. Acta Neurol Scand. 2013;127(3):161–9. https://doi.org/10.1111/j.1600-0404.2012.01692.x.
42. Sicking J, Voß KM, Spille DC, Schipmann S, Holling M, Paulus W, et al. The evolution of cranial meningioma surgery-a single-center 25-year experience. Acta Neurochir. 2018;160(9):1801–12. https://doi.org/10.1007/s00701-018-3617-6.
43. Nittby HR, Maltese A, Ståhl N. Early postoperative haematomas in neurosurgery. Acta Neurochir. 2016;158(5):837–46. https://doi.org/10.1007/s00701-016-2778-4.
44. Connolly ID, Cole T, Veeravagu A, Popat R, Ratliff J, Li G. Craniotomy for resection of meningioma: an age-stratified analysis of the MarketScan longitudinal database. World Neurosurg. 2015;84(6):1864–70. https://doi.org/10.1016/j.wneu.2015.08.018.
45. Chan RC, Thompson GB. Morbidity, mortality, and quality of life following surgery for intracranial meningiomas. A retrospective study in 257 cases. J Neurosurg. 1984;60(1):52–60. https://doi.org/10.3171/jns.1984.60.1.0052.
46. Chen WC, Magill ST, Wu A, Vasudevan HN, Morin O, Aghi MK, et al. Histopathological features predictive of local control of atypical meningioma after surgery and adjuvant radiotherapy. J Neurosurg. 2018;130(2):443–50. https://doi.org/10.3171/2017.9.Jns171609.
47. Rogers L, Zhang P, Vogelbaum MA, Perry A, Ashby LS, Modi JM, et al. Intermediate-risk meningioma: initial outcomes from NRG Oncology RTOG 0539. J Neurosurg. 2018;129(1):35–47. https://doi.org/10.3171/2016.11.Jns161170.
48. Becker G, Jeremic B, Pitz S, Buchgeister M, Wilhelm H, Schiefer U, et al. Stereotactic fractionated radiotherapy in patients with optic nerve sheath meningioma. Int J Radiat Oncol Biol Phys. 2002;54(5):1422–9. https://doi.org/10.1016/s0360-3016(02)03753-7.
49. Kim M, Lee DH, Kim Rn HJ, Cho YH, Kim JH, Kwon DH. Analysis of the results of recurrent intracranial meningiomas treated with re-radiosurgery. Clin Neurol Neurosurg. 2017;153:93–101. https://doi.org/10.1016/j.clineuro.2016.12.014.
50. Sheehan JP, Lee C-C, Xu Z, Przybylowski CJ, Melmer PD, Schlesinger D. Edema following Gamma Knife radiosurgery for parasagittal and parafalcine meningiomas. J Neurosurg JNS. 2015;123(5):1287–93. https://doi.org/10.3171/2014.12.JNS142159.

51. Milano MT, Sharma M, Soltys SG, Sahgal A, Usuki KY, Saenz JM, et al. Radiation-induced edema after single-fraction or multifraction stereotactic radiosurgery for meningioma: a critical review. Int J Radiat Oncol Biol Phys. 2018;101(2):344–57. https://doi.org/10.1016/j.ijrobp.2018.03.026.
52. Meniai-Merzouki F, Bernier-Chastagner V, Geffrelot J, Tresch E, Lacornerie T, Coche-Dequeant B, et al. Hypofractionated stereotactic radiotherapy for patients with intracranial meningiomas: impact of radiotherapy regimen on local control. Sci Rep. 2018;8(1):13666. https://doi.org/10.1038/s41598-018-32124-8.
53. Magill ST, Lau D, Raleigh DR, Sneed PK, Fogh SE, McDermott MW. Surgical resection and interstitial Iodine-125 brachytherapy for high-grade meningiomas: a 25-year series. Neurosurgery. 2017;80(3):409–16. https://doi.org/10.1227/neu.0000000000001262.
54. El Shafie RA, Czech M, Kessel KA, Habermehl D, Weber D, Rieken S, et al. Evaluation of particle radiotherapy for the re-irradiation of recurrent intracranial meningioma. Radiat Oncol. 2018;13(1):86. https://doi.org/10.1186/s13014-018-1026-x.
55. Weber DC, Grau C, Lim PS, Georg D, Lievens Y. Bringing Europe together in building clinical evidence for proton therapy—the EPTN-ESTRO-EORTC endeavor. Acta Oncol. 2019;58(10):1340–2. https://doi.org/10.1080/0284186x.2019.1624820.
56. Imber BS, Neal B, Casey DL, Darwish H, Lin AL, Cahlon O, et al. Clinical outcomes of recurrent intracranial meningiomas treated with proton beam reirradiation. Int J Part Ther. 2019;5(4):11–22. https://doi.org/10.14338/IJPT-18-00045.1.
57. Lomax A. Intensity modulation methods for proton radiotherapy. Phys Med Biol. 1999;44(1):185. https://doi.org/10.1088/0031-9155/44/1/014.
58. Lambrecht M, Eekers DBP, Alapetite C, Burnet NG, Calugaru V, Coremans IEM, et al. Radiation dose constraints for organs at risk in neuro-oncology; the European Particle Therapy Network consensus. Radiother Oncol. 2018;128(1):26–36. https://doi.org/10.1016/j.radonc.2018.05.001.
59. Florijn MA, Sharfo AWM, Wiggenraad RGJ, van Santvoort JPC, Petoukhova AL, Hoogeman MS, et al. Lower doses to hippocampi and other brain structures for skull-base meningiomas with intensity modulated proton therapy compared to photon therapy. Radiother Oncol. 2020;142:147–53. https://doi.org/10.1016/j.radonc.2019.08.019.
60. Schrell UMH, Rittig MG, Anders M, Kiesewetter F, Marschalek R, Koch UH, et al. Hydroxyurea for treatment of unresectable and recurrent meningiomas. I. Inhibition of primary human meningioma cells in culture and in meningioma transplants by induction of the apoptotic pathway. J Neurosurg. 1997;86(5):845–52. https://doi.org/10.3171/jns.1997.86.5.0845.
61. Fuentes S, Chinot O, Dufour H, Paz-Paredes A, Métellus P, Barrie-Attarian M, et al. [Hydroxyurea treatment for unresectable meningioma]. Neurochirurgie 2004;50(4):461–7. https://doi.org/10.1016/s0028-3770(04)98326-9.
62. Chamberlain MC, Tsao-Wei DD, Groshen S. Temozolomide for treatment-resistant recurrent meningioma. Neurology. 2004;62(7):1210–2. https://doi.org/10.1212/01.wnl.0000118300.82017.f4.
63. Chamberlain MC, Tsao-Wei DD, Groshen S. Salvage chemotherapy with CPT-11 for recurrent meningioma. J Neuro-Oncol. 2006;78(3):271–6. https://doi.org/10.1007/s11060-005-9093-x.
64. Norden AD, Ligon KL, Hammond SN, Muzikansky A, Reardon DA, Kaley TJ, et al. Phase II study of monthly pasireotide LAR (SOM230C) for recurrent or progressive meningioma. Neurology. 2015;84(3):280–6. https://doi.org/10.1212/wnl.0000000000001153.
65. Cossu G, Levivier M, Daniel RT, Messerer M. The role of mifepristone in meningiomas management: a systematic review of the literature. Biomed Res Int. 2015;2015:267831. https://doi.org/10.1155/2015/267831.
66. Germano G, Frapolli R, Simone M, Tavecchio M, Erba E, Pesce S, et al. Antitumor and anti-inflammatory effects of trabectedin on human myxoid liposarcoma cells. Cancer Res. 2010;70(6):2235–44. https://doi.org/10.1158/0008-5472.Can-09-2335.
67. Preusser M, Silvani A, Le Rhun E, Soffietti R, Lombardi G, Sepulveda JM, et al. Trabectedin for recurrent WHO grade 2 or 3 meningioma: a randomized phase II study of the EORTC Brain Tumor Group (EORTC-1320-BTG). Neuro-Oncology. 2022;24(5):755–67. https://doi.org/10.1093/neuonc/noab243.
68. Chamberlain MC. The role of chemotherapy and targeted therapy in the treatment of intracranial meningioma. Curr Opin Oncol. 2012;24(6):666–71. https://doi.org/10.1097/CCO.0b013e328356364d.
69. Raizer JJ, Grimm SA, Rademaker A, Chandler JP, Muro K, Helenowski I, et al. A phase II trial of PTK787/ZK 222584 in recurrent or progressive radiation and surgery refractory meningiomas. J Neuro-Oncol. 2014;117(1):93–101. https://doi.org/10.1007/s11060-014-1358-9.
70. Shih KC, Chowdhary S, Rosenblatt P, Weir AB 3rd, Shepard GC, Williams JT, et al. A phase II trial of bevacizumab and everolimus as treatment for patients with refractory, progressive intracranial meningioma. J Neuro-Oncol. 2016;129(2):281–8. https://doi.org/10.1007/s11060-016-2172-3.
71. Graillon T, Sanson M, Campello C, Idbaih A, Peyre M, Peyrière H, et al. Everolimus and octreotide for patients with recurrent meningioma: results from the phase II CEVOREM Trial. Clin Cancer Res. 2020;26(3):552–7. https://doi.org/10.1158/1078-0432.Ccr-19-2109.
72. Norden AD, Raizer JJ, Abrey LE, Lamborn KR, Lassman AB, Chang SM, et al. Phase II trials of erlo-

tinib or gefitinib in patients with recurrent meningioma. J Neuro-Oncol. 2010;96(2):211–7. https://doi.org/10.1007/s11060-009-9948-7.

73. Wen PY, Yung WK, Lamborn KR, Norden AD, Cloughesy TF, Abrey LE, et al. Phase II study of imatinib mesylate for recurrent meningiomas (North American Brain Tumor Consortium study 01-08). Neuro-Oncology. 2009;11(6):853–60. https://doi.org/10.1215/15228517-2009-010.
74. Kaley TJ, Wen P, Schiff D, Ligon K, Haidar S, Karimi S, et al. Phase II trial of sunitinib for recurrent and progressive atypical and anaplastic meningioma. Neuro-Oncology. 2015;17(1):116–21. https://doi.org/10.1093/neuonc/nou148.
75. Lamszus K, Lengler U, Schmidt NO, Stavrou D, Ergün S, Westphal M. Vascular endothelial growth factor, hepatocyte growth factor/scatter factor, basic fibroblast growth factor, and placenta growth factor in human meningiomas and their relation to angiogenesis and malignancy. Neurosurgery. 2000;46(4):938–47. ; discussion 47–8. https://doi.org/10.1097/00006123-200004000-00033.
76. Kumthekar P, Grimm SA, Aleman RT, Chamberlain MC, Schiff D, Wen PY, et al. A multi-institutional phase II trial of bevacizumab for recurrent and refractory meningioma. Neurooncol Adv. 2022;4(1):vdac123. https://doi.org/10.1093/noajnl/vdac123.
77. Graillon T, Defilles C, Mohamed A, Lisbonis C, Germanetti AL, Chinot O, et al. Combined treatment by octreotide and everolimus: octreotide enhances inhibitory effect of everolimus in aggressive meningiomas. J Neuro-Oncol. 2015;124(1):33–43. https://doi.org/10.1007/s11060-015-1812-3.
78. Koper JW, Zwarthoff EC, Hagemeijer A, Braakman R, Avezaat CJ, Bergström M, et al. Inhibition of the growth of cultured human meningioma cells by recombinant interferon-alpha. Eur J Cancer. 1991;27(4):416–9. https://doi.org/10.1016/0277-5379(91)90375-n.
79. Chamberlain MC, Glantz MJ. Interferon-alpha for recurrent World Health Organization grade 1 intracranial meningiomas. Cancer. 2008;113(8):2146–51. https://doi.org/10.1002/cncr.23803.

Management of Recurrent Skull Base Meningiomas

Sachin Govind, Khizar R. Nandoliya, Rushmin Khazanchi, Rishi Jain, Timothy L. Sita, Priya Kumthekar, and Stephen T. Magill

1 Meningioma Recurrence

Meningiomas are the most prevalent primary CNS neoplasms, representing roughly 40% of all intracranial tumors [1], 20–30% of which occur along the skull base [2]. While most meningiomas are histologically benign lesions, they have the potential to recur after initial treatment [3]. Several studies have investigated the risk factors associated with meningioma recurrence after initial treatment, finding that:

- Age is unrelated to recurrence risk [4–6].
- Ionizing radiation to the head and neck is associated with the risk of meningioma development and is also associated in a dose-dependent manner with increased recurrence risk [7–9].
- Higher World Health Organization (WHO) grade tumors (grade 2 and 3) exhibit significantly higher rates of recurrence [10, 11]. Elevated cellularity, an increased mitotic rate, and the presence of necrotic lesions are indicative of a higher risk of recurrence [12]. Skull base meningiomas are more likely to have lower histological grade than non-skull base meningiomas [13].
- Surgical factors such as extent of initial resection are strongly associated with meningioma recurrence, with incomplete resection (higher Simpson grade) being associated with increased risk of recurrence [5, 6, 14]. Most recurrent lesions will progress within the initial surgical bed or within 1.5 cm of the prior resection cavity [15, 16]. This is particularly true for skull base meningiomas, where performing gross-total resection is often not possible due to unacceptable morbidity, so there is an increased risk of progression of residual tumor.

S. Govind · K. R. Nandoliya · R. Khazanchi · R. Jain
Feinberg School of Medicine, Northwestern University, Chicago, IL, USA

Department of Neurological Surgery, Northwestern University, Chicago, IL, USA

T. L. Sita
Department of Radiation Oncology, Northwestern University, Chicago, IL, USA

P. Kumthekar
Department of Neurology, Northwestern University, Chicago, IL, USA

Lou & Jean Malnati Brain Tumor Institute, Northwestern University, Chicago, IL, USA

S. T. Magill (✉)
Feinberg School of Medicine, Northwestern University, Chicago, IL, USA

Department of Neurological Surgery, Northwestern University, Chicago, IL, USA

Lou & Jean Malnati Brain Tumor Institute, Northwestern University, Chicago, IL, USA
e-mail: stephen.magill@nm.org

F. Maiuri, M. Del Basso De Caro (eds.), *Meningiomas*,
https://doi.org/10.1007/978-3-031-76680-0_21

- Non-skull base tumors are more likely to be WHO grade 2 [17], which have a high rate of recurrence, although other studies have suggested tumor location is not an independent predictor of recurrence risk [5].
- Skull base lesions do have a longer median time to recurrence [18].
- Molecular changes in meningioma are strongly associated with recurrence; in particular, loss of Chr 1p, loss of CDKN2A/B, and other high risk CNVs and gene expression profiles [19–23].

Considering these risk factors, meningioma recurrence—particularly lesions in difficult-to-resect regions—is a challenging clinical entity. Few specific differences between skull base and non-skull base meningiomas lead to changes in recurrence management, as the options are essentially the same regardless of tumor location. For high grade meningioma with aggressive molecular features, recurrence rates can be as high as 94% and patients should be counseled on this [10]. Thus, understanding risk factors associated with progression or recurrence of skull base meningiomas and the available treatment modalities is important for management of recurrent skull base lesions. In this chapter, we will discuss considerations for repeat surgery, adjuvant radiotherapy, and a discussion of medical treatments that may be considered or are being actively explored as options.

2 Repeat Surgery

The decision to undergo repeat surgery for recurrent skull base meningiomas may be informed by several factors, such as recurrent tumor size, symptomology, patient age, and history of prior radiation therapy. While the literature on approaches and outcomes is robust for primary meningiomas, there is a relative paucity of studies evaluating the outcomes of repeat surgery for recurrent tumors. Repeat surgery for skull base meningiomas is associated with good long-term survival, but with an increased risk of postoperative complications [24–26]. Rates of complications such as hydrocephalus, CSF leak, wound infections, and postoperative hematoma have been reported to increase from 32% after the first reoperation to 41% after the second reoperation [24]. Additionally, time to retreatment may decrease significantly between surgeries in patients requiring repeated resections [25], suggesting that in patients with more aggressive meningiomas, subsequent reoperations may be less effective for oncologic control.

Posterior fossa tumor location has been shown to be a strong predictor of perioperative complications following reoperation [24]. Given the strong correlation between extent of resection and meningioma recurrence [27, 28], it is unsurprising that reoperation for skull base meningiomas, particularly those in the posterior fossa, may have high complication rates given the presence of critical anatomic structures in the region, which hinder gross-total resection with acceptable morbidity.

Achieving gross-total resection of recurrent meningiomas can be challenging, particularly in situations with bone invasion. Cranioplasties with customized implants can be a useful adjunct for the management of recurrent meningiomas with bone involvement [29–31]. Given the importance of gross-total resection for oncologic control, the neurosurgeon may consider removing involved bone fragments and use of a custom implant to achieve greater extent of resection and optimal cosmetic outcome (Fig. 1). A variety of cranioplasty materials are available for custom cranioplasties, including 3D printed and designed polyether ether ketone (PEEK), pre-fabricated polymethyl methacrylate (pPMMA), titanium custom implants or mesh, and hand-molded polymethyl methacrylate cement (hPMMA). Pre-fabricated materials have been associated with fewer postoperative complications [30].

Another emerging tool in the management of recurrent meningiomas is 1,4,7,10-tetraazacyclo dodecane-1,4,7,10-tetraacetic acid (DOTA)-octreotate positron emission tomography (DOTATATE-PET). DOTATATE, a somatostatin analog radionuclide, binds somatostatin receptors (SSTRs) and is widely used to improve imaging and surgical management of neuroendocrine tumors [32–34]. SSTRs are expressed in nearly 100% of meningiomas [35–38], and in

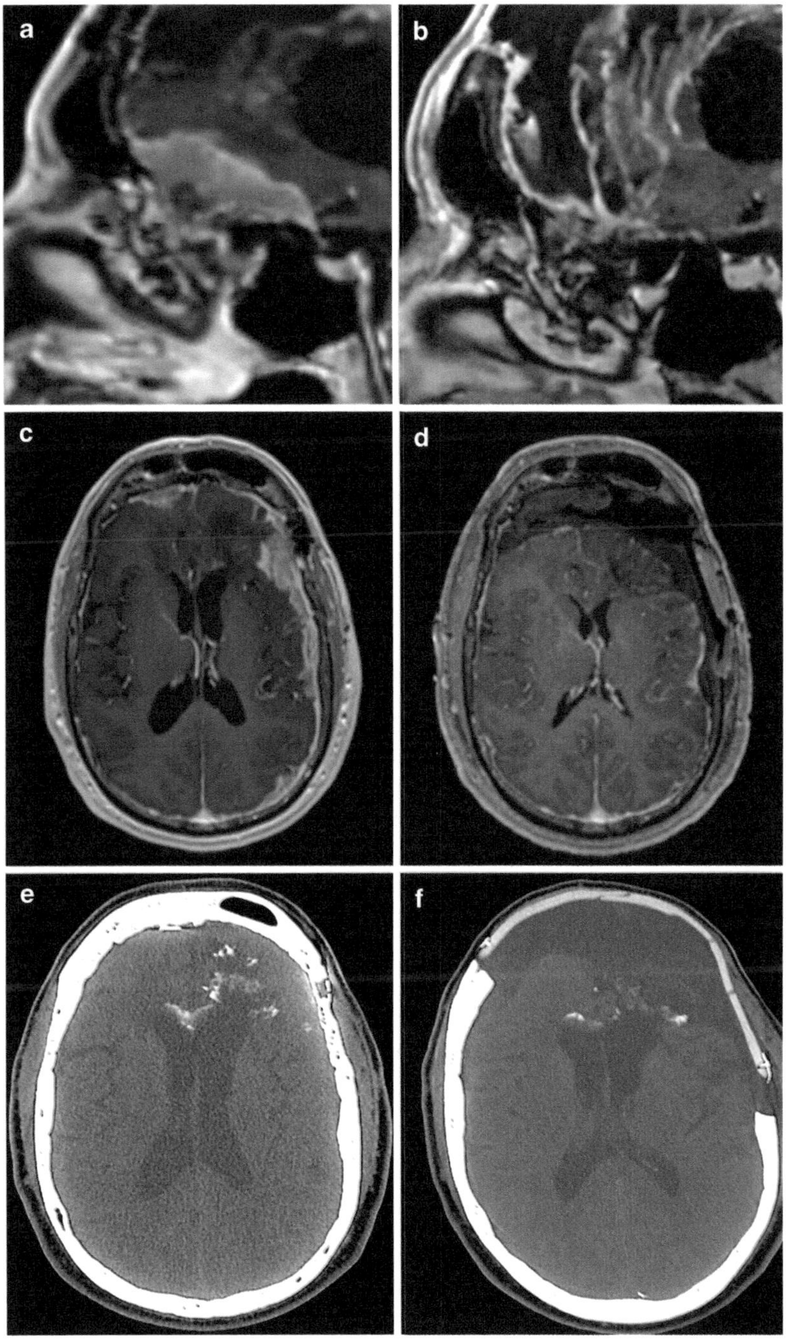

Fig. 1 A 64-year-old patient with multifocal three times recurrent atypical meningioma underwent a bifrontal and left temporo-orbital craniectomy to remove progressive skull base and convexity disease. (**a**, **c**) Preoperative and (**b**, **d**) postoperative T1 MRI show resection of a planum sphenoidale meningioma. (**e**) Preoperative and (**f**) postoperative computed tomography images show a bilateral custom cranioplasty implant. (Published with permission)

recent years, its use has been reported for surgical planning of intraosseous [39] and parasagittal [40] meningiomas (Fig. 2). For recurrent high grade tumors, it is advised that they undergo immunohistochemistry for SSTR2A to ensure that the tumor continues to express SSTR2A, which can occasionally be lost in high grade meningiomas [41, 42].

In cases of symptomatic meningiomas, surgical resection is the primary treatment choice, with the goal of achieving symptom relief and maximal tumor removal to enhance the patient's

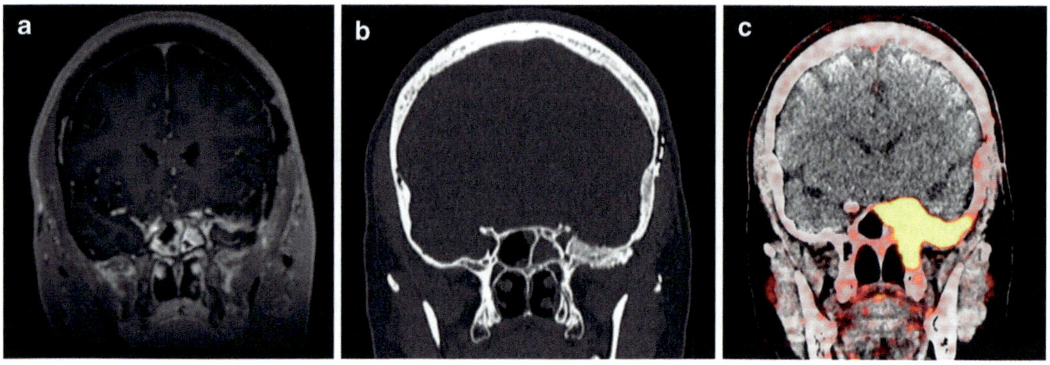

Fig. 2 A 46-year-old patient with a history of pterional craniotomy for subtotal resection of a left hyperostosing sphenoid wing WHO grade 2 meningioma presented with severe vision deficits and proptosis in her left eye secondary to compression of the optic nerve, orbit, and orbital apex due to recurrent skull base meningioma. Preoperative (**a**) MRI and (**b**) CT demonstrate hyperostotic bone in the middle fossa. (**c**) DOTATATE-PET imaging shows extension of tumor into the pterygoid plates that is not appreciated easily on the MRI or CT, which was important in resection planning and postoperative RT planning. (Published with permission)

quality of life and reduce the risk of recurrence. Higher Simpson grades (III-V) are associated with shorter disease-free survival, diminished quality of life, and an elevated likelihood of early tumor recurrence [5]. There is a rare risk of seeding metastasis along the surgical tract, particularly in deep-seated meningiomas [43]. In general, systemic metastases are rare, however, for meningiomas that have undergone at least two surgeries and radiation and are being considered for a third resection, systemic screening for distal extracranial metastases should be considered with whole body DOTATATE imaging or CT of the chest/abdomen/pelvis, as the number needed to screen in that scenario is only three patients to find one with distant metastases [44].

3 Radiation

Radiation, whether delivered through fractionated radiotherapy or radiosurgery, has a long history of use for meningiomas. It provides excellent tumor control for residual WHO grade 1 meningiomas or as an adjuvant treatment to residual or recurrent meningiomas [40, 45]. Radiation treatment tends to stabilize residual or recurrent meningiomas, whether delivered as radiotherapy or radiosurgery; however, 30–35% of meningiomas will see some decrease in size within 2–3 years of treatment, and neurologic deficits may improve in 23–89% of patients [40]. Given the frequency of subtotal resection along the skull base, there is debate about whether to radiate residual tumor upfront or simply observe and radiate if it grows. This is often guided by the size of the residual, with larger residuals being radiated earlier, and small residuals with separation from the brainstem or cranial nerves often being observed initially if the tumor is low grade. Multiple studies have shown efficacy of radiosurgery for recurrent skull base tumors across all locations [40, 45, 46].

The role of repeat radiation therapy for recurrent skull base meningiomas remains less studied. A small institutional series attempted to answer this question by exploring the role of salvage surgery with or without reirradiation for recurrent skull base meningiomas that had been radiated once before. The results suggested that the impact on progression-free survival for these tumors was dependent on WHO grading. Grade 1 tumors did not benefit from repeat radiotherapy regardless of extent of resection, Grade 2 tumors had significantly better outcomes (specifically when STR was paired with repeat adjuvant radiation), and Grade 3 tumors benefited from radiation regardless of resection status [47].

DOTATATE-PET imaging has shown to be quite beneficial in radiation planning, as it can often highlight tumor extension into bone, dural tails, or soft tissue that is not always clear on MRI or CT (Fig. 2). Indeed, when DOTATATE-PET is used to guide planning for radiotherapy, there is an improvement in disease control for cavernous sinus meningiomas, suggesting that failures in the past may be due to incomplete tumor coverage [48]. Recent series have demonstrated that the DOTATATE imaging can alter the treatment plan in up to 40% of patients, consistent with the chapter author's experience [49]. Finally, a prospective study comparing pre-radiation to post-radiation DOTATATE-PET imaging revealed that the PET signal correlates with meningioma response to radiation and may be useful in guiding patient care [50].

4 Brachytherapy

Brachytherapy is considered for specific cases following the resection of recurrent meningiomas [51]. The primary indications for brachytherapy are cases of recurrent atypical and malignant meningiomas that have already undergone external beam irradiation in prior treatments and are undergoing additional resection [52, 53]. It may be considered for recurrent meningiomas that are too large for post-surgery radiosurgery and for cases where a thin residual tumor remains unresectable, such as against the superior sagittal sinus or eloquent brain. Patients with recurrent meningiomas who are being considered for reoperation with brachytherapy implants should undergo systemic staging as discussed above, as distant metastases are possible in this setting [44].

Brachytherapy for recurrent tumors is associated with relatively significant complication rates, although are decreased with Cs-131 sources compared to I-125 sources that were used historically [54]. Wound complications are the most common early complications after tumor resection and brachytherapy implants, and can occur in up to 40% of patients [52]. Radiation necrosis is a prevalent delayed side effect, occurring in up to 19% of patients due to the high local radiation doses from the radioactive implant, particularly in brain tissue that has already received maximum radiation doses externally [55]. Management of radiation necrosis typically involves dexamethasone treatment, with surgical reoperation or the use of Bevacizumab considered if the condition persists or worsens. Despite the challenges and complications associated with brachytherapy in recurrent aggressive meningiomas, numerous case series have demonstrated good long-term survival with its use, although no randomized data exists, in part due to the complicated heavily pre-treated nature of the patients in whom it is used [56].

5 Medical Management

There are no highly effective chemotherapies available for meningioma, however, many treatments have been trialed and are used in cases where meningiomas fail surgery and radiation. While most studies assessing drug therapies are not specific to skull base meningiomas, they continue to shape mainstay treatment paradigms and are reviewed below:

5.1 Mifepristone

Mifepristone (RU-486) has garnered attention for its potential to inhibit meningioma growth and proliferation by disrupting progesterone-mediated signaling pathways [57]. While initially developed as an anti-glucocorticoid and anti-progestin with abortive properties, its applications have been extended to the management of uterine fibroids, Cushing's Disease, and luminal breast carcinomas [58].

Prevalent progesterone receptor (PR) expression poises meningiomas as potent Mifepristone targets; however, clinical data in vivo suggests that, while treatment is well tolerated, freedom from progression is limited in practice. In one cohort, only 30% of patients with unresectable meningiomas—mixed non-skull base and skull base (sphenoid, parasellar, suprasellar, and tuberculum sellae)—completed a 2-year Mifepristone course without disease progression; no significant difference was observed between placebo and Mifepristone-prescribed arms in progression-free or overall survival (OS) [59]. Another group noted that 58% of mixed, unresectable cases over a 12-month treatment span displayed volumetric regressions or stabilizations; however, 17% of these tumors proceeded to increase in size post-treatment [60]. Percent changes in tumor volume and long-term outcomes were hence not encouraging.

Heterogenous outcomes reflect an incomplete understanding of the role of progesterone in meningioma progression. Although 70% of meningiomas express PRs, their functional status can vary greatly depending on isoforms present and the involvement of adjacent (e.g. estrogen) hormonal receptors [61]; clinical data suggests that distinct receptor genotypes can confer variable anti-progestin resistance [62]. Other evidence indicates that Mifepristone does not sufficiently stifle non-hormonal signal transduction pathways—namely EGF, integrin, and Ras driven sequences—integral to tumor pathogenesis [63]. A deeper understanding of the interplay between different hormone receptors, individual variability in hormonal sensitivity, and the potential for hormone-independent growth will be required before determining if hormonally targeted treatments will have a role in recurrent meningiomas.

5.2 Octreotide/Everolimus

Octreotide/Everolimus has emerged as a commonly used option for recurrent WHO grade 2–3 meningiomas. Octreotide binds somatostatin receptors (SSTRs; predominantly SSTR2A) and Everolimus inhibits the PI3K/Akt/mTOR pathway. This combination better stabilizes meningioma progression compared to each individual drug treatment [64]. While standalone Octreotide treatment inhibits cell proliferation by interfering with the GH/IGF-1 axes and adjacent hormonal systems [65], it induces an additive effect with Everolimus on the inhibition of downstream proteins involved in abnormal translation and control of the cell cycle.

Unlike Mifepristone-based therapies, targeted cooperativity is maintained in vivo. A successful phase II, multi-center clinical trial observed a 6-month progression-free survival rate (PFS) of 55% (compared to the current standard of 11–15%) and >50% decrease in growth rate in 78% of tumors within 3 months [66]. While the study population included patients with skull base meningiomas, results were not stratified by tumor location or characteristics. A separate single-institution study demonstrated that these outcomes persist longitudinally and despite complex pathologies: a 12-month PFS rate of 61.1% and OS rate of 83.3% were noted [67]. The optimization of somatostatin agonist-mTOR inhibitor treatment by changing dosage, varying choice of analog (e.g., Pasireotide), and targeting different SST subtypes is an area of interest that may improve outcomes for meningioma patients [68].

5.3 Immunotherapy

Immunotherapies have emerged as another possible medical treatment for refractory meningioma. Evidence of increased load of immunogenic neoepitopes, programmed death-ligand (PDL)-1/2 signaling dysfunction by regulatory T-cells, and reductions in CD4+/CD8+ T-cell count are recognized as core components of the general immunosuppressive meningioma niche [69]. These experimental findings have raised the possibility that systemic treatments that repopulate antitumor cells such as immune check point inhibition (ICI) or even potentially chimeric antigen receptor T-cell therapies may hold promise [70].

A recent phase 2 study evaluated pembrolizumab (anti-PDL-1 monoclonal antibody) for management of WHO grade 2–3 recurrent meningiomas. Stabilization in growth rate was observed in 40% (6/15) of patients relative to pre-treatment trajectories with a small decrease in tumor size seen in two patients, and PFS-6 was achieved by 48% (12/25) of the overall cohort, albeit higher in patients who had not received prior systemic therapy (56.3% vs. 33.3%) [69].

Other anti-PDL-1 antibody treatments such as avelumab have also demonstrated significant antitumor responses pre-clinically, with one group citing the efficacy of the treatment in tandem with the introduction of high-affinity natural killer T-cell lines [71]. While dramatic reports of meningioma response to immunotherapy continue to offer hope, it remains to be determined how to identify meningiomas that may respond to this strategy.

5.4 Lutathera

Lutathera is an intravenous peptide receptor radionuclide therapy (PRRT) that combines radionuclide Lutetium-177 with somatostatin analog DOTATATE to target SSTRs. With high binding affinity, selective retention, and long-term beta-particle emission, Lutathera can induce tumor cell apoptosis [72].

The treatment achieved FDA approval for management of SSTR positive gastroenteropancreatic neuroendocrine tumors (GEP-NETs) in 2018 and has since been tested in limited, off-label cases for treatment-refractory meningiomas [73]. While more clinical evidence is necessary, preliminary analyses have shown that with minimal complications—beyond possible grade 1 hematotoxicities and asthenia—Lutathera achieves notable 6-month PFS for WHO grade 1–2 recurrent meningiomas, with PFS rates ranging from 40–85% [73–75]. Two ongoing phase 2 studies are in process that will better reveal the efficacy of this modality.

6 Clinical Trials

An emerging area of research for treatment of skull base meningiomas involves subclassification using molecular biomarkers. Analysis of meningiomas for the presence of specific germline or somatic mutations, cytogenetic copy number variations, and methylation profiling has been shown to provide significant prognostic value as well as stratification that could inform specific treatments [76]. For skull base meningiomas specifically, AKT1 gain-of-function mutations have been shown to upregulate the mTOR and ERK signaling pathways, leading to a shorter time to recurrence. More broadly, CDKN2A/B loss of function, SMO/SUFU gain of function, TERT promoter regional mutations, and 1p or 14q chromosomal arm deletions have been associated with increased risk of meningioma recurrence in the literature [77–81].

One clinical trial of note is NCT02523014 sponsored by the Alliance for Clinical Trials in Oncology. This phase II trial is studying the effect of Vismodegib, FAK Inhibitor GSK2256098, Capivasertib, and Abemaciclib in the treatment of progressive or recurrent intracranial meningioma [82]. Patients enrolled were assigned to one of four treatment arms defined by the mutational status of the tumor. Initial results from the trial are promising, with results from a subgroup of NF2 mutation positive meningiomas treated with GSK2256098 demonstrating an improved 6-month progression-free survival rate compared to historical controls [82].

7 Conclusion

In conclusion, recurrent meningiomas are challenging tumors to manage and lead to significant morbidity and mortality for patients. These patients are best managed with a multi-disciplinary team including neurosurgery, radiation oncology, and neuro-oncology. Management primarily involves repeat surgery and radiation with either radiotherapy, radiosurgery, and in select cases, brachytherapy. For tumors that are refractory to surgery and radiotherapy, there are no proven medical treatments, but commonly used treatments with some benefit include SSTR analogs with mTOR inhibitors, immunotherapy, and VEGF inhibitors. Clinical trials are needed for molecularly-guided therapies that are targeted to the specific molecular changes within a tumor to offer hope for better treatments for these difficult-to-manage tumors.

References

1. Ostrom QT, et al. CBTRUS statistical report: primary brain and other central nervous system tumors diagnosed in the United States in 2016–2020. Neuro-Oncology. 2023;25(Supplement_4):iv1–iv99.
2. Shrivastava RK, et al. Harvey Cushing's Meningiomas text and the historical origin of resectability criteria

for the anterior one third of the superior sagittal sinus. J Neurosurg. 2003;99(4):787–91.
3. Hortobágyi T, et al. Meningioma recurrence. Open Med (Wars). 2016;11(1):168–73.
4. Adegbite AB, et al. The recurrence of intracranial meningiomas after surgical treatment. J Neurosurg. 1983;58(1):51–6.
5. Ildan F, et al. Predicting the probability of meningioma recurrence in the preoperative and early postoperative period: a multivariate analysis in the midterm follow-up. Skull Base. 2007;17(3):157–71.
6. Mahmood A, Qureshi NH, Malik GM. Intracranial meningiomas: analysis of recurrence after surgical treatment. Acta Neurochir. 1994;126(2–4):53–8.
7. Mansouri A, et al. The evolving role of radiosurgery in the management of radiation-induced meningiomas: a review of current advances and future directions. Biomed Res Int. 2014;2014:107526.
8. Shoshan Y, et al. Radiation-induced meningioma: historical perspective, presentation, management, and genetics. In: Lee JH, editor. Meningiomas. London: Springer; 2009. p. 163–76.
9. Umansky F, et al. Radiation-induced meningioma. Neurosurg Focus. 2008;24(5):E7.
10. Marciscano AE, et al. Benign meningiomas (WHO Grade I) with atypical histological features: correlation of histopathological features with clinical outcomes. J Neurosurg. 2016;124(1):106–14.
11. Mirian C, et al. The importance of considering competing risks in recurrence analysis of intracranial meningioma. J Neuro-Oncol. 2024;166(3):503–11.
12. Böker DK, Meurer H, Gullotta F. Recurring intracranial meningiomas. Evaluation of some factors predisposing for tumor recurrence. J Neurosurg Sci. 1985;29(1):11–7.
13. Hashimoto N, et al. Slower growth of skull base meningiomas compared with non-skull base meningiomas based on volumetric and biological studies. J Neurosurg. 2012;116(3):574–80.
14. Nakasu S, et al. Preoperative identification of meningiomas that are highly likely to recur. J Neurosurg. 1999;90(3):455–62.
15. Ong K, Rizzuto M, Makarenko S. Location pattern of recurrence of fully resected grade 1 meningiomas. Acta Neurochir. 2023;165(10):2865–71.
16. Susko MS, et al. Letter: patterns of intermediate- and high-risk meningioma recurrence after treatment with postoperative external beam radiotherapy. Neurosurgery. 2021;89(1):E99–e101.
17. Magill ST, et al. Relationship between tumor location, size, and WHO grade in meningioma. Neurosurg Focus. 2018;44(4):E4.
18. Maiuri F, et al. Time to recurrence of intracranial meningiomas from a monoinstitutional surgical series. World Neurosurg. 2024;185:e612.
19. Choudhury A, et al. Meningioma DNA methylation groups identify biological drivers and therapeutic vulnerabilities. Nat Genet. 2022;54(5):649–59.
20. Chen WC, et al. Targeted gene expression profiling predicts meningioma outcomes and radiotherapy responses. Nat Med. 2023;29(12):3067–76.
21. Driver J, et al. A molecularly integrated grade for meningioma. Neuro-Oncology. 2022;24(5):796–808.
22. Nassiri F, et al. A clinically applicable integrative molecular classification of meningiomas. Nature. 2021;597(7874):119–25.
23. Maas SLN, et al. Integrated molecular-morphologic meningioma classification: a multicenter retrospective analysis, retrospectively and prospectively validated. J Clin Oncol. 2021;39(34):3839–52.
24. Magill ST, et al. Surgical outcomes after reoperation for recurrent skull base meningiomas. J Neurosurg. 2018;130(3):876–83.
25. Lemée JM, Corniola MV, Meling TR. Benefits of re-do surgery for recurrent intracranial meningiomas. Sci Rep. 2020;10(1):303.
26. Richardson GE, et al. Clinical outcomes following re-operations for intracranial meningioma. Cancers (Basel). 2021;13(19)
27. Lemée JM, et al. Extent of resection in meningioma: predictive factors and clinical implications. Sci Rep. 2019;9(1):5944.
28. Materi J, et al. Predictors of recurrence and high growth rate of residual meningiomas after subtotal resection. J Neurosurg. 2020;134(2):410–6.
29. Barros A, et al. One-step customized PEEK cranioplasty after 3D printed resection template assisted surgery for a frontal intraosseous meningioma: a case report. Turk Neurosurg. 2021;31(1):142–7.
30. Clynch AL, et al. Cranial meningioma with bone involvement: surgical strategies and clinical considerations. Acta Neurochir. 2023;165(5):1355–63.
31. Sundblom J, et al. Removal of giant intraosseous meningioma followed by cranioplasty using a custom-made bioceramic implant: case report. J Neurosurg. 2018;131(3):735–9.
32. Buchmann I, et al. Comparison of 68Ga-DOTATOC PET and 111In-DTPAOC (Octreoscan) SPECT in patients with neuroendocrine tumours. Eur J Nucl Med Mol Imaging. 2007;34(10):1617–26.
33. Deppen SA, et al. Safety and efficacy of 68Ga-DOTATATE PET/CT for diagnosis, staging, and treatment management of neuroendocrine tumors. J Nucl Med. 2016;57(5):708–14.
34. Sanli Y, et al. Neuroendocrine tumor diagnosis and management: (68)Ga-DOTATATE PET/CT. AJR Am J Roentgenol. 2018;211(2):267–77.
35. Menke JR, et al. Somatostatin receptor 2a is a more sensitive diagnostic marker of meningioma than epithelial membrane antigen. Acta Neuropathol. 2015;130(3):441–3.
36. Reubi JC, et al. High incidence of somatostatin receptors in human meningiomas: biochemical characterization. J Clin Endocrinol Metab. 1986;63(2):433–8.
37. Schulz S, et al. Immunohistochemical determination of five somatostatin receptors in meningioma reveals frequent overexpression of somatostatin receptor subtype sst2A. Clin Cancer Res. 2000;6(5):1865–74.

38. Wu W, et al. Clinical Significance of Somatostatin Receptor (SSTR) 2 in meningioma. Front Oncol. 2020;10:1633.
39. Guinto-Nishimura GY, et al. 68Ga-DOTATOC-PET/CT-guided resection of a primary intraosseous meningioma: technical note. Neurosurg Focus. 2021;50(1):E6.
40. Chen WC, et al. Radiotherapy for meningiomas. J Neuro-Oncol. 2022;160(2):505–15.
41. Roytman M, et al. Somatostatin receptor-2 negative meningioma: pathologic correlation and imaging implications. Clin Imaging. 2020;66:18–22.
42. Behling F, et al. Differences in the expression of SSTR1-5 in meningiomas and its therapeutic potential. Neurosurg Rev. 2022;45(1):467–78.
43. Maddah G, et al. Iatrogenic seeding of tumor cells in thigh soft tissue upon surgical removal of intracranial meningioma. Basic Clin Neurosci. 2016;7(2):159–64.
44. Dalle Ore CL, et al. Meningioma metastases: incidence and proposed screening paradigm. J Neurosurg. 2019;132(5):1447–55.
45. Chen WC, et al. Radiotherapy and radiosurgery for meningiomas. Neurooncol Adv. 2023;5(Suppl 1):i67–83.
46. Vera E, et al. A review of stereotactic radiosurgery practice in the management of skull base meningiomas. J Neurol Surg B Skull Base. 2014;75(3):152–8.
47. Rubino F, et al. Impact of salvage surgery with or without reirradiation for skull base meningiomas recurring after prior radiotherapy. J Neurosurg. 2023;139(3):798–809.
48. Hadi I, et al. Multimodal therapy of cavernous sinus meningioma: impact of surgery and (68) Ga-DOTATATE PET-guided radiation therapy on tumor control and functional outcome. Neurooncol Adv. 2021;3(1):vdab114.
49. Hintz EB, et al. Using 68 Ga-DOTATATE PET for postoperative radiosurgery and radiotherapy planning in patients with meningioma: a case series. Neurosurgery. 2023;93(1):95–101.
50. Perlow HK, et al. A prospective registry study of (68) Ga-DOTATATE PET/CT incorporation into treatment planning of intracranial meningiomas. Int J Radiat Oncol Biol Phys. 2024;118(4):979–85.
51. Gutin PH, et al. Brachytherapy of recurrent tumors of the skull base and spine with iodine-125 sources. Neurosurgery. 1987;20(6):938–45.
52. Magill ST, et al. Brachytherapy for meningiomas. Handb Clin Neurol. 2020;170:303–7.
53. Ware, M.L., et al., Surgical resection and permanent brachytherapy for recurrent atypical and malignant meningioma. Neurosurgery, 2004.54(1): p. 55–63; discussion 63–4.
54. Chen WC, et al. Resection with intraoperative cesium-131 brachytherapy as salvage therapy for recurrent brain tumors. J Neurosurg. 2022;137:1–7.
55. Kerschbaumer J, et al. Risk factors for radiation necrosis in patients undergoing cranial stereotactic radiosurgery. Cancers (Basel). 2021;13(19)
56. Koch MJ, et al. Brachytherapy as an adjuvant for recurrent atypical and malignant meningiomas. Neurosurgery. 2019;85(5):E910–e916.
57. Islam MS, et al. Selective progesterone receptor modulators-mechanisms and therapeutic utility. Endocr Rev. 2020;41(5)
58. Elía A, et al. Beneficial effects of mifepristone treatment in patients with breast cancer selected by the progesterone receptor isoform ratio: results from the MIPRA trial. Clin Cancer Res. 2023;29(5):866–77.
59. Ji Y, et al. Double-blind phase III randomized trial of the antiprogestin agent mifepristone in the treatment of unresectable meningioma: SWOG S9005. J Clin Oncol. 2015;33(34):4093–8.
60. Lamberts SW, et al. Mifepristone (RU 486) treatment of meningiomas. J Neurol Neurosurg Psychiatry. 1992;55(6):486–90.
61. Donnell MS, Meyer GA, Donegan WL. Estrogen-receptor protein in intracranial meningiomas. J Neurosurg. 1979;50(4):499–502.
62. Cossu G, et al. The role of mifepristone in meningiomas management: a systematic review of the literature. Biomed Res Int. 2015;2015:267831.
63. Koper JW, et al. Effects of progesterone on the response to epidermal growth factor and other growth factors in cultured human meningioma cells. Cancer Res. 1990;50(9):2604–7.
64. El-Habr EA, et al. Complex interactions between the components of the PI3K/AKT/mTOR pathway, and with components of MAPK, JAK/STAT and Notch-1 pathways, indicate their involvement in meningioma development. Virchows Arch. 2014;465(4):473–85.
65. Graillon T, et al. Combined treatment by octreotide and everolimus: octreotide enhances inhibitory effect of everolimus in aggressive meningiomas. J Neuro-Oncol. 2015;124(1):33–43.
66. Graillon T, et al. Everolimus and octreotide for patients with recurrent meningioma: results from the phase II CEVOREM trial. Clin Cancer Res. 2020;26(3):552–7.
67. Youssef M, et al. CTNI-54. long-term use of everolimus plus octreotide for recurrent intracranial or spinal meningioma: a single institution experience. Neuro Oncol. 2020;22(Suppl 2):ii55.
68. Graillon T, et al. P11.11 aggressive meningiomas: in vitro study of the combination pasireotide-everolimus vs. octreotide everolimus. Neuro Oncol. 2016;18(Suppl 4):iv68.
69. Brastianos PK, et al. Phase 2 study of pembrolizumab in patients with recurrent and residual high-grade meningiomas. Nat Commun. 2022;13(1):1325.

70. Tang X, et al. Bioactivity and safety of B7-H3-targeted chimeric antigen receptor T cells against anaplastic meningioma. Clin Transl Immunol. 2020;9(6):e1137.
71. Giles AJ, et al. Efficient ADCC killing of meningioma by avelumab and a high-affinity natural killer cell line, haNK. JCI Insight. 2019;4(20)
72. Hennrich U, Kopka K. Lutathera(®): the first FDA- and EMA-approved radiopharmaceutical for peptide receptor radionuclide therapy. Pharmaceuticals (Basel). 2019;12(3):114.
73. Zahid A, Johnson DR, Kizilbash SH. Efficacy of (177)Lu-Dotatate therapy in the treatment of recurrent meningioma. Mayo Clin Proc Innov Qual Outcomes. 2021;5(1):236–40.
74. Salgues B, et al. Somatostatin receptor theranostics for refractory meningiomas. Curr Oncol. 2022;29(8):5550–65.
75. Braat A, et al. Will (177)Lu-DOTATATE treatment become more effective in salvage meningioma patients, when boosting somatostatin receptor saturation? A promising case on intra-arterial administration. Cardiovasc Intervent Radiol. 2019;42(11):1649–52.
76. Wang EJ, et al. Recent advances in the molecular prognostication of meningiomas. Front Oncol. 2022;12:910199.
77. Guyot A, et al. Analysis of CDKN2A gene alterations in recurrent and non-recurrent meningioma. J Neuro-Oncol. 2019;145(3):449–59.
78. Pawloski JA, et al. Genomic biomarkers of meningioma: a focused review. Int J Mol Sci. 2021;22(19)
79. Sahin B, et al. Influence of human telomerase reverse transcriptase mutation on the aggressiveness and recurrence in meningiomas. Cureus. 2021;13(5):e15342.
80. Och W, et al. The correlation of clinical and chromosomal alterations of benign meningiomas and their recurrences. Neurol Neurochir Pol. 2016;50(6):395–402.
81. Peng W, et al. Potential molecular mechanisms of recurrent and progressive meningiomas: a review of the latest literature. Front Oncol. 2022;12:850463.
82. Brastianos PK, et al. Alliance A071401: phase II trial of focal adhesion kinase inhibition in meningiomas with somatic NF2 mutations. J Clin Oncol. 2023;41(3):618–28.

Multiple Reoperations for Recurrent Meningiomas

Francesco Maiuri and Sergio Corvino

1 Introduction

Intracranial meningiomas often recur even after gross-total resection with and without adjuvant radiation treatment, with a rate ranging from 10% to 32% at 10 years [1–3]. The main risk factors of recurrence include the WHO grade [4–7], the proliferation index Ki67-MIB1 [6, 8–11] and mitotic index [12], the biomolecular profile, the extent of resection (EOR) according to Simpson [13–17] and the postoperative adjuvant treatments [1, 18, 19]. Other investigated factors include patient age and sex [4, 6, 20, 21], tumor size [20, 22, 23], location [24–28] and morphology [22, 24, 29], brain invasion, progesterone receptor (PR) expression [6, 30–34].

Most patients with recurrent not anaplastic meningiomas are cured after one reoperation and adjuvant radiation therapy; nevertheless, a lesser percentage experiences further recurrences even after many years. Although the extent of resection plays a significant role, the factors correlated to multiple recurrences have scarcely been investigated. This chapter discusses the main risk factors for multiple recurrences and reoperations of meningiomas and their management.

F. Maiuri (✉) · S. Corvino
Department of Neurosciences and Reproductive and Odontostomatological Sciences, Neurosurgical Division, University of Naples "Federico II", Naples, Italy
e-mail: frmaiuri@unina.it

2 Incidence and Clinic-Pathological Factors of Multiple Recurrences

The occurrence of multiple recurrences requiring multiple (two or more) reoperations is a rather frequent event (from 22% up to 40% of all recurrences reoperated on) [35–38]. In an own study [38], patients with multiple recurrences, as compared to those with single recurrence, were mainly males; this agrees with the known more aggressive behavior of meningiomas in male sex [39]. The role of the age at initial diagnosis is less defined. Lemeé et al. [37] report significantly higher rate of older patients in the group with one recurrence, but no age difference in those with multiple recurrences. In our study [38], younger patients were found to be significantly more numerous in the multiple recurrence group. This may be due to the longer follow-up because of the longer life expectancy; conversely, elderly patients have a shorter follow-up period since they are more at risk of death for other causes.

The meningioma location is variably reported; meningiomas in patients with multiple recurrences were significantly more frequent in non-skull base location [36], at parasagittal-falx region [38] or at skull base [37]. The higher recurrence rates of non-skull base and lateral skull base meningiomas [24, 26, 40–43] may reflect the embryological origin of the meninges of these region from the mesoderm and the different biomolecular expression [44–47].

F. Maiuri, M. Del Basso De Caro (eds.), *Meningiomas*,
https://doi.org/10.1007/978-3-031-76680-0_22

Recurrences of spinal meningiomas are rare (0–10%) and occur at the previous site; thus, they require a single reoperation [48]. Only rare extradural or extra-intradural "en-plaque" meningiomas may show multiple recurrences requiring more than one reoperations because of the large circumferential and unresectable dural invasion [49].

In the four reviewed studies [35–38], 91 patients had at least two reoperations; among them, 41 (45%) were cured after surgery and irradiation. Concerning the other patients with further recurrence, 36% required a third reoperation, 13% a fourth reoperation, and 5% more than four reoperations and up to eight.

The time to reoperation along the further recurrences is variably reported. Lemeé et al. [37] found that it decreases progressively between surgeries in patients requiring repeated operations. In our report [38] and in the studies by Magill et al. [35, 36], decrease of the time to retreatment was also observed, although late recurrences occurred even after many years [50]. This may result from the entity of resection at reoperations and the only late histological progression.

The histological WHO grade and Ki67MiB1 more often show progression between the recurrences, with increase of WHO grades II and III [37, 38]. However, in many cases, the tumor was atypical WHO grade II also after two or more reoperations with no or only late anaplastic transformation. This suggests the need for reoperating on the further recurrences early, before the tumor becomes anaplastic.

3 Predictive Factors for Multiple Reoperations

Several findings, both at initial surgery and at the first reoperation, are at risk of multiple recurrences.

3.1 Findings at Initial Surgery

The flat-shaped aspect of the meningioma and the surrounding dural enhancement are at risk of multicentric-diffuse and often multiple recurrences [51]. Flat-shaped meningiomas are characterized by prevalent dural involvement as compared to round-shaped ones. Thus, it is likely that they may be associated to various degree of even distant microscopic dural infiltration [52], sometimes resulting in more diffuse and late recurrences. On the other hand, "the dural tail" does not reflect the dural infiltration and is not correlated to the recurrence [2, 15].

Other tumor features, such as size, margins, brain-tumor interface, perilesional edema, are not correlated to multiple recurrences.

The extent of surgical resection at initial surgery according to the Simpson grade is a well-recognized factor correlated to the recurrence of meningiomas [1, 13, 15, 22, 41, 53, 54]. Our recent study [38] shows that it is a risk factor also for multiple recurrences. Grades III and IV resections are more at risk than grades I and II ones; residual intrasinusal, bone and intradural tumor is more like to progress and escape to resection at the first reoperation. Although the term "gross-total resection" is often used to define Simpson grades I-II-III together [55], most cases of grade III resection have different prognosis than grades I and II.

The histological WHO grade I versus II and the proliferation index at initial diagnosis are significantly correlated to the recurrence rate and the recurrence-free survival of meningiomas [6, 10–12, 31, 56]. A Ki67 cut-off of 4% is mostly used to differentiate between high and low risk of recurrence. One study on multiple recurrences [38] did not find significant difference of WHO grade (I versus II) and Ki67-MIB1 at initial diagnosis between patients who develop single or multiple reoperations; three other studies on multiple reoperations do not include data of Ki67-MIB1 [35–37].

Many studies focusing on PR expression and recurrence of meningiomas found an inverse correlation, with higher recurrence rates in meningiomas with low PR expression at initial surgery [6, 30–33, 57–59]. On the other hand, no significant difference of PR expression at initial diagnosis was evidenced between patients with single and multiple recurrences.

3.2 Findings at the First Reoperation

The pattern of regrowth and the extent of resection at the first recurrence are important risk factors for multiple reoperations.

Meningiomas may recur with different patterns of growth [51, 60–62]. Most recur at the previous dural site ("localized") or inside and outside the original tumor margins within 1 cm ("peripheral"); some may recur in multicentric and diffuse forms, with multiple nodules, even distant, and interposed seemly normal ("multicentric") or infiltrated ("diffuse") dura. Multicentric and diffuse recurrences represent the progressive growth of multiple distant dural tumor nodules with different potential of growth [52, 63, 64]. Patients with multicentric-diffuse recurrences, when compared to those with localized-peripheral recurrences, show significantly higher rates of flat-shaped tumors and values of ki67-MIB1 > 4% and lower rates of gross-total resections [51]. A recent study [38] demonstrated that multicentric and diffuse patterns of regrowth at the first recurrence are significantly associated to multiple recurrences and reoperations.

The significantly higher rate of multicentric-diffuse patterns of regrowth results in significantly lower rate of gross-total resection in patients who later develop multiple recurrences. This finding confirms that the obtaining gross-total resection at the first reoperation is the most important factor to achieve definitive cure and no further regrowth.

The increase of the WHO grade from I to II between the initial surgery and first reoperation, occurring in about 20% of the patients who experience recurrence, is not a risk factor if the resection was complete; on the other hand, the diagnosis of anaplastic WHO III tumor at reoperation is associated to multiple recurrences.

4 Management of Multiple Recurrences

4.1 Indication to Repeated Surgeries

The neurosurgeon must often decide how to treat meningiomas which further recur after the first reoperation. In almost all cases, the radiotherapy was already administered after the initial surgery or at the first recurrence. Thus, surgery remains the best option. The decision must be discussed with the neuro-oncologist and radiation oncologist and is based on several factors, including patient age and symptoms, tumor size and location, pattern of regrowth, histology of initial and recurrent tumor, time lap after the first reoperation.

A new re-surgery is indicated in symptomatic patients with resectable tumors (Fig. 1).

Asymptomatic patients should be reoperated on when they harbor large recurrences; however, re-surgery should be avoided in elderly individuals with comorbidities and recurrent meningiomas involving critical functional structures (Fig. 2).

The pattern of regrowth at the first recurrence is also important: patients with local or peripheral recurrences are more easily treated by surgery; on the other hand, multicentric and diffuse recurrences with prevalent and extensive dural infiltration and lesser intradural mass escape to attempts of complete resection.

Although the WHO grade and proliferation index often increase in further recurrences [37], in many patients, they remain almost unchanged [38, 50]. This reflects the different time to further recurrences and is important for the indication to surgery. Meningiomas with unchanged WHO grade (or progression form I to II) and unchanged or slightly increased proliferation index at the second recurrence, may present further recurrences even after several years; in those cases, repeated

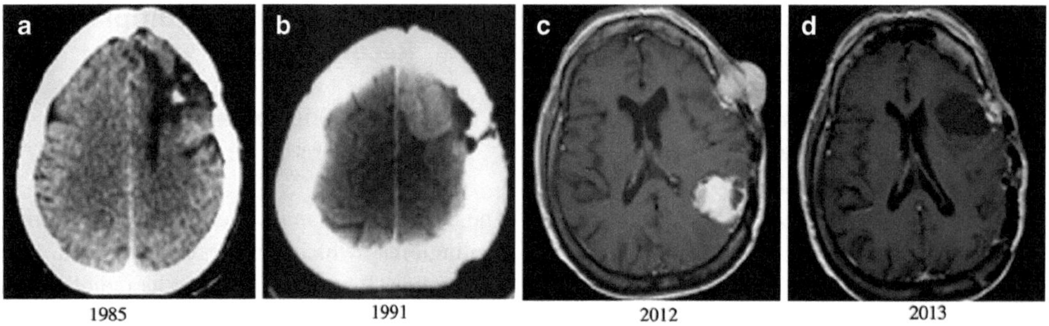

Fig. 1 (**a**) CT scan of a young female patient who underwent resection of the frontal convexity meningioma few years ago: peripheral recurrence posterior to the initial site treated by re-surgery. (**b**) CT scan 6 years after re-surgery: large recurrence at the initial left frontal site. (**c**) Magnetic resonance imaging 21 years after second surgery: recurrence in the left frontal and occipito-temporal regions. (**d**) MRI 1-years after the third surgery: good tumor control. No further follow-up

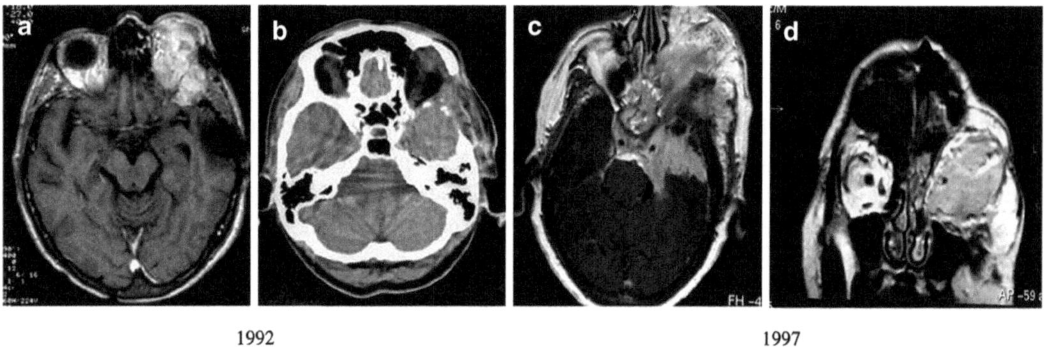

Fig. 2 (**a**) Contrast-enhanced MRI. Recurrence of left spheno-orbital atypical meningioma 4 years after initial surgery, treated by reoperation and irradiation. (**b**) Postoperative CT scan showing good tumor resection with tumor-free orbital cavity. (**c–d**) MRI at 5 years: large and diffuse recurrence involving the orbital cavity, suprasellar region, ambient cistern, and sphenoid bone. No further surgery performed

reoperations are justified, if the other conditions (age, extent of the recurrent tumor, location) are favorable. On the other hand, further short-term recurrences with anaplastic WHO III transformation and significantly higher proliferation index, particularly if with different growth patterns, are not an indication to further surgeries; these patients may be treated by repeated low dose radiotherapy and medicine therapy.

4.2 Postoperative Complications

Postoperative complications after repeated surgeries for intracranial meningiomas include wound dehiscence and infection, cerebrospinal fluid leak, hydrocephalus, hematoma, neurological deficits. Although superficial wound infections are not uncommon, cases requiring a reoperation for wound closure are not more frequent after multiple recurrences than at the first recurrence [37, 65, 66]. Prior radiation therapy has been implicated in the wound-related complications [67], but it is mostly administered at the first recurrence. Also, the rate of postoperative hematomas does not increase after the reoperations [35–37, 66, 68].

Neurological complications include the worsening of a previous deficit and the occurrence of new transient or permanent deficits. The rate of neurological complications increases after repeated surgeries [35, 37, 66]. Older age, presence of cognitive changes and meningioma location at skull base, posterior fossa, and middle third parasagittal region are at higher risk of neurological deficits [35, 37].

5 Outcome After Multiple Meningioma Surgeries

No perioperative mortality was reported in four studies focusing on multiple reoperations for meningiomas [35–38]. The median overall survival after the second reoperation is still high (7–8 years to 12.2 years according to the location in two studies [35, 36]), but it decreases progressively after further surgeries. This is correlated to the increased difficulty to achieve complete resection and the increased WHO grade and proliferation index in most cases.

6 Conclusion

The pattern of regrowth and the extent of resection at the first recurrence are the most important risk factors for multiple reoperations of intracranial meningiomas. The extent of resection at initial surgery is also significant. On the other hand, the pathological findings at initial diagnosis, although well-recognized risk factors of recurrence, are not correlated to the number of recurrences. Biomolecular studies need to better define this aspect. Multiple reoperations are often indicated in selected patients with multiple recurrences according to the above discussed criteria. Even subtotal or partial resections may allow long survivals with good quality of life, at least up to the tumor becomes anaplastic.

References

1. Aghi MK, Carter BS, Cosgrove GR, Ojemann RG, Amin-Hanjani S, Martuza RL, et al. Long-term recurrence rates of atypical meningiomas after gross total resection with or without postoperative adjuvant radiation. Neurosurgery. 2009;64(1):56–60; discussion. https://doi.org/10.1227/01.NEU.0000330399.55586.63.
2. Nakasu S, Fukami T, Jito J, Nozaki K. Recurrence and regrowth of benign meningiomas. Brain Tumor Pathol. 2009;26(2):69–72. https://doi.org/10.1007/s10014-009-0251-2.
3. van Alkemade H, de Leau M, Dieleman EM, Kardaun JW, van Os R, Vandertop WP, et al. Impaired survival and long-term neurological problems in benign meningioma. Neuro-Oncology. 2012;14(5):658–66. https://doi.org/10.1093/neuonc/nos013.
4. Matsuno A, Fujimaki T, Sasaki T, Nagashima T, Ide T, Asai A, et al. Clinical and histopathological analysis of proliferative potentials of recurrent and non-recurrent meningiomas. Acta Neuropathol. 1996;91(5):504–10.
5. Yamasaki F, Yoshioka H, Hama S, Sugiyama K, Arita K, Kurisu K. Recurrence of meningiomas. Cancer. 2000;89(5):1102–10.
6. Maiuri F, De Caro MB, Esposito F, Cappabianca P, Strazzullo V, Pettinato G, et al. Recurrences of meningiomas: predictive value of pathological features and hormonal and growth factors. J Neuro-Oncol. 2007;82(1):63–8. https://doi.org/10.1007/s11060-005-9078-9.
7. Franca RA, Della Monica R, Corvino S, Chiariotti L, Del Basso De Caro M. WHO grade and pathological markers of meningiomas: clinical and prognostic role. Pathol Res Pract. 2023;243:154340. https://doi.org/10.1016/j.prp.2023.154340.
8. Oya S, Kawai K, Nakatomi H, Saito N. Significance of Simpson grading system in modern meningioma surgery: integration of the grade with MIB-1 labeling index as a key to predict the recurrence of WHO Grade I meningiomas. J Neurosurg. 2012;117(1):121–8. https://doi.org/10.3171/2012.3.JNS111945.
9. Abramovich CM, Prayson RA. MIB-1 labeling indices in benign, aggressive, and malignant meningiomas: a study of 90 tumors. Hum Pathol. 1998;29(12):1420–7.
10. Liu N, Song SY, Jiang JB, Wang TJ, Yan CX. The prognostic role of Ki-67/MIB-1 in meningioma: a systematic review with meta-analysis. Medicine (Baltimore). 2020;99(9):e18644. https://doi.org/10.1097/MD.0000000000018644.
11. Mirian C, Skyrman S, Bartek J, Jensen LR, Kihlström L, Förander P, et al. The Ki-67 proliferation index as a marker of time to recurrence in intracranial meningioma. Neurosurgery. 2020;87(6):1289–98. https://doi.org/10.1093/neuros/nyaa226.
12. Olar A, Wani KM, Sulman EP, Mansouri A, Zadeh G, Wilson CD, et al. Mitotic index is an independent predictor of recurrence-free survival in meningioma. Brain Pathol. 2015;25(3):266–75. https://doi.org/10.1111/bpa.12174.
13. Simpson D. The recurrence of intracranial meningiomas after surgical treatment. J Neurol Neurosurg Psychiatry. 1957;20(1):22–39.
14. Champeaux C, Dunn L. World Health Organization Grade II meningioma: a 10-year retrospective study for recurrence and prognostic factor assessment. World Neurosurg. 2016;89:180–6. https://doi.org/10.1016/j.wneu.2016.01.055.
15. Hwang WL, Marciscano AE, Niemierko A, Kim DW, Stemmer-Rachamimov AO, Curry WT, et al. Imaging and extent of surgical resection predict risk of meningioma recurrence better than WHO histopathological grade. Neuro-Oncology. 2016;18(6):863–72. https://doi.org/10.1093/neuonc/nov285.

16. Stafford SL, Perry A, Suman VJ, Meyer FB, Scheithauer BW, Lohse CM, et al. Primarily resected meningiomas: outcome and prognostic factors in 581 Mayo Clinic patients, 1978 through 1988. Mayo Clin Proc. 1998;73(10):936–42. https://doi.org/10.4065/73.10.936.
17. de Carvalho GTC, da Silva-Martins WC, de Magalhães KCSF, Nunes CB, Soares AN, Tafuri LSA, et al. Recurrence/regrowth in grade I meningioma: how to predict? Front Oncol. 2020;10:1144. https://doi.org/10.3389/fonc.2020.01144.
18. Rogers L, Barani I, Chamberlain M, Kaley TJ, McDermott M, Raizer J, et al. Meningiomas: knowledge base, treatment outcomes, and uncertainties. A RANO review. J Neurosurg. 2015;122(1):4–23. https://doi.org/10.3171/2014.7.JNS131644.
19. Sun SQ, Kim AH, Cai C, Murphy RK, DeWees T, Sylvester P, et al. Management of atypical cranial meningiomas, part 1: predictors of recurrence and the role of adjuvant radiation after gross total resection. Neurosurgery. 2014;75(4):347–54.; discussion 54–5; quiz 55. https://doi.org/10.1227/NEU.0000000000000461.
20. Domingues PH, Sousa P, Otero Á, Gonçalves JM, Ruiz L, de Oliveira C, et al. Proposal for a new risk stratification classification for meningioma based on patient age, WHO tumor grade, size, localization, and karyotype. Neuro-Oncology. 2014;16(5):735–47. https://doi.org/10.1093/neuonc/not325.
21. Maiuri F, Corvino S, Lorenzetti M, Franca RA, Esposito F, Del Basso De Caro M. Intracranial meningiomas in patients aged ≥80 years: pathological features and surgical problems. World Neurosurg. 2023;173:e498–508. https://doi.org/10.1016/j.wneu.2023.02.086.
22. Ildan F, Erman T, Göçer AI, Tuna M, Bağdatoğlu H, Cetinalp E, et al. Predicting the probability of meningioma recurrence in the preoperative and early postoperative period: a multivariate analysis in the midterm follow-up. Skull Base. 2007;17(3):157–71. https://doi.org/10.1055/s-2007-970554.
23. Magill ST, Young JS, Chae R, Aghi MK, Theodosopoulos PV, McDermott MW. Relationship between tumor location, size, and WHO grade in meningioma. Neurosurg Focus. 2018;44(4):E4. https://doi.org/10.3171/2018.1.FOCUS17752.
24. Nakasu S, Nakasu Y, Nakajima M, Matsuda M, Handa J. Preoperative identification of meningiomas that are highly likely to recur. J Neurosurg. 1999;90(3):455–62. https://doi.org/10.3171/jns.1999.90.3.0455.
25. McGovern SL, Aldape KD, Munsell MF, Mahajan A, DeMonte F, Woo SY. A comparison of World Health Organization tumor grades at recurrence in patients with non-skull base and skull base meningiomas. J Neurosurg. 2010;112(5):925–33. https://doi.org/10.3171/2009.9.JNS09617.
26. Maiuri F, Mariniello G, Guadagno E, Barbato M, Corvino S, Del Basso De Caro M. WHO grade, proliferation index, and progesterone receptor expression are different according to the location of meningioma. Acta Neurochir. 2019;161(12):2553–61. https://doi.org/10.1007/s00701-019-04084-z.
27. Maiuri F, Corvino S. Meningiomas of the rolandic region: risk factors for motor deficit and role of intra-operative monitoring. Acta Neurochir. 2023;165(7):1707–16. https://doi.org/10.1007/s00701-023-05630-6.
28. Maiuri F, Mariniello G, Barbato M, Corvino S, Guadagno E, Chiariotti L, et al. Malignant intraventricular meningioma: literature review and case report. Neurosurg Rev. 2022;45(1):151–66. https://doi.org/10.1007/s10143-021-01585-3.
29. Nawashiro H. Tumor shape and recurrence. J Neurosurg. 2000;93(3):528.
30. Fewings PE, Battersby RD, Timperley WR. Long-term follow up of progesterone receptor status in benign meningioma: a prognostic indicator of recurrence? J Neurosurg. 2000;92(3):401–5. https://doi.org/10.3171/jns.2000.92.3.0401.
31. Roser F, Nakamura M, Bellinzona M, Rosahl SK, Ostertag H, Samii M. The prognostic value of progesterone receptor status in meningiomas. J Clin Pathol. 2004;57(10):1033–7. https://doi.org/10.1136/jcp.2004.018333.
32. Pravdenkova S, Al-Mefty O, Sawyer J, Husain M. Progesterone and estrogen receptors: opposing prognostic indicators in meningiomas. J Neurosurg. 2006;105(2):163–73. https://doi.org/10.3171/jns.2006.105.2.163.
33. Maiuri F, Mariniello G, de Divitiis O, Esposito F, Guadagno E, Teodonno G, et al. Progesterone receptor expression in meningiomas: pathological and prognostic implications. Front Oncol. 2021;11:611218. https://doi.org/10.3389/fonc.2021.611218.
34. Maiuri F, Mariniello G, Somma T, Guadagno E, Corvino S, Pagano S, et al. Meningiomas in premenopausal women: role of the hormone related conditions. Front Oncol. 2020;10:556701. https://doi.org/10.3389/fonc.2020.556701.
35. Magill ST, Lee DS, Yen AJ, Lucas CG, Raleigh DR, Aghi MK, et al. Surgical outcomes after reoperation for recurrent skull base meningiomas. J Neurosurg. 2018;130(3):876–83. https://doi.org/10.3171/2017.11.JNS172278.
36. Magill ST, Dalle Ore CL, Diaz MA, Jalili DD, Raleigh DR, Aghi MK, et al. Surgical outcomes after reoperation for recurrent non-skull base meningiomas. J Neurosurg. 2018;131:1–9. https://doi.org/10.3171/2018.6.JNS18118.
37. Lemée JM, Corniola MV, Meling TR. Benefits of re-do surgery for recurrent intracranial meningiomas. Sci Rep. 2020;10(1):303. https://doi.org/10.1038/s41598-019-57254-5.
38. Maiuri F, Corvino S, Corazzelli G, Del Basso De Caro M. Single versus multiple reoperations for recurrent intracranial meningiomas. J Neuro-Oncol. 2024;168:527.
39. Brokinkel B, Holling M, Spille DC, Heß K, Sauerland C, Bleimüller C, et al. Surgery for meningioma in the elderly and long-term survival: comparison with an

age- and sex-matched general population and with younger patients. J Neurosurg. 2017;126(4):1201–11. https://doi.org/10.3171/2016.2.JNS152611.
40. Mansouri A, Klironomos G, Taslimi S, Kilian A, Gentili F, Khan OH, et al. Surgically resected skull base meningiomas demonstrate a divergent postoperative recurrence pattern compared with non-skull base meningiomas. J Neurosurg. 2016;125(2):431–40. https://doi.org/10.3171/2015.7.JNS15546.
41. Lam Shin Cheung V, Kim A, Sahgal A, Das S. Meningioma recurrence rates following treatment: a systematic analysis. J Neuro-Oncol. 2018;136(2):351–61. https://doi.org/10.1007/s11060-017-2659-6.
42. Materi J, Mampre D, Ehresman J, Rincon-Torroella J, Chaichana KL. Predictors of recurrence and high growth rate of residual meningiomas after subtotal resection. J Neurosurg. 2020;134(2):410–6. https://doi.org/10.3171/2019.10.JNS192466.
43. Mariniello G, de Divitiis O, Corvino S, Strianese D, Iuliano A, Bonavolontà G, et al. Recurrences of spheno-orbital meningiomas: risk factors and management. World Neurosurg. 2022;161:e514. https://doi.org/10.1016/j.wneu.2022.02.048.
44. Clark VE, Erson-Omay EZ, Serin A, Yin J, Cotney J, Ozduman K, et al. Genomic analysis of non-NF2 meningiomas reveals mutations in TRAF7, KLF4, AKT1, and SMO. Science. 2013;339(6123):1077–80. https://doi.org/10.1126/science.1233009.
45. Bi WL, Abedalthagafi M, Horowitz P, Agarwalla PK, Mei Y, Aizer AA, et al. Genomic landscape of intracranial meningiomas. J Neurosurg. 2016;125(3):525–35. https://doi.org/10.3171/2015.6.JNS15591.
46. Boetto J, Peyre M, Kalamarides M. Meningiomas from a developmental perspective: exploring the crossroads between meningeal embryology and tumorigenesis. Acta Neurochir. 2021;163(1):57–66. https://doi.org/10.1007/s00701-020-04650-w.
47. Okano A, Miyawaki S, Hongo H, Dofuku S, Teranishi Y, Mitsui J, et al. Associations of pathological diagnosis and genetic abnormalities in meningiomas with the embryological origins of the meninges. Sci Rep. 2021;11(1):6987. https://doi.org/10.1038/s41598-021-86298-9.
48. Maiuri F, Del Basso De Caro M, de Divitiis O, Guadagno E, Mariniello G. Recurrence of spinal meningiomas: analysis of the risk factors. Br J Neurosurg. 2019;34:1–6. https://doi.org/10.1080/02688697.2019.1638886.
49. Mariniello G, Briganti F, De Caro ML, Maiuri F. Cervical extradural "en-plaque" meningioma. J Neurol Surg A Cent Eur Neurosurg. 2012;73(5):330–3. https://doi.org/10.1055/s-0032-1304222.
50. Maiuri F, Corvino S, Corazzelli G, Berardinelli J, Di Crescenzo RM, Del Basso De Caro M. Time to recurrence of intracranial meningiomas from a monoinstitutional surgical series. World Neurosurg. 2024;185:e612. https://doi.org/10.1016/j.wneu.2024.02.087.
51. Maiuri F, Mariniello G, Peca C, Guadagno E, Corvino S, d'Avanzo S, et al. Multicentric and diffuse recurrences of meningiomas. Br J Neurosurg. 2020;34:1–8. https://doi.org/10.1080/02688697.2020.1754335.
52. Borovich B, Doron Y. Recurrence of intracranial meningiomas: the role played by regional multicentricity. J Neurosurg. 1986;64(1):58–63. https://doi.org/10.3171/jns.1986.64.1.0058.
53. Gousias K, Schramm J, Simon M. The Simpson grading revisited: aggressive surgery and its place in modern meningioma management. J Neurosurg. 2016;125(3):551–60. https://doi.org/10.3171/2015.9.JNS15754.
54. Nanda A, Bir SC, Maiti TK, Konar SK, Missios S, Guthikonda B. Relevance of Simpson grading system and recurrence-free survival after surgery for World Health Organization Grade I meningioma. J Neurosurg. 2017;126(1):201–11. https://doi.org/10.3171/2016.1.JNS151842.
55. Goldbrunner R, Minniti G, Preusser M, Jenkinson MD, Sallabanda K, Houdart E, et al. EANO guidelines for the diagnosis and treatment of meningiomas. Lancet Oncol. 2016;17(9):e383–91. https://doi.org/10.1016/S1470-2045(16)30321-7.
56. Abry E, Thomassen I. The significance of Ki-67/MIB-1 labeling index in human meningiomas: a literature study. Pathol Res Pract. 2010;206:810.
57. Strik HM, Strobelt I, Pietsch-Breitfeld B, Iglesias-Rozas JR, Will B, Meyermann R. The impact of progesterone receptor expression on relapse in the long-term clinical course of 93 benign meningiomas. In Vivo. 2002;16(4):265–70.
58. Metellus P, Nanni I, Dussert C, Trinkhaus M, Fuentes S, Chinot O, et al. [Prognostic implications of biologic markers in intracranial meningiomas: 120 cases]. Neurochirurgie 2008;54(6):750–6. https://doi.org/10.1016/j.neuchi.2008.07.005.
59. Abdelzaher E, El-Gendi SM, Yehya A, Gowil AG. Recurrence of benign meningiomas: predictive value of proliferative index, BCL2, p53, hormonal receptors and HER2 expression. Br J Neurosurg. 2011;25(6):707–13. https://doi.org/10.3109/02688697.2010.522743.
60. Maiuri F, Corvino S, Guadagno E, Del Basso De Caro M. Topographic distribution of intracranial meningioma's recurrences: localized versus diffuse-multicentric [Internet]. Front Clin Neurosurg IntechOpen. 2021; https://doi.org/10.5772/intechopen.97120:2021.
61. Obiri-Yeboah D, Soni P, Oyem PC, Almeida JP, Murayi R, Recinos PF, et al. Beyond the surgical margin: patterns of recurrence in World Health Organization Grade 2 intracranial meningiomas. World Neurosurg. 2023;177:e593. https://doi.org/10.1016/j.wneu.2023.06.100.
62. Ong K, Rizzuto M, Makarenko S. Location pattern of recurrence of fully resected grade 1 meningiomas. Acta Neurochir. 2023;165(10):2865–71. https://doi.org/10.1007/s00701-023-05758-5.
63. Kinjo T, al-Mefty O, Kanaan I. Grade zero removal of supratentorial convexity meningiomas. Neurosurgery.

1993;33(3):394.–9; discussion 9. https://doi.org/10.1227/00006123-199309000-00007.
64. Qi ST, Liu Y, Pan J, Chotai S, Fang LX. A radiopathological classification of dural tail sign of meningiomas. J Neurosurg. 2012;117(4):645–53. https://doi.org/10.3171/2012.6.JNS111987.
65. Klinger DR, Flores BC, Lewis JJ, Hatanpaa K, Choe K, Mickey B, et al. Atypical meningiomas: recurrence, reoperation, and radiotherapy. World Neurosurg. 2015;84(3):839–45. https://doi.org/10.1016/j.wneu.2015.04.033.
66. Sicking J, Voß KM, Spille DC, Schipmann S, Holling M, Paulus W, et al. The evolution of cranial meningioma surgery-a single-center 25-year experience. Acta Neurochir. 2018;160(9):1801–12. https://doi.org/10.1007/s00701-018-3617-6.
67. Haubner F, Ohmann E, Pohl F, Strutz J, Gassner HG. Wound healing after radiation therapy: review of the literature. Radiat Oncol. 2012;7:162. https://doi.org/10.1186/1748-717X-7-162.
68. Connolly ID, Cole T, Veeravagu A, Popat R, Ratliff J, Li G. Craniotomy for resection of meningioma: an age-stratified analysis of the MarketScan longitudinal database. World Neurosurg. 2015;84(6):1864–70. https://doi.org/10.1016/j.wneu.2015.08.018.

GPSR Compliance

The European Union's (EU) General Product Safety Regulation (GPSR) is a set of rules that requires consumer products to be safe and our obligations to ensure this.

If you have any concerns about our products, you can contact us on ProductSafety@springernature.com

In case Publisher is established outside the EU, the EU authorized representative is:

Springer Nature Customer Service Center GmbH
Europaplatz 3
69115 Heidelberg, Germany

Batch number: 10731385

Printed by Printforce, the Netherlands